13th International Conference on Numerical Modeling of Space Plasma Flows (ASTRONUM 2018)

Journal of Physics: Conference Series Volume 1225

Panama City Beach, Florida, USA
25 - 29 June 2018

ISBN: 978-1-5108-8945-3
ISSN: 1742-6588

Printed from e-media with permission by:

Curran Associates, Inc.
57 Morehouse Lane
Red Hook, NY 12571

Some format issues inherent in the e-media version may also appear in this print version.

Institute of Physics
Dirac House, Temple Back
Bristol BS1 6BE UK

Phone: 44 1 17 929 7481
Fax: 44 1 17 920 0979

techtracking@iop.org

Additional copies of this publication are available from:

Curran Associates, Inc.
57 Morehouse Lane
Red Hook, NY 12571 USA
Phone: 845-758-0400
Fax: 845-758-2633
Email: curran@proceedings.com
Web: www.proceedings.com

TABLE OF CONTENTS

ASTROPHYSICAL FLOWS

SPACE PLASMA FLOWS

NUMERICAL METHODS

IOP Conf. Series: Journal of Physics: Conf. Series **1225** (2019) 011001 doi:10.1088/1742-6596/1225/1/011001

PREFACE

The Center for Space Plasma and Aeronomic Research (CSPAR) at the University of Alabama in Huntsville (UAH) and Maison de la Simulation, a joint laboratory at the French Alternative Energies and Atomic Energy Commission (CEA), National Center for Scientific Research (CNRS), University of Paris-Sud, and University of Versailles, organized the 13th International Conference on Numerical Modeling of Space Plasma Flows (ASTRONUM-2018) on June 25—29, 2018 in Panama City Beach, Florida, USA.

The Program Committee consisted of Tahar Amari (CNRS Ecole Polytechnique, France), Edouard Audit (CEA/CNRS Maison de la Simulation, Gif-sur-Yvette, France, co-chair), Amitava Bhattacharjee (Princeton University, USA), Phillip Colella (Lawrence Berkeley National Laboratory, USA), Anthony Mezzacappa (University of Tennessee, Knoxville, USA), Ewald Müller (Max-Planck-Institute for Astrophysics, Garching, Germany), Nikolai Pogorelov (University of Alabama in Huntsville/CSPAR, USA, chair), Kazunari Shibata (Kyoto University, Japan), James Stone (Princeton University, USA), Jon Linker (Predictive Science, Inc., USA), and Gary P. Zank (University of Alabama in Huntsville, USA).

The conference attracted 67 scientists from 14 countries representing different branches of the plasma simulation community. The distinctive feature of this conference is a combination of diverse research topics, all of which are essential for performing high-resolution, continuum mechanics and particle, simulations of physical phenomena in space physics and astrophysics. Among such topics were software packages for modeling and analyzing plasma flows; advanced numerical methods for space and astrophysical flows; large-scale fluid-based, kinetic, and hybrid simulations; turbulence and cosmic ray transport; and magnetohydrodynamics. The applications discussed included cosmology and galaxy formation, supernova explosions, physics of the Sun-heliosphere-magnetosphere environments, the interstellar medium and star formation, stellar physics, experimental plasma physics, astrophysical accretion, numerical methods for ideal and non-ideal, relativistic and nonrelativistic MHD, etc. The proceedings volume is structured so that it covers all of these topics.

The structure of the ASTRONUM conference series is based on the idea that modelers working in seemingly distant fields should have an opportunity to share their scientific achievements with the broad community of computational scientists performing numerical experiments. As in previous ASTRONUM meetings, we were interested in physical systems that are coupled across a multiplicity of spatial and temporal that incorporate diverse physical processes.

The contributors to this volume are both young researchers and renowned experts in space physics and astrophysics, applied mathematics, and computational physics. This volume describes the application of numerical methods and the algorithms themselves, allowing us to discuss the challenges that theory imposes on numerical schemes for solving partial differential equations describing collisional and collisionless processes in space and astrophysical plasmas.

We would like to thank the participants who submitted their papers to Proceedings of ASTRONUM-2017 and especially to those who reviewed manuscripts thus ensuring the high quality of this publication. We also are grateful to Adele Corona (International Conferencing and Networking Solutions, icnsmeetings.com) for the excellent management of the conference.

IOP Conf. Series: Journal of Physics: Conf. Series **1225** (2019) 011001 doi:10.1088/1742-6596/1225/1/011001

The volume will be useful to graduate and postgraduate students majoring in space physics, astrophysics, numerical, engineering, and applied mathematics. It is also aimed at specialists in applied mathematics, and various fields of physics that involve flows of partially ionized plasmas at both the collisional and collisionless levels.

Nikolai V. Pogorelov, Edouard Audit, and Gary P. Zank

April 19, 2019

IOP Conf. Series: Journal of Physics: Conf. Series **1225** (2019) 011002 doi:10.1088/1742-6596/1225/1/011002

Peer review statement

All papers published in this volume of *Journal of Physics: Conference Series* have been peer reviewed through processes administered by the proceedings Editors. Reviews were conducted by expert referees to the professional and scientific standards expected of a proceedings journal published by IOP Publishing.

IOP Conf. Series: Journal of Physics: Conf. Series **1225** (2019) 012001 doi:10.1088/1742-6596/1225/1/012001

3D relativistic MHD simulations of bow-shock Pulsar Wind Nebulae with highly asymmetric geometry

Maxim V. Barkov[1,2], Maxim Lyutikov[1] and Dmitry Khangulyan[3]
[1]Department of Physics and Astronomy, Purdue University, West Lafayette, IN 47907-2036, USA
[2] Astrophysical Big Bang Laboratory, RIKEN, 351-0198 Saitama, Japan
[3]Department of Physics, Rikkyo University, Nishi-Ikebukuro 3-34-1, Toshima-ku, Tokyo 171-8501, Japan
E-mail: mbarkov@purdue.edu

Abstract. Bow-shock pulsar wind nebulae (PWNe) show a variety of morphological shapes. We attribute this diversity to the geometrical factors: relative orientations of the pulsar rotation axis, proper velocity, and the line of sight. Here we report study of morphology in the most complicated geometry, when the pulsar rotation axis and its proper velocity makes an arbitrary angle (in the simulations we adopt a value of $\sim \pi/4$). Using 3D RMHD simulations we obtain the MHD structure of PWN and compute synthetic synchrotron emissivity maps.

1. Introduction

Pulsars produce relativistic magnetized winds that create pulsar wind nebulae (PWNe) ([1] and references therein). A distinct type of PWNe is produced by runaway pulsar, *i.e.* fast moving pulsars that quickly escape from the supernova remnant (for a recent review see Ref.[2]). Runaway pulsars can have velocities of hundreds kilometers per second, significantly exceeding the typical sound speeds in the interstellar medium (ISM), $c_{s,\rm ISM} = 10 - 100\,{\rm km\,s^{-1}}$. The interaction of such pulsars with the ISM produces a bow-shock nebula with an extended tail.

Pulsar proper velocity and rotation axis are often aligned allowing axial symmetric two dimensional treatment of these systems. A number of analytical and numerical models were developed under this assumption (*e.g.* [3, 4]). Also this configuration was considered in three dimensional (3D) hydrodynamic limit [5] and recently with relativistic magnetohydrodynamics (RMHD) approach [6]. However, the alignment does not always take place, and such not-aligned systems require a 3D MHD simulations in complex geometries.

The structure of magnetic fields in the bow-shock PWNe is, generally, a complicated mixture of the wind-structure produces by the pulsar, and the interaction of the wind with the ISM. The formation of wind produced on the magnetosphere scale, which is much smaller than the termination shock (TS) radius. Thus, the formation of pulsar wind in the bow-shock PWNe is not affected by the processes taking place in the PWN. Thus, one can expect that the properties of pulsar winds in PWNe formed by fast- and slow-moving pulsars are similar. Thus, we utilize the model of the pulsar wind that is used in RMHD modeling of the Crab Nebula. Namely, we assume that on scales larger than the light cylinder radius, the energy flux in the wind (which is mostly Poynting power) changes as $L(\theta_p) \propto \sin^2\theta_p$, where θ_p is the angle to the pulsar rotation

IOP Conf. Series: Journal of Physics: Conf. Series **1225** (2019) 012001 doi:10.1088/1742-6596/1225/1/012001

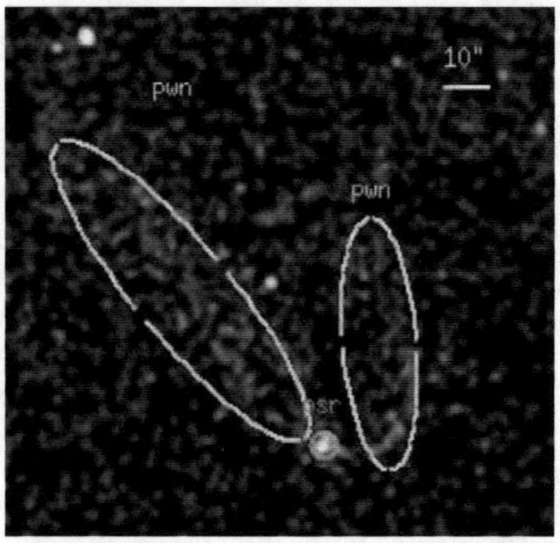

Figure 1. PSR J1135−6055 0.3-10 keV Chandra Imaging from Ref.[7]. The two yellow ellipses highlight two distinct features of the nebular: a single-sided bent jet and an extended tail.

axis [8, 9, 10]. In the equatorial region, occupying a section of the polar angles of the order of the pulsar magnetic inclination angle, the magnetic field is reversing polarity each half a period, while at larger latitudes it is unidirectional [9, 11]. As the wind propagates from the light cylinder, the magnetic field is dissipated in the region of field reversals [12, 13]. Thus, the magnetization σ in the wind is small near the equator and can be significant in the polar region.

In addition to the fairly complected structure of the winds of slow-moving pulsars, for bow-shock pulsars there is an additional complication due to the pulsar motion. This introduces another special direction - along the pulsar velocity - and generally makes the whole structure non-axisymmetric. Thus, the structure of the magnetic field depends on two geometrical factors: (i) the angle between the pulsar rotation axis and the magnetic moment; (ii) the angle between the pulsar rotation axis and the direction of motion. Obviously, the observational appearance of the formed complex 3D structure depends strongly on the line-of-sight direction.

Three distinct geometric cases were considered (see Ref. [6] and in Fig. 2): (i) "Rifle Bullet" geometry – when the rotation axis is aligned with the direction of motion. In this case the whole system has a cylindrical symmetry with concentric areas of toroidal fields of changing polarity. (ii) "Frisbee" geometry – when the rotation axis is perpendicular to the direction of motion and is in the plane of the sky; (iii) "Cart Wheel" geometry – when the rotation axis is perpendicular both to the direction of motion and the plane of the sky. The two physically distinct geometries, rifle-bullet and frisbee – cart-wheel, will have very different magnetic fields in the tail, Fig. 3. Here we consider a mixed case when the angle between the pulsar rotation axis and proper velocity is $\pi/4$. We aim at qualitatively explaining the morphology of the PWN formed around PSR J1135−6055 (see in Fig. 1). A detailed consideration of different configurations is presented in Ref. [6].

2. 3D RMHD simulations of bow-shock PWNe

We performed 3D RMHD simulations of the interaction of magnetized pulsar wind with magnetized external medium using a 3D geometry in Cartesian coordinates using the *PLUTO* code[1] [14]. Spatial parabolic interpolation, a 3$^{\rm rd}$ order Runge-Kutta approximation in time, and an HLL Riemann solver are utilized [15]. *PLUTO* is a modular Godunov-type code entirely written in C and intended mainly for astrophysical applications and high Mach number flows in multiple spatial dimensions. The simulations were performed on CFCA XC30 cluster of National Astronomical Observatory of Japan (NAOJ). The flow has been approximated as an

[1] Link http://plutocode.ph.unito.it/index.html

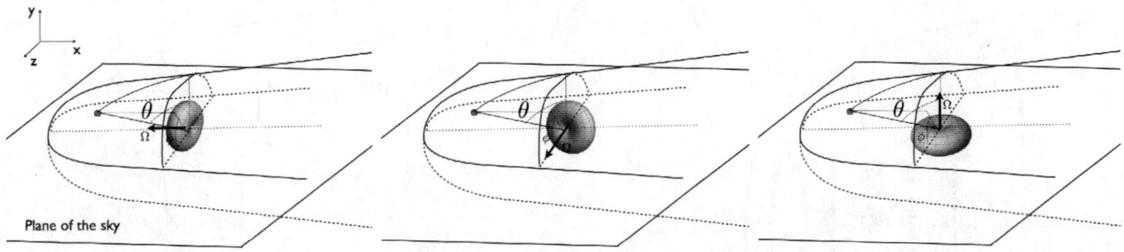

Figure 2. Basic geometries: rifle-bullet (with the spin of the NS aligned with the velocity), frisbee (with the spin of the NS perpendicular to the velocity but lying in the plane of the sky), and cart-wheel (with the spin of the NS perpendicular both to the velocity and the plane of the sky). The central doughnut-like structure indicates the distribution of wind power, $\propto \sin^2\theta_p$, where θ_p is the pulsar rotation axis; θ is a polar angle with respect to the velocity.

Table 1. Parameters of the Grid

Coordinates	Left	N_l	Left-center	N_c	Right-center	N_r	Right
X	-4	72	-1	144	1	252	10
Y	-5	96	-1	144	1	96	5
Z	-5	96	-1	144	1	96	5

ideal, relativistic adiabatic gas, one particle species, and polytropic index of 4/3. The simulations were performed in a domain of size $x \in [-4, 10]$, y and $z \in [-5, 5]$. The initial ISM velocity is directed along x-axis. To achieve a good resolution in the central region and long tail zone, which are the mostly relevant for studying the PWN morphology, we use non uniform resolution stretched grid with the total number of cells $N_X = 468$, and $N_Y = N_Z = 336$, see Table 1 for details.

In this work we use the prescription of pulsar wind similar to what was used by [16]. The pulsar with radius 0.2 is placed at the origin. The pulsar produces supersonic magnetized pulsar wind with a toroidal magnetic field that changes its polarity in northern an southern hemispheres. For the total energy flux density of the wind we adopt the monopole model [8, 9]

$$f_{\rm tot}(r, \theta_p) = L_0 \left(\frac{1}{r}\right)^2 \left(\sin^2\theta_p + g\right). \tag{1}$$

where we added $g = 0.03$ to avoid vanishing energy flux at the poles.

In the wind the energy is distributed between the magnetic f_m component

$$f_m(r, \theta_p) = \frac{\sigma(\theta_p) f_{tot}(r, \theta_p)}{1 + \sigma(\theta_p)} \tag{2}$$

and kinetic f_k one

$$f_k(r, \theta_p) = \frac{f_{tot}(r, \theta_p)}{1 + \sigma(\theta_p)}, \tag{3}$$

where $\sigma(\theta_p)$ is the wind magnetization, which depends on the co-latitude, θ_p measured relative to the pulsar spin axis.

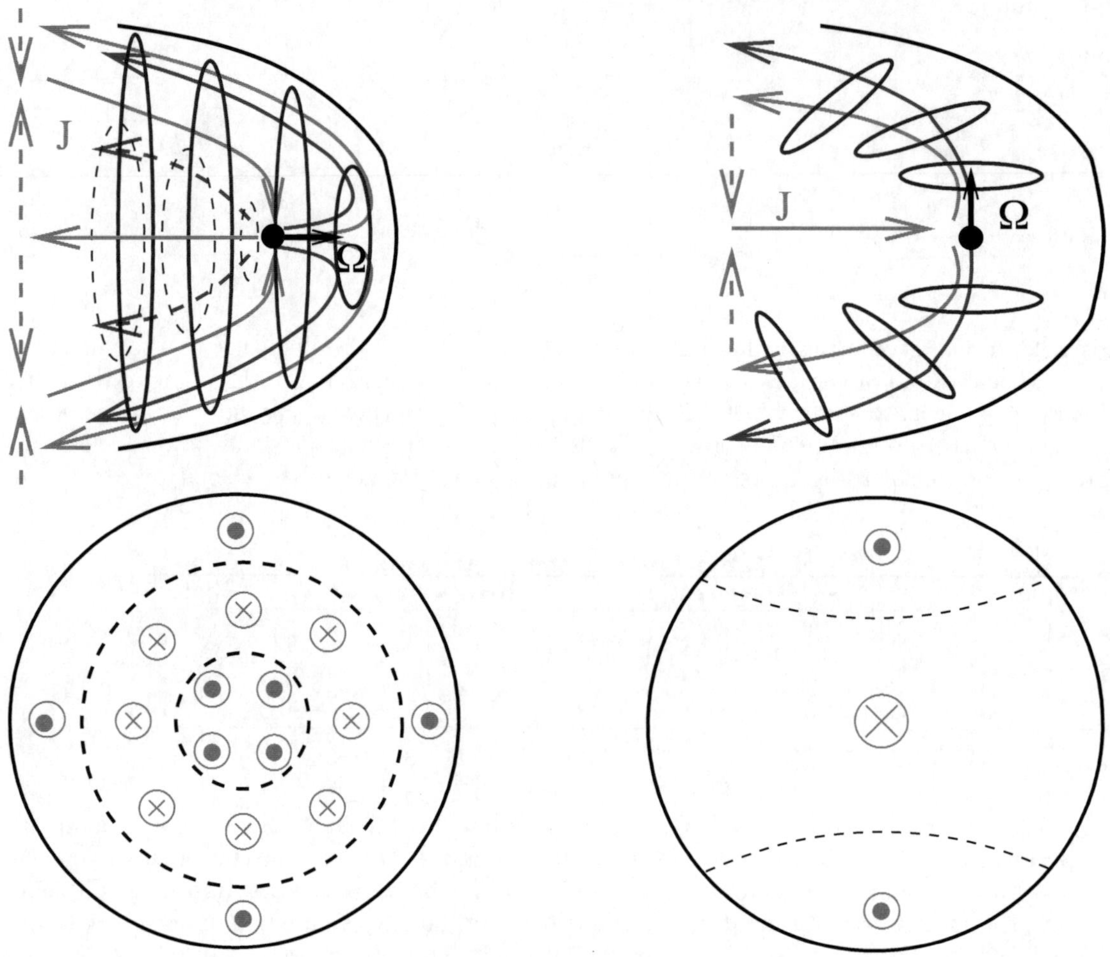

Figure 3. Magnetic field in the rifle-bullet (left) and frisbee/cart-wheel (right) geometries. In case of a rifle-bullet geometry (NS spin oriented along the velocity), the pulsar produces axisymmetric current flows and magnetic fields. Far in the tail the electric currents form a triple sequence of concentric regions with oppositely directed current flows (low left). For frisbee, (NS spin is orthogonal to the velocity; Sun is in a frisbee/cart-wheel configuration), the pulsar produces three adjacent regions of opposite current flows (low right).

As one adopts a torroidal geometry of the magnetic field, the numerical stability of the code requires vanishing of the magnetic field close to the polar axis. This is achieved by introducing the following dependence of the pulsar wind magnetization:

$$\sigma_0(\theta_p) = \sigma_0 \min\left(1, \theta_p^2/\theta_0^2, (\pi - \theta_p)^2/\theta_0^2\right) \tag{4}$$

where θ_0 is a small parameter, which was set to 0.2.

Near the equator the alternating components of magnetic field are assumed to annihilate, leaving a low-magnetized equatorial sector with magnetization varying according to

$$\sigma(\theta_p) = \frac{\sigma_0(\theta_p)\chi_\alpha(\theta_p)}{1 + \sigma_0(\theta_p)(1 - \chi_\alpha(\theta_p))}, \tag{5}$$

IOP Conf. Series: Journal of Physics: Conf. Series **1225** (2019) 012001 doi:10.1088/1742-6596/1225/1/012001

Table 2. Parameters of the model

θ	ψ	σ_0	α
$\pi/4$	$\pi/4$	1	$\pi/4$

where

$$\chi_\alpha(\theta_p) = \begin{cases} (2\phi_\alpha(\theta_p)/\pi - 1)^2, & |\pi/2 - \theta_p| < \alpha \\ 1 & \text{otherwise} \end{cases}, \tag{6}$$

and $\cos\phi_\alpha(\theta_p) = -\cot(\theta_p)\cot(\alpha)$. The angle α is an angle between magnetic axis and pulsar rotation axis (see Ref. [17] for detail). The pulsar wind was injected with the initial Lorentz factor, $\Gamma = 2.9$.

We start our simulation with a non-equilibrium configuration and evolve it until a quasi-stationary solution is settled. From the left edge ($X = -4$) we inject ISM. To reduce computational expenses we set the initial ISM speed to $v_{\rm ISM} = 0.1c$, which corresponds to the Mach number of $M = 85$. The density of the ISM was adopted so that in the case of non-magnetized spherical pulsar wind the TS radius equals 1. The adopted initial ISM speed is not realistic, but it does not affect significantly the region inside the contact discontinuity [18]. The ISM flow carries a weak magnetic field with magnetization $\sigma_{\rm ISM} = 0.01$; the ISM magnetic field is directed along the z-axis.

We choose the intermediate configuration, frisbee – rifle-bullet. The orientation is determined by two angles θ (clockwise turn around Y axis) and angle ψ (clockwise turn around axis Z), The intermediate case has $\theta = \pi/4$ and $\psi = \pi/4$. The parameters of the models are presented in Table 2. The magnetization of pulsar wind is $\sigma_0 = 1$, and pulsar magnetic inclination is $\alpha = \pi/4$.

Results are shown in Fig. 4. As it was expected the whole structures is highly asymmetric. This is due to the effects of magnetic hoop stresses near the rotational axis: in the head part the wind is slowed down and efficiently confined by the ISM ram pressure. This allows magnetic stresses to get accumulated and produce larger distortion than in the tailward part. Thus, the front outflow forms a narrow jet-like structure. The tail is formed by the back "jet" and partially by a matter and magnetic field from front "jet" which was turned backwards near the head of the bow shock. Turned back flow has a magnetic field directed differently as compared to the back "jet". This provides sites suitable for the magnetic field reconnection in the tail. A similar configuration is formed in the pure rifle-bullet configuration (see Ref. [6]). Finally, we note that the shape of pulsar wind TS is complicated, similar to the one revealed with 3D hydrodynamic simulations in Ref. [5] (see also Ref. [6]).

3. Synchrotron synthetic maps

The observed X-ray emission from PWNe is generated via synchrotron radiation by non-thermal particles, which are presumably accelerated at shocks and/or in the reconnection events within the PWNe. Conventional ideal RMHD simulations produce only hydrodynamic quantities – density, thermal pressure, velocity, and magnetic field. Thus, we have no direct information about energy distribution and density of non-thermal particles. To obtain this additional information one needs to perform dedicated simulation of evolution of non-thermal particles (see *e.g.* [19, 20]). However, if the particle cooling is dominated by adiabatic losses, one can use a simplified approach and reconstruct the spectrum of non-thermal particles based on MHD

IOP Conf. Series: Journal of Physics: Conf. Series **1225** (2019) 012001 doi:10.1088/1742-6596/1225/1/012001

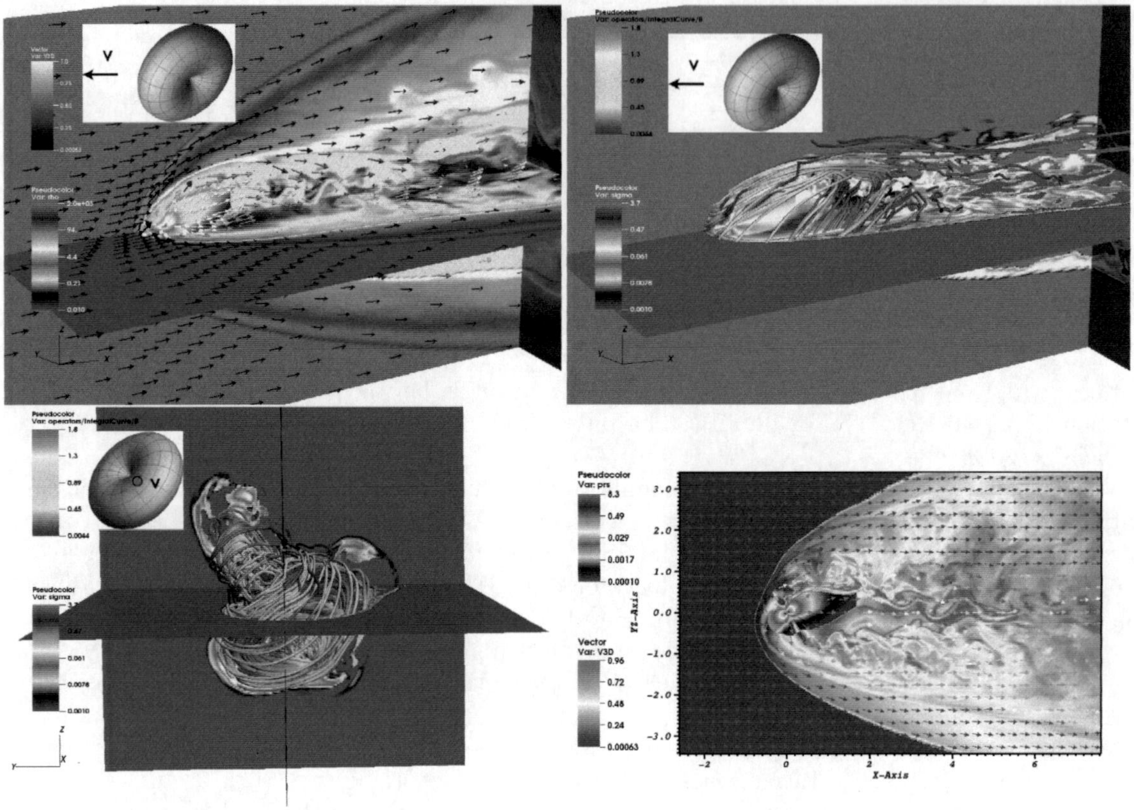

Figure 4. 3D rendering in the mixed rifle-bullet – frisbee configuration. Top row: Plasma density (bottom colour scale, logarithmic) and the velocity field (arrows, upper colour scale) and Pressure/velocity plot at bottom right.

parameters only (see [21]).

In bow-shock PWNe, the strength of the magnetic field might be quite high, exceeding the field inferred in PWNe around slow-moving pulsars. Although the structure of the magnetic field in PWNe is quite complicated, the characteristic magnetic field can be obtained from the pressure, $\sim \rho_{\rm ISM} V_{NS}^2$, required to support the nebula. Thus, one obtains

$$B \sim 2 \times 10^{-4} n_{\rm ISM,0}^{1/2} V_{NS,7.5}\,{\rm G}\,. \tag{7}$$

For the magnetic field of strength B' and photon energy ϵ' (both in flow co-moving frame) the required Lorentz factor of the radiating particles is

$$\gamma_{\rm SYN} \simeq \sqrt{\frac{2}{3}\frac{m_e c \epsilon'_\gamma}{e\hbar B'}} = 7 \times 10^6 {\epsilon'_{\gamma,1\,\rm keV}}^{1/2} {B'}_{-3}^{-1/2}\,, \tag{8}$$

which should be easily attainable for non-thermal particles in PWNe (*e.g.* in Crab Nebula one expects particle acceleration to PeV energies, see [19, 22]).

The corresponding synchrotron cooling time is

$$t_{\rm SYN} \simeq 10^8\, {\epsilon'}_{\gamma,1\,\rm keV}^{-1/2} {B'}_{-3}^{-3/2}\ {\rm s}. \tag{9}$$

IOP Conf. Series: Journal of Physics: Conf. Series **1225** (2019) 012001 doi:10.1088/1742-6596/1225/1/012001

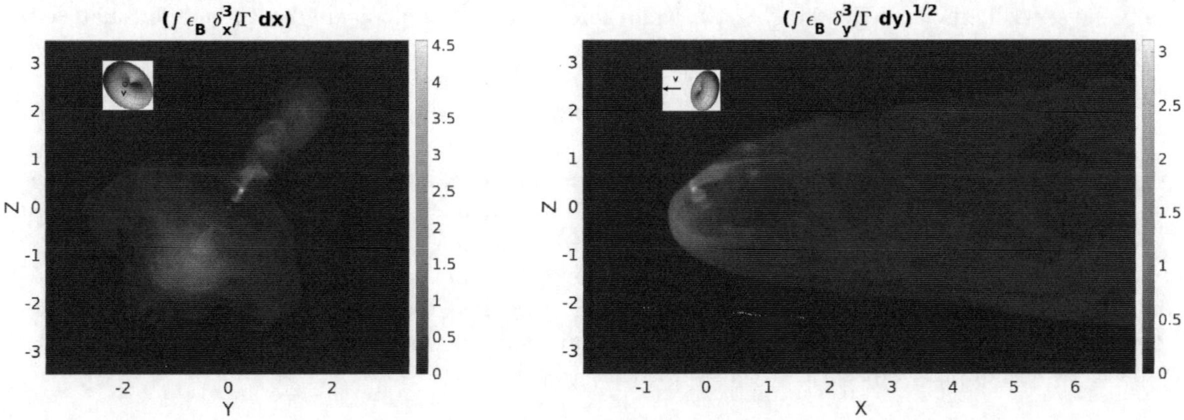

Figure 5. Synchrotron emissivity map projected along X (top) and Y (bottom) axis.

This cooling time should be compared to adiabatic cooling time, which can be estimated as the time required for the flow to cross the characteristic hydrodynamic scale, *e.g.* r_s:

$$t_{\rm AD} \simeq \frac{r_s}{c} \simeq 10^6 n_{\rm ISM,0}^{-1/2} V_{NS,7.5}^{-1} L_{w,36}^{1/2}\ {\rm s}\,. \tag{10}$$

The ratio of synchrotron and adiabatic cooling times is

$$\frac{t_{\rm SYN}}{t_{\rm AD}} \simeq 5\times 10^2 \epsilon'^{-1/2}_{\gamma,1\,{\rm keV}} n_{\rm ISM,0}^{-1/4} V_{NS,7.5}^{-1/2} L_{w,36}^{-1/2}\,, \tag{11}$$

which implies that in a bow-shock PWNe the cooling of particles, responsible for X-ray emission, proceeds predominately due to adiabatic losses. Thus, one can utilize the simple approach for computing synchrotron radiation [21, 6].

To compute synthetic synchrotron emission maps, we follow a procedure outlined below.

- Our simulations produce 3D distribution of pressure, density, velocity and magnetic field.
- Using proper p and $\boldsymbol{B}'$ we calculate the synchrotron emissivity according to Ref. [6].
- For given local synchrotron emissivity we then choose a given line of sight and integrate the emissivity, assuming optically thin regime and taking into account the local velocity and the corresponding Doppler factor.

To make clearly visible the X-ray morphology, we use different quantities to produce the synthetic maps, depending on the orientation of the line-of-sight. Namely, we found that the emission intensity maps are more illustrative in the case if the pulsar moves toward the observer. If the pulsar moves side way, then we plot the square-root of the emission intensity. The latter quantity is somewhat arbitrary, chosen to allow a better highlighting of faint X-ray features.

In Fig. 5 we show synthetic synchrotron maps for the considered configuration. If such a system is seen along the axis, then the morphology is similar to the frisbee case [6]. If the PWN is seen side way, then the PWN appears as an asymmetric bow-shock structure (Fig. 5, right panel) - the front/up "jet" is brighter and narrower as compared to the down/back "jet". Thus, this simulations allow us to reproduce the X-ray structure of PWN formed by PSR J1135−6055: a single-sided bent jet and an extended tail (see Fig. 1) [7]).

The considered RMHD model also explains the apparent low X-ray efficiency of the bow-shock PWNe [23, 2]. Bow-shock PWNes are very inefficient in converting spin-down luminosity in X-rays, with efficiencies $\sim 10^{-3} - 10^{-5}$ [2]. These values are typically much smaller than for slow-moving pulsars, where the conversion efficiency can be as high as tens of per cents *e.g.* [1].

Such low effectiveness can be explained by strong adiabatic loses $L_{\rm X-ray}/\dot{E}_{\rm sd} \sim t_{\rm AD}/t_{\rm SYN} < 1/1000$ in the head of PWNe. In bow-shock PWNe the crossing time of the relativistic plasma through the tail is shorter than the synchrotron cooling time. As a results, particles are able to emit only small fraction of the energy that they acquired during acceleration at the reverse shock or in reconnection sites - adiabatic cooling dominates and most of the wind luminosity is spent on pdV work inflating the bubble at the large distance from a pulsar.

This model may also explain the apparent disagreement between the estimate of the magnetic field from equipartition arguments (even initially weakly magnetized flow after the shock transition reaches approximate equipartition) and the observed length of the X-ray tail [23] - particles are quickly adiabatically cooled.

4. Conclusion

In this work we present analytical and numerical 3D MHD calculations of the interaction of relativistic wind produced by fast moving pulsars relative to the interstellar medium in highly asymmetric geometry. We capture both the flow dynamics in the head part of the resulting bow-shock PWNe, as well as the evolution of the flow in the tail part. Our results indicate that magnetic fields play the most important role in shaping the morphology of the bow-shock PWNe. We also compute synthetic synchrotron intensity maps for frisbee – rifle-bullet geometry. Our synchrotron emissivity maps can reproduce the key features revealed with Chandra X-ray observation of PSR J1135−6055: a single-sided bent jet and an extended tail. Finally, our model naturally addresses the small X-ray efficiencies seen from many bow-shock PWNe. The rate of adiabatic losses appeared to be significantly shorter than the synchrotron cooling time, thus the bulk of the spin-down losses is spent on mechanical work needed for PWN expansion.

Acknowledgments

We would like to thank Anatoliy Spitkovsky, Joseph Gelfand, Oleg Kargaltsev, Victoria Kaspi, Andrey Bykov and Mallory Roberts for numerous enlightening discussions.

The calculations were carried out in the CFCA cluster of National Astronomical Observatory of Japan. We thank the *PLUTO* team for the possibility to use the *PLUTO* code and for technical support. The visualization of the results performed in the VisIt package [24]. This work had been supported by NSF grant AST-1306672, DoE grant DE-SC0016369 and NASA grant 80NSSC17K0757. D.K. acknowledges support by JSPS KAKENHI Grant Number JP18H03722.

References

[1] Kargaltsev O, Cerutti B, Lyubarsky Y and Striani E 2015 SSRv **191** 391–439 (*Preprint* 1507.03662)
[2] Kargaltsev O, Pavlov G G, Klingler N and Rangelov B 2017 Journal of Plasma Physics **83** 635830501 (*Preprint* 1708.00456)
[3] Wilkin F P 1996 ApJ **459** L31
[4] Bucciantini N 2002 A&A **387** 1066–1073 (*Preprint* astro-ph/0203504)
[5] Vigelius M, Melatos A, Chatterjee S, Gaensler B M and Ghavamian P 2007 MNRAS **374** 793–808 (*Preprint* astro-ph/0610454)
[6] Barkov M V and Lyutikov M 2018 ArXiv e-prints (*Preprint* 1804.07327)
[7] Marelli M 2012 ArXiv e-prints (*Preprint* 1205.1748)
[8] Michel F C 1973 ApJ **180** 207–226
[9] Bogovalov S V 1999 A&A **349** 1017–1026 (*Preprint* arXiv:astro-ph/9907051)
[10] Bogovalov S V and Khangoulian D V 2002 MNRAS **336** L53–L55 (*Preprint* astro-ph/0209269)
[11] Komissarov S S and Lyubarsky Y E 2004 MNRAS **349** 779–792
[12] Coroniti F V 1990 ApJ **349** 538–545
[13] Lyubarsky Y E 2003 MNRAS **345** 153–160 (*Preprint* astro-ph/0306435)

IOP Conf. Series: Journal of Physics: Conf. Series **1225** (2019) 012001 doi:10.1088/1742-6596/1225/1/012001

[14] Mignone A, Bodo G, Massaglia S, Matsakos T, Tesileanu O, Zanni C and Ferrari A 2007 ApJS **170** 228–242 (*Preprint* arXiv:astro-ph/0701854)
[15] Harten A 1983 Journal of Computational Physics **49** 357393 ISSN 0021-9991 URL http://www.sciencedirect.com/science/article/pii/0021999183901365
[16] Porth O, Komissarov S S and Keppens R 2014 MNRAS **438** 278–306 (*Preprint* 1310.2531)
[17] Komissarov S S 2013 MNRAS **428** 2459–2466 (*Preprint* 1207.3192)
[18] Barkov M, Lyutikov M, Klingler N and Bordas P 2018 in preparation
[19] Kennel C F and Coroniti F V 1984 ApJ **283** 710–730
[20] Vaidya B, Mignone A, Bodo G, Rossi P and Massaglia S 2018 ApJ **865** 144 (*Preprint* 1808.08960)
[21] Barkov M and Bosch-Ramon V 2018 in preparation
[22] Atoyan A M and Aharonian F A 1996 MNRAS **278** 525–541
[23] Kaspi V M, Gotthelf E V, Gaensler B M and Lyutikov M 2001 ApJ **562** L163–L166 (*Preprint* astro-ph/0110188)
[24] Hank Childs H, Brugger E, Whitlock B and et al 2012 VisIt: An End-User Tool For Visualizing and Analyzing Very Large Data High Performance Visualization–Enabling Extreme-Scale Scientific Insight pp 357–372

IOP Conf. Series: Journal of Physics: Conf. Series **1225** (2019) 012002 doi:10.1088/1742-6596/1225/1/012002

Thermonuclear (Type Ia) Supernovae and Progenitor Evolution

A C Calder[1,2], D E Willcox[1,3], C J DeGrendele[1], D Shangase[1,4], M Zingale[1], and D M Townsley[5]

[1] Department of Physics and Astronomy, Stony Brook University, Stony Brook, NY 11794-3800, USA

[2] Institute for Advanced Computational Science, Stony Brook University, Stony Brook, NY 11794-5250, USA

[3] Center for Computational Sciences and Engineering, Lawrence Berkeley National Lab, Berkeley, CA 94720 USA

[4] Department of Physics, University of Michigan, Ann Arbor, MI 48109-1040, USA

[5] Department of Physics and Astronomy, University of Alabama, Tuscaloosa, AL 35487-0324, USA

E-mail: alan.calder@stonybrook.edu

Abstract. Thermonuclear (type Ia) supernovae are bright stellar explosions with the unique property that the light curves can be standardized, allowing them to be used as distance indicators for cosmological studies. Many fundamental questions bout these events remain, however. We provide a critique of our present understanding of these and present results of simulations assuming the single-degenerate progenitor model consisting of a white dwarf that has gained mass from a stellar companion. We present results from full three-dimensional simulations of convection with weak reactions comprising the A=23 Urca process in the progenitor white dwarf.

1. Introduction

Thermonuclear (type Ia) supernovae are bright stellar explosions thought to occur when approximately one solar mass of stellar material composed principally of C and O burns under degenerate conditions. The light curves of these events have the unique property that the rate of decline from maximum light is inversely proportional to the overall brightness. This "brighter is broader" relationship allows these events to be standardized and used as distance indicators [1]. Despite highly successful use of these for cosmology [2–4], the setting of these events is still subject to debate.

At present, there are three progenitor systems widely under study: the "single degenerate" scenario in which a C/O white dwarf (WD) gains mass from a companion and explodes as it approaches a limiting mass, the "double degenerate" scenario in which two WDs merge and subsequently explode, and the "double detonation" model in which a detonation in an accreted He layer triggers a detonation in the underlying C/O WD [5–8, and references therein]. In this work, we explore the evolution of the progenitor white dwarf as it approaches the limiting mass in the single degenerate scenario.

IOP Conf. Series: Journal of Physics: Conf. Series **1225** (2019) 012002 doi:10.1088/1742-6596/1225/1/012002

2. The Effects of the History and Composition of the WD Progenitor

The capability of modern computing allows development of models with a vast amount of included physics and thus simulations of thermonuclear supernovae with unprecedented realism. Correspondingly, advances in observational technology allow a synthesis between observation and theory that allows study of the connections between progenitors and the observed light curves and spectra [9, 10]. Both approaches allow studying the efficacy of proposed progenitor systems, and within a given progenitor system allow the study of systematic effects on the brightness of an event due to age and composition.

Considerable research on systematic effects has been performed in the single degenerate paradigm. Research has addressed how properties of the progenitor WD that should follow from the age and metallicity of the host galaxy influence the outcome of explosions, including the composition of progenitor white dwarf [11–15] and the internal structure of the progenitor white dwarf [12, 16–18]. Also, as we will illustrate immediately below, research has applied the latest in stellar evolution theory to produce new types of progenitors and explored the results of explosions occurring in these new progenitors [19–21].

As an example of contemporary research, we present highlights from a study of explosions from a new class of progenitors. These progenitors arise from recent results in stellar evolution indicating that convection in the interior of a late-time AGB star can strongly influence the nuclear burning. Studies suggest that convective boundary layers may inhibit burning in regions of the interior with the result that supernova progenitors can be a "hybrid" between traditional C/O and O/Ne white dwarfs [22–24]. Our group investigated supernova explosions from these progenitors using the hybrid C/O/Ne progenitor models of Denissenkov [22, 23] and compared the results to an earlier study of explosions from traditional C/O progenitors [16, 17].

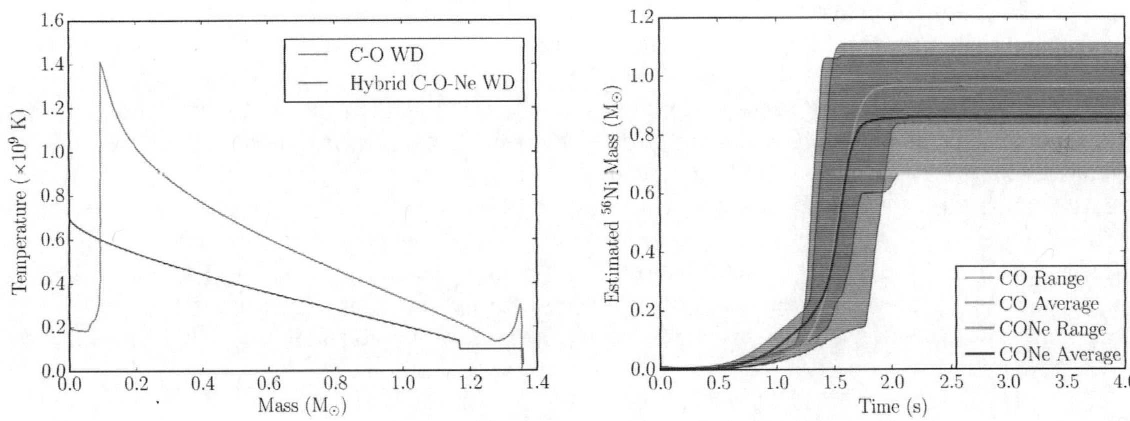

Figure 1. Left panel: Temperature profiles of the hybrid C-O-Ne white dwarf (red) and a reference C-O white dwarf (blue). Right panel: Evolution of the estimated ^{56}Ni yields for C-O and Hybrid C-O-Ne WD realizations. The ensemble-averaged value among C-O realizations is shown in yellow, with the full range of values at any point in time for the C-O realizations shown in red. Likewise, the ensemble-averaged value among C-O-Ne realizations is shown in blue and their range of values shown in green. Panels from [21] © AAS. Reproduced with permission.

The left panel of Figure 1 shows the initial temperature profiles of the hybrid progenitors and ^{56}Ni explosion yields. A unique feature of these hybrid progenitors is that the temperature peak is predicted to lie near a mass coordinate of 0.15 $M_{\odot}$ due to the composition and energy generation structures during the accretion phase. Thus while the overall structure of these hybrid progenitors is similar to traditional progenitors, the ignition occurs farther out in the star and leads to a slightly different way of burning with the inner core even being consumed last in some simulations. The right panel of Figure 1 presents the evolution of estimated ^{56}Ni yields from a traditional progenitor and a hybrid progenitor. The results of the study indicated that explosions from these progenitors are largely consistent with explosions from traditional progenitors, but indicated less kinetic energy in ejecta and slightly less ^{56}Ni is synthesized on average, which suggests slightly dimmer events. Complete results from the study may be found in [21].

Motivated by these earlier studies of systematic effects on the brightness of explosions, we initiated a study of conditions in the progenitor as it approaches ignition. At this point in the evolution, relatively slow fusion reactions occur that change the composition and drive convection. If conditions are right, reactions can occur and energy can be lost by a process known as the convective Urca process. Below we present an initial look at simulations of this process in the progenitor white dwarf.

3. The Convective Urca Process

The Urca process is a nuclear decay cycle of alternating electron capture and beta decay on the same nucleus with the net effect of producing neutrinos. An electron capture

$$e^- + (Z, A) - \quad (Z-1, A) + \nu_e \tag{1}$$

is followed by a beta decay

$$(Z-1, A) - \quad (Z, A) + e^- + \bar{\nu}_e \ , \tag{2}$$

both of which emit a neutrino or an anti-neutrino [25]. The rates of these depend on the local density, temperature, and composition, with electron captures occurring at densities high enough for the electron energy to exceed the threshold for electron capture and beta decays occurring at densities low enough that there is significant phase space for the emitted electron to occupy.

The convective Urca process, introduced by Paczynski [26], occurs in the single degenerate scenario when the WD has gained enough mass to compress the core and start slow C fusion reactions that drive a period of simmering prior to the runaway. Convection carries a warm blob of material up to lower densities were it cools and beta decays and then descends to capture electrons and heat. Neutrinos emitted during this cycle escape the stellar material ($\rho \lesssim 5 \times 10^9$ g cm^{-3}) and transport energy away, cooling the star. Several pairs of nuclei can experience the convective Urca process, but in this study we consider only the A = 23 pair, ^{23}Na/^{23}Ne.

Paczynski introduced the Urca process as an energy loss mechanism to delay the thermonuclear runaway of the WD. Subsequent analytical work by Bruenn accounted for heating due to the relaxation of the electron distribution in regions of higher density than the electron capture threshold and argued that the convective Urca process would not stabilize the WD core against thermonuclear runaway [27]. Subsequent studies using analytic results and one-dimensional stellar evolution models [28–34] showed that the convective Urca process is complex and stressed the need for multidimensional simulations.

Stein and Wheeler performed two-dimensional simulations of the convective Urca process with the DWARF implicit hydrodynamics code in order to evaluate the interaction between electron capture, β^--decay, and convection, but the simulations were limited to a wedge of the star [35]. The study found that under some conditions, the A = 23 Urca process limited the extent of the convection zone to the region of the star interior to the density at which the electron capture and beta decay rates balance. This density occurs at a particular radius of the star and defines the Urca shell. The result of Stein and Wheeler was important because the extent of the convection zone across the Urca shell influences the energy evolution of the core and it is not possible to constrain with one-dimensional models.

Contemporary studies performed with detailed stellar evolution simulation codes must make assumptions about the extent of convection into the Urca shell. Denissenkov et al., for example, address the problem by performing simulations with different assumptions about the structure of the convection zone around the A = 23 Urca shell and compare the results to the accepted evolution path (c.f. Figure 9) [23]. Considerable work in stellar evolution included the convective Urca process in one-dimensional calculations [22, 23, 36–39], but the restriction of one dimension requires assumptions about turbulent convection and the interaction with the Urca shell, an inherently three-dimensional phenomenon. The results we present here, a preliminary investigation of the convective A = 23 Urca process with full three-dimensional simulations, is meant to better our understanding of this process and hence better characterize the conditions in the simmering white dwarf progenitors of a thermonuclear supernovae.

4. Simulation Instrument

The simulation instrument we used for this study was the Maestro code, which is part of the AMReX Astrophysics Suite of adaptive mesh refinement hydrodynamics codes for reactive astrophysical flows [40]. The hydrodynamics in Maestro is designed for low Mach number flows as are found in applications like convection. Maestro uses a projection method to filter out acoustic waves but accounts for compressibility effects due to energy generation and gravitational stratification [41–43]. The method thus allows the time step to be set by the advection speed and not the sound speed, which allows for considerably larger time steps than traditional hydrodynamics methods.

Maestro has been applied to a host of astrophysics problems including an earlier study of WD convective simmering on which this study of the Urca process is based [44]. Other applications include type I X-ray bursts [45, 46] and convection in helium shells of sub-Chandrasekhar mass WDs in the double-detonation progenitor model for SNIa [47, 48]. In the simulations of the convective Urca process we present here, we use a velocity "sponge" to damp the velocities in the low-density gas above the star [44] so such regions do not restrict the time step. Also, in these simulations the electron chemical potential contribution to energy generation is included as a heating source after each Strang-split reaction integration but not as a heating source for temperature evolution during the reaction integration. Because MAESTRO calculates local temperature from density and either pressure or enthalpy, we expect this arrangement to have only a slight effect compared to the case of including the contribution to the temperature evolution. We will address this discrepancy with additional simulations.

5. Results

The simulations of the convective Urca process were performed in progenitor WDs near the Chandrasekhar mass. As the progenitor star simmers, the temperature in the core increases, which changes the extent of the convective zone. These preliminary simulations were meant to characterize the extent of the convective zone and allow us to determine the central temperatures and densities of the progenitor models that allow the convective zone to reach the A = 23 Urca shell. We show results from a simulation with central temperature 5.5×10^8 K, central density $4.5 \times 10^9\ \mathrm{g\,cm^{-3}}$, and total mass of $1.395\ M_\odot$.

We constructed hydrostatic models by choosing the central density, temperature, and species mass fractions and integrating the equation of hydrostatic equilibrium radially outward to obtain the initial spherical base state for Maestro. In each zone of the initial model, we solve for the mass fractions of ^{23}Na and ^{23}Ne such that the electron capture and beta decay contributions to $\dot{\mathrm{X}}(^{23}\mathrm{Na})$ and $\dot{\mathrm{X}}(^{23}\mathrm{Ne})$ summed to zero while maintaining a constant value of $\mathrm{X}(^{23}\mathrm{Na}) + \mathrm{X}(^{23}\mathrm{Ne})$. This approach to setting the initial mass fractions ensured that the A = 23 Urca species were initially in equilibrium with each other in the hydrostatic background state. We also supplement the equation of hydrostatic equilibrium by requiring the initial model have an isentropic region interior to mass coordinate 0.5 $M_\odot$, with the outer region isothermal. This condition allows us to capture the effects of a core convection zone in the initial model thermodynamics. For the central conditions described above, we used a composition consisting of $X(^{12}C) = 0.39975$, $X(^{16}O) = 0.59975$, and $X(^{23}Na) + X(^{23}Ne) = 5 \times 10^{-4}$.

We initialized this model in Maestro using the multipole initialization method of [44] for the core velocity field and ran the two simulations at 5 km and 2.5 km effective resolution (3 and 4 spatial levels) to obtain the results shown here. We evolved these simulations for 842 and 627 seconds, respectively, corresponding to 5–10 convective eddy turnover times. In Figure 2, we show representative radially binned isotope production rates for ^{23}Na and ^{23}Ne in our two models, indicating radial zones for which $\dot{X} > 0$. In these plots, we attribute the spikes at small radii to binning artifacts and note that at both resolutions the production rates for both ^{23}Na and ^{23}Ne are large within the core of the WD out to around 100 km in radius, with $\dot{X}(^{23}Na) > \dot{X}(^{23}Ne)$. In this region within 100 km, ^{12}C-burning produces ^{23}Na, which subsequently captures an electron to produce ^{23}Ne. As the ^{12}C burning rate drops between 100 km and 200 km in radius, we see the corresponding sharp decline in ^{23}Na production. The region between 160 km and 1000 km is divided into two shells at 400 km by the A = 23 Urca shell, where for r < 400 km, electron captures onto ^{23}Na produce ^{23}Ne. The ^{23}Na participating in this reaction is produced in the core from ^{12}C fusion and transported into the core from above the Urca shell via convection. For the outer shell where r > 400 km, ^{23}Ne brought upwards by convection undergoes beta decay to ^{23}Na. The bulk features presented in this radial binning appear qualitatively identical at the two resolutions, suggesting that the interaction between convection and reactions across the Urca shell is likely a physical effect and not an artifact of resolution in these simulations.

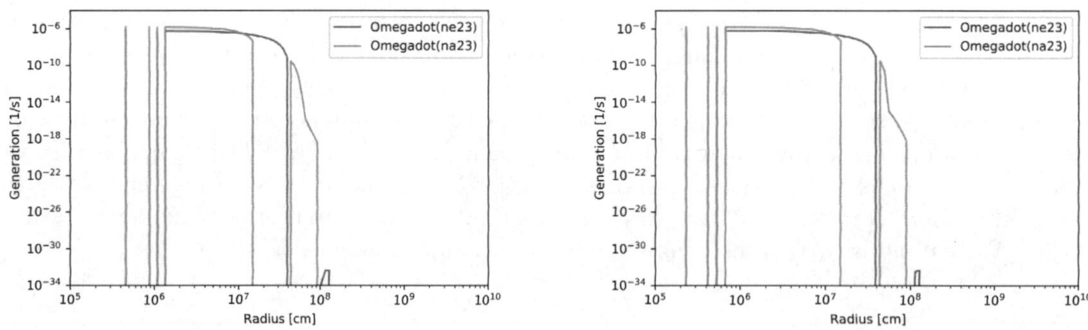

Figure 2. Plots of the production of ^{23}Ne and ^{23}Na within a white dwarf star as a function of radius. Left panel: Results at 3 levels of refinement, corresponding to an effective resolution of 5 km at a simulation time of 443.9 seconds. Right panel: Results at 4 levels of refinement, corresponding to an effective resolution of 2.5 km at a simulation time of 445.0 seconds.

The energy generation structure arising from these reactions is shown in Figure 3 for the 2.5 km-resolution model. The region of electron capture onto ^{23}Na within approximately 400 km in radius is evident and includes the central ^{12}C burning region where the sum of reaction energy is exothermic. Between the exothermic burning core and the Urca shell lies an endothermic shell region where in the absence of burning electron captures occur either onto ^{23}Na produced in the core and transported upwards or on ^{23}Na transported downwards via convection from above the Urca shell. Outside the Urca shell the beta decay of ^{23}Ne creates a thin exothermic shell structure

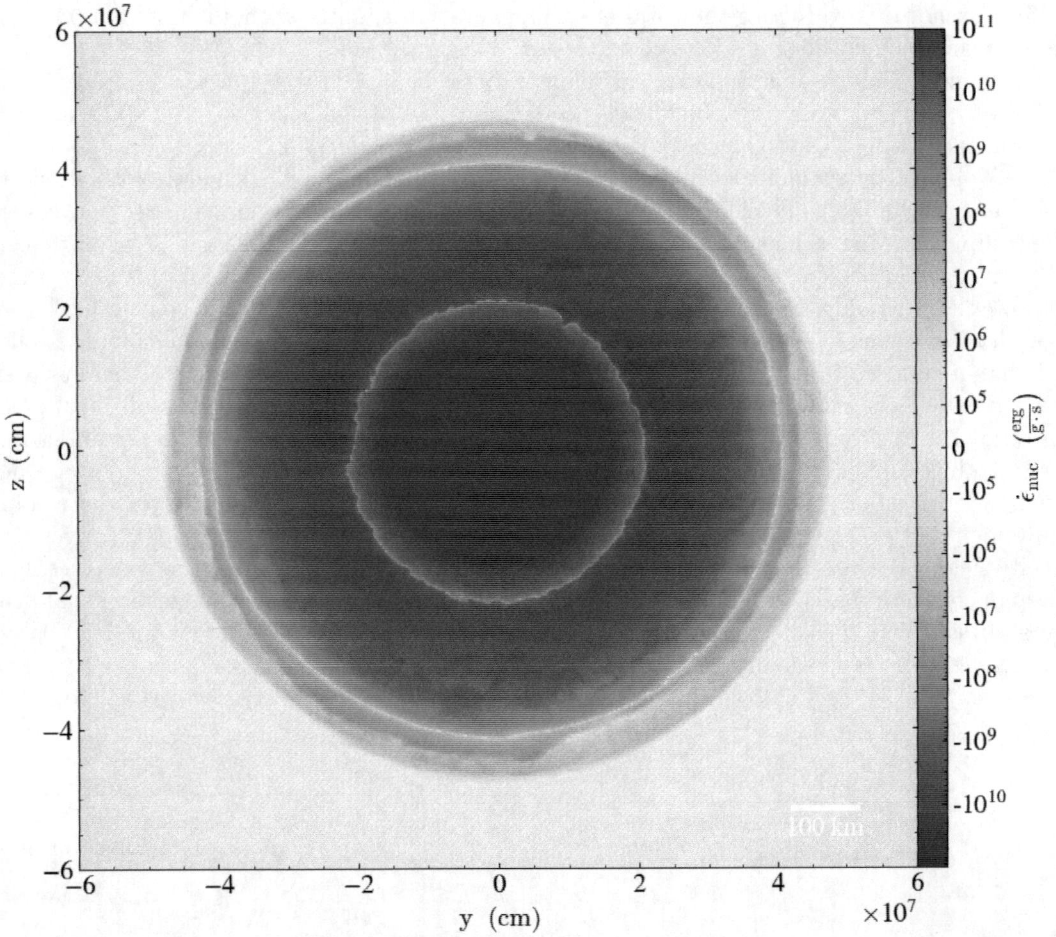

Figure 3. Slice through the core of a WD at $x = 0$ showing the specific energy generation rate as a result of ^{12}C burning and the A = 23 Urca reactions. The WD shown has a central density of 4.5×10^9 g cm^{-3}, central temperature of 5.5×10^8 K, and is spatially resolved to 2.5 km in the energy generating regions shown. Positive specific energy generation is shown in green, and negative specific energy generation is shown in magenta.

at the upper edge of the convection zone. We note that while the bulk shell structure is spherical, the local energy generation rate can vary significantly depending on the local convective flow. This is to be expected from the convective Urca process, in which species transport across the Urca shell and the ensuing reactions are tightly coupled to convection.

The global energy generation arising from this shell structure is shown in Figure 4 (left), where we plot the energy generation rate (in ergs s^{-1}) integrated from $r = 0$ to $r = R$ as a function of radius R. The resolutions we explored yield consistent energy generation rates and we see that the peak energy generation integral occurs just outside the central ^{12}C burning region and decreases due to the endothermic electron captures between the core and the Urca shell. Exothermic beta decays outside the Urca shell do not appear significant to the overall energy generation rate of the WD. We attribute this result to both the limited extent of the convection zone above the Urca shell and the energy loss to neutrinos during both the electron capture and beta decay reactions. We also note that the net effect of the Urca process under the conditions shown is to reduce the total energy generation slightly without having a net cooling effect on the WD, consistent with the results of [35].

We show in Figure 4 (right) the turbulent kinetic energy power spectra computed for the models at 5 km and 2.5 km resolution at simulation times of 445.0 seconds and 443.9 seconds in blue and red, respectively. Over a large range of wavenumbers the 2.5 km resolution simulation follows Kolmogorov scaling better than the coarser 5 km simulation, although both simulations depart from Kolmogorov scaling at the largest wavenumbers. These kinetic energy power spectra were computed using a density weighting of $\rho^{1/3}$ for the spatial velocities and a cutoff density of 10^9 g cm^{-3} below which the velocity field is set to zero. The motivation for this is to compute the power spectra only over the portion of the domain corresponding to the convective WD core since the low density, non-convective regions of the WD exhibit spatial noise in the velocity field that may arise from unresolved gravity waves. For the 5 km simulation, there is a small region between the convective core and the cutoff density containing such unphysical velocity noise that contributes to the excess energy at high wavenumber. In the 2.5 km simulation, the corresponding region contains mostly gravity waves, which likely account for the energy deficit at high wavenumber. We are currently exploring this issue to understand better the resolution requirements of gravity waves in this model and the effect of unresolved gravity waves on the velocity field. We note that energy generation appears consistent between these two resolutions although convection is coupled to energy generation via the Urca process, suggesting that gravity waves outside the convective zone are not critical to the Urca process in this model. That will not in general be true for different central conditions and convection zone sizes, where it may be necessary for models with a smaller convection zone to resolve gravity waves above the Urca shell.

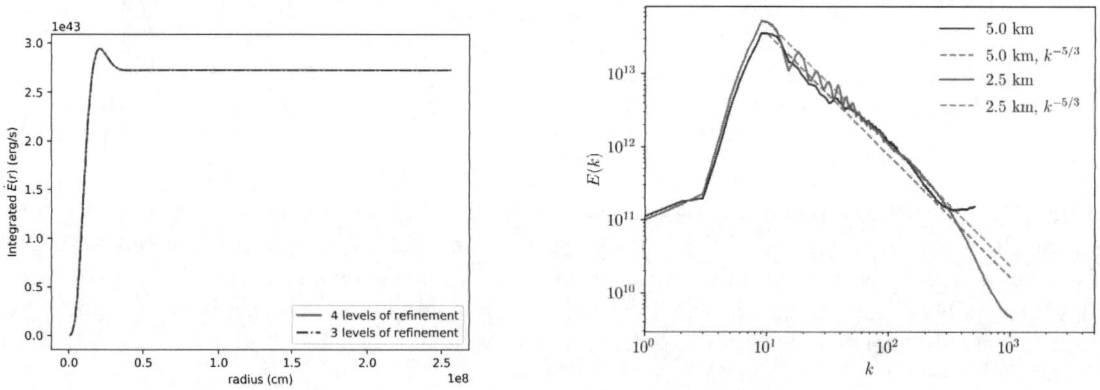

Figure 4. Left panel: Plot of integrated energy generation as a function of radius. Right panel: Plot of the turbulent kinetic energy power spectrum of the white dwarf. Red represents a simulation that was run with 4 levels of refinement corresponding to an effective resolution of 2.5 km. Blue represents a simulation run with 3 levels of refinement corresponding to an effective resolution of 5 km. The Kolmogorov scalings are plotted for these resolutions as dashed lines only to guide the eye and were matched to the simulation results at the tenth spectral point.

6. Conclusions

With the results presented here, we have working models for further investigation into the convective Urca process. These models produce energy and react as we expect from previous studies and, importantly, we were able to produce progenitors in which the convective region reaches the Urca shell, thus ensuring an active convective Urca process. The models passed basic tests such as showing consistency between the two resolutions even though the power spectra show the simulations may not fully capture the effects of turbulence at high wavenumbers, and the energy generation and cooling are consistent with earlier work. We conclude, therefore, that these models are reasonable for further study of the convective Urca process. Future work will include other pairs, e.g. A = 25.

Acknowledgments
Maestro is freely available on GitHub (`https://github.com/AMReX-Astro/`), and all problem setup files for the calculations presented here are in the code repository. This work was supported in part by the US Department of Energy under grant DE-FG02-87ER40317. The reaction networks were generated using the pynucastro library [49]. Results in this paper were obtained using the high-performance computing system at the Institute for Advanced Computational Science at Stony Brook University. Support also came from the Data + Computing = Discovery summer program at the Institute for Advanced Computational Science at Stony Brook University, supported in part by the NSF under a supplement to award AST-1211563. An award of computer time was provided by the Innovative and Novel Computational Impact on Theory and Experiment (INCITE) program. This research used resources of the Oak Ridge Leadership Computing Facility at the Oak Ridge National Laboratory, which is supported by the Office of Science of the U.S. Department of Energy under Contract No. DE-AC05-00OR22725. This research used resources of the National Energy Research Scientific Computing Center, which is supported by the Office of Science of the U.S. Department of Energy under Contract No. DE-AC02-05CH11231. Visualizations and part of this analysis made use of yt [50]. The authors thank Ilana Bromberg and Brianna Isola for previewing this manuscript.

References

[1] Phillips M M 1993 *ApJ* **413** L105

[2] Riess A G, Filippenko A V, Challis P, Clocchiatti A, Diercks A, Garnavich P M, Gilliland R L, Hogan C J, Jha S, Kirshner R P, Leibundgut B, Phillips M M, Reiss D, Schmidt B P, Schommer R A, Smith R C, Spyromilio J, Stubbs C, Suntzeff N B and Tonry J 1998 *AJ* **116** 1009 (*Preprint* `astro-ph/9805201`)

[3] Perlmutter S, Aldering G, Goldhaber G, Knop R A, Nugent P, Castro P G, Deustua S, Fabbro S, Goobar A, Groom D E, Hook I M, Kim A G, Kim M Y, Lee J C, Nunes N J, Pain R, Pennypacker C R, Quimby R, Lidman C, Ellis R S, Irwin M, McMahon R G, Ruiz-Lapuente P, Walton N, Schaefer B, Boyle B J, Filippenko A V, Matheson T, Fruchter A S, Panagia N, Newberg H J M, Couch W J and The Supernova Cosmology Project 1999 *ApJ* **517** 565 (*Preprint* `astro-ph/9812133`)

[4] Leibundgut B 2001 *Annu. Rev. Astron. Astrophys* **39** 67–98

[5] Calder A C, Krueger B K, Jackson A P and Townsley D M 2013 *Frontiers of Physics* **8** 168–188 (*Preprint* `1303.2207`)

[6] Maoz D, Mannucci F and Nelemans G 2014 *Annual Reviews of Astronomy and Astrophysics* **52** 107–170 (*Preprint* `1312.0628`)

[7] Seitenzahl I R and Townsley D M 2017 *Handbook of Supernovae* ed Alsabti A W and Murdin P (Springer International Publishing AG) p 1955

[8] Röpke F K and Sim S A 2018 *Space Sci. Rev.* **214** 72 (*Preprint* `1805.07268`)

[9] Höflich P, Dragulin P, Mitchell J, Penney B, Sadler B, Diamond T and Gerardy C 2013 *Frontiers of Physics* **8** 144–167

[10] Hillebrandt W, Kromer M, Röpke F K and Ruiter A J 2013 *Frontiers of Physics* **8** 116–143 (*Preprint* `1302.6420`)

[11] Jackson A P, Calder A C, Townsley D M, Chamulak D A, Brown E F and Timmes F X 2010 *ApJ* **720** 99–113 (*Preprint* 1007.1138)

[12] Bravo E, Althaus L G, García-Berro E and Domínguez I 2011 *A&A* **526** A26 (*Preprint* 1010.5098)

[13] Ohlmann S T, Kromer M, Fink M, Pakmor R, Seitenzahl I R, Sim S A and Röpke F K 2014 *A&A* **572** A57 (*Preprint* 1409.2866)

[14] Miles B J, van Rossum D R, Townsley D M, Timmes F X, Jackson A P, Calder A C and Brown E F 2016 *ApJ* **824** 59 (*Preprint* 1508.05961)

[15] Leung S C and Nomoto K 2018 *ApJ* **861** 143 (*Preprint* 1710.04254)

[16] Krueger B K, Jackson A P, Townsley D M, Calder A C, Brown E F and Timmes F X 2010 *ApJ* **719** L5–L9 (*Preprint* 1007.0910)

[17] Krueger B K, Jackson A P, Calder A C, Townsley D M, Brown E F and Timmes F X 2012 *ApJ* **757** 175 (*Preprint* 1208.1986)

[18] Seitenzahl I R, Ciaraldi-Schoolmann F and Röpke F K 2011 *MNRAS* **414** 2709–2715 (*Preprint* 1012.4929)

[19] Kromer M, Ohlmann S T, Pakmor R, Ruiter A J, Hillebrandt W, Marquardt K S, Röpke F K, Seitenzahl I R, Sim S A and Taubenberger S 2015 *MNRAS* **450** 3045–3053 (*Preprint* 1503.04292)

[20] Bravo E, Gil-Pons P, Gutiérrez J L and Doherty C L 2016 *A&A* **589** A38 (*Preprint* 1603.00641)

[21] Willcox D E, Townsley D M, Calder A C, Denissenkov P A and Herwig F 2016 *ApJ* **832** 13 (*Preprint* 1602.06356)

[22] Denissenkov P A, Herwig F, Truran J W and Paxton B 2013 *ApJ* **772** 37 (*Preprint* 1305.2649)

[23] Denissenkov P A, Truran J W, Herwig F, Jones S, Paxton B, Nomoto K, Suzuki T and Toki H 2015 *MNRAS* **447** 2696–2705 (*Preprint* 1407.0248)

[24] Brooks J, Schwab J, Bildsten L, Quataert E and Paxton B 2017 *ApJ* **834** L9 (*Preprint* 1611.03061)

[25] Clayton D 1983 *Principles of Stellar Evolution and Nucleosynthesis* (University of Chicago Press) ISBN 9780226109534

[26] Paczyński B 1972 *Astrophysical Letters* **11** 53 ISSN 0004-6388

[27] Bruenn S W 1973 *The Astrophysical Journal Letters* **183** L125 ISSN 0004-637X

[28] Couch R G and Arnett W D 1973 *ApJL* **180** L101

[29] Couch R G and Arnett D W 1974 *ApJ* **194** 537–542

[30] Couch R G and Arnett W D 1975 *The Astrophysical Journal* **196** 791–803 ISSN 0004-637X

[31] Lazareff B 1975 *Astronomy and Astrophysics* **45** 141–143 ISSN 0004-6361

[32] Iben Jr I 1978 *ApJ* **219** 213–225

[33] Iben Jr I 1978 *ApJ* **226** 996–1033

[34] Iben Jr I 1982 *ApJ* **253** 248–259

[35] Stein J and Wheeler J C 2006 *ApJ* **643** 1190–1197 (*Preprint* astro-ph/0512580)

[36] Lesaffre P, Podsiadlowski P and Tout C A 2005 *MNRAS* **356** 131–144

[37] Martínez-Rodríguez H, Piro A L, Schwab J and Badenes C 2016 *The Astrophysical Journal* **825** 57 ISSN 0004-637X

[38] Schwab J, Bildsten L and Quataert E 2017 *Monthly Notices of the Royal Astronomical Society* **472** 3390–3406 ISSN 0035-8711, 1365-2966 arXiv: 1708.07514

[39] Schwab J, Martínez-Rodríguez H, Piro A L and Badenes C 2017 *The Astrophysical Journal* **851** 105 ISSN 1538-4357 arXiv: 1711.04780

[40] 2018 Amrex `https://amrex-codes.github.io/` freely available

[41] Almgren A S, Bell J B, Rendleman C A and Zingale M 2006 *ApJ* **637** 922–936 (*Preprint* `arXiv:astro-ph/0509892`)

[42] Almgren A S, Bell J B, Rendleman C A and Zingale M 2006 *ApJ* **649** 927–938 (*Preprint* `arXiv:astro-ph/0606496`)

[43] Almgren A S, Bell J B, Nonaka A and Zingale M 2008 *ApJ* **684** 449–470

[44] Zingale M, Almgren A S, Bell J B, Nonaka A and Woosley S E 2009 *ApJ* **704** 196–210 (*Preprint* `0908.2668`)

[45] Malone C M, Nonaka A, Almgren A S, Bell J B and Zingale M 2011 *The Astrophysical Journal* **728** 118 ISSN 0004-637X

[46] Malone C M, Zingale M, Nonaka A, Almgren A S and Bell J B 2014 *The Astrophysical Journal* **788** 115 ISSN 0004-637X

[47] Zingale M, Nonaka A, Almgren A S, Bell J B, Malone C M and Orvedahl R J 2013 *The Astrophysical Journal* **764** 97 ISSN 0004-637X

[48] Jacobs A M, Zingale M, Nonaka A, Almgren A S and Bell J B 2016 *The Astrophysical Journal* **827** 84 ISSN 0004-637X

[49] Willcox D E and Zingale M 2018 *Journal of Open Source Software* **3** 588 URL `https://doi.org/10.21105/joss.00588`

[50] Turk M J, Smith B D, Oishi J S, Skory S, Skillman S W, Abel T and Norman M L 2011 *ApJS* **192** 9 (*Preprint* `1011.3514`)

IOP Conf. Series: Journal of Physics: Conf. Series **1225** (2019) 012003 doi:10.1088/1742-6596/1225/1/012003

Radiation-Hydrodynamic Simulations of Core-collapse Supernovae with 6 Dimensional Boltzmann Neutrino Transport

Hiroki Nagakura[1,2], Shun Furusawa[3], Hajime Togashi[4], Kohsuke Sumiyoshi[5] and Shoichi Yamada[6]

[1]Department of Astrophysical Sciences, Princeton University, Princeton NJ 08544
[2]TAPIR, Walter Burke Institute for Theoretical Physics, Mailcode 350-17, California Institute of Technology, Pasadena, CA 91125, USA
[3]Interdisciplinary Theoretical and Mathematical Sciences Program (iTHEMS), RIKEN 2-1 Hirosawa, Wako, Saitama 351-0198, Japan
[4]Nishina Center for Accelerator-based Science, RIKEN 2-1 Hirosawa, Wako, Saitama 351-0198, Japan
[5]Numazu College of Technology, Ooka 3600, Numazu, Shizuoka 410-8501, Japan
[6]Advanced Research Institute for Science & Engineering, Waseda University, 3-4-1 Okubo, Shinjuku, Tokyo 169-8555, Japan

E-mail: [1]`hirokin@caltech.edu`

Abstract. Core-collapse supernovae (CCSNe) are intrinsically multi-scale, multi-physics and multi-dimensional phenomena. Because of the enormous complexity, the first-principles numerical simulations under realistic input physics are strongly required to uncover the explosion mechanism, predict observational signals (neutrinos, gravitational waves and electromagnetic waves) and prove physical state in extremely hot and dense matter of supernova core. We have tackled the development of multi-dimensional radiation-hydrodynamic code with full Boltzmann neutrino transport and performed several scientific CCSNe simulations in the last few years. In this article, we report the recent progress of our CCSNe numerical simulations with the most up-to-date equation-of-state (EOS) and nuclear weak interactions. We also present preliminary results of non-rotating CCSNe simulations in spatial axisymmetry.

1. Introduction

The explosion mechanism of core-collapse supernova (CCSN) is one of long-standing problems in our universe. CCSNe are intrinsically multi-scale, multi-physics and multi-dimensional phenomena with there being a complex interplay of all four fundamental forces in nature. First-principle numerical simulations are the most promising approach to reveal the nature of the explosion mechanism. They have made rapid progress with ever increasing computational resources in the last few years. Indeed, nowadays sophisticated three-dimensional CCSNe simulations can be carried out with sufficient long-term in post-bounce phase [1, 2, 3, 4, 5, 6, 7] and some of these simulations demonstrated the success of shock revival. In addition to this, under axisymmetry in space, CCSNe simulations with full Boltzmann neutrino transport are also possible on $\sim$ 10 PFLOPS supercomputer facilities and our group achieved such simulations [8]. We keep improving our code towards general relativistic three-dimensional

CCSNe simulations with full Boltzmann neutrino transport, which will be performed in the exa-scale era of computational facilities.

Nuclear equation of states (EOSs) and weak interactions in hot and densed matter are other key ingredients for core-collapse supernovae (CCSNe). Thermodynamical quantities given by EOS play a primary role for hydrodynamics. Weak interactions also dictate exchanging energy, momentum and leptons between neutrinos and matter, which affect the dynamics of CCSNe. On the other hand, accurate treatments of EOS and weak interactions on CCSNe simulations are very challenging issues, for instance, we have not yet fully undersood the property at supranuclear densities at the moment due to many difficulties on theories, observations and experiments (see a recent review by [9]); the description of nuclear abundances relevant to CCSNe has not yet been well established due to the appearance of extremely heavy nuclei where no experimental data are available. The ambiguity of nuclear state obscures the effect of neutrino-nucleus interactions, which implicitly but significantly influence on CCSNe.

Very recently, we improved the input physics by developing self-consistent treatments of nuclear EOS and weak interactions. We employ an up-to-date EOS whose uniform matter is computed based on the variational method with including realistic nuclear force. For inhomogeneous nuclear matter, the full ensemble nuclei are taken into account with various finite-density and thermal effects. The abundance of nuclei is used to compute reaction rates of nuclear weak interactions such as electron captures of heavy nuclei and electron and positron captures of light nuclei. In this article, we briefly report the essence of these improvements and then present recent results of CCSNe simulations with incorporating these improvements.

2. Multi-nuclear VM EOS

In this section, we describe the essence of physics in multi-nuclear variational method (VM) EOS. The free energy density is calculated with the variational many-body theory. The Hamiltonian consists of the kinetic, the AV18 two-body potential [10] and the UIX three-body potential [11, 12] terms. This formulation is extended to non-uniform nuclear matter at sub-nuclear densities to compute the free energy density of unbound nucleons that drip from heavy nuclei, $f_{n,p}$, as well as the bulk energies of heavy nuclei, E_{AZ}^{bulk}. The free energy density of non-uniform nuclear matter is given by

$$\begin{aligned} f &= \eta f_{n,p} + \sum_{AZ} n_{AZ} \times \\ &\left\{ \kappa T \left[\ln\left(\frac{n_{AZ}}{g_{AZ}(M_{AZ}T/2\pi\hbar)^{3/2}} \right) - 1 \right] + M_{AZ} \right\}, \end{aligned} \tag{1}$$

where n_{AZ}, M_{AZ}, and g_{AZ} are the number density, mass, and internal degrees of freedom for nucleus with the mass number A and atomic number Z. The excluded volume effects for nucleons and nuclei are defined as $\eta = 1/[1 - \sum_{AZ}(n_{AZ}A/n_{sAZ})]$ and $\kappa = n_B/n_0$, respectively. The quantities n_0 is the nuclear saturation density of symmetric matter. We also take into account the dependencies of iso-spin of each nucleus, Z/A, and temperature on the saturation densities of heavy nuclei, n_{sAZ}.

The mass model for heavy nuclei with atomic numbers $6 \leq Z \leq 1000$ and mass numbers $A \leq 2000$ is expressed as

$$M_{AZ} = E_{AZ}^{bulk} + E_{AZ}^{surf} + E_{AZ}^{Coul} + E_{AZ}^{shell}. \tag{2}$$

The surface energy E_{AZ}^{surf} and Coulomb energy E_{AZ}^{Coul} depend on number densities of uniformly-distributed dense electrons, dripped nucleons and on shape changes of heavy nuclei from normal droplets to bubbles just below nuclear normal density. The shell energies of heavy nuclei E_{AZ}^{shell},

IOP Conf. Series: Journal of Physics: Conf. Series **1225** (2019) 012003 doi:10.1088/1742-6596/1225/1/012003

which represent microscopic effects such as neutron- and proton-magic numbers and pairing, are taken from the experimental and theoretical mass data [13, 14]. The temperature dependencies of E_{AZ}^{shell} and g_{AZ} are also phenomenologically taken into account. The mass model of the light nuclei with $Z < 6$ is given by $M_{AZ} = M_{data} + \Delta E_{AZ}^{Coul} + \Delta E_{AZ}^{Pauli} + \Delta E_{AZ}^{self}$ where M_{data} is the experimental mass, ΔE_{AZ}^{Coul} is Coulomb energy shift, and a quantum approach is incorporated to evaluate Pauli-energy shift ΔE_{AZ}^{Pauli} and self-energy shift ΔE_{AZ}^{self}.

3. Electron and positron captures

We calculate weak interaction rates of nuclei which are evaluated under the consistent treatment with the nuclear abundances provided by the EOS. For some nuclei, we use the reaction data in [15, 16, 17, 18], which are based on the shell model or its extension. It should be noted, however, that these theoretical computations do not cover the full nuclei which appear in CCSNe. We adopt an analytical formula as a function of the Q-value [16] for the neutron-rich and/or heavy nuclei with the unavailable data:

$$\lambda_{AZ} = \frac{(\ln 2)B}{K}\left(\frac{T}{m_e c^2}\right)^5 \times \left[F_4(\eta_{AZ}) - 2\chi_{AZ}F_3(\eta_{AZ}) + \chi_{AZ}^2 F_2(\eta_{AZ})\right], \quad (3)$$

where $K = 6146$ sec, $\chi_{AZ} = (Q_{AZ} - \Delta E)/T$, $\eta_{AZ} = (\mu_e + Q_{AZ} - \Delta E)/T$ with the electron chemical potential μ_e, and F_k is the relativistic Fermi integrals of order k. The Q-value is defined as $Q_{AZ} = M_{AZ} - M_{A,Z-1}$, where the finite temperature and density effects on the nuclear masses given by Eq. (2) are included with being consistent with the nuclear abundances on EOS tables. We use the detailed balance relation to evaluate the rate of neutrino absorption on heavy nuclei.

As for light nuclei, we include the following weak interactions in our simulation, referring to [19],

$$\text{(elpp)}: \quad \nu_e + {}^2\text{H} \longleftrightarrow e^- + p + p, \quad (4)$$
$$\text{(ponn)}: \quad \bar{\nu_e} + {}^2\text{H} \longleftrightarrow e^+ + n + n, \quad (5)$$
$$\text{(el2h)}: \quad \nu_e + n + n \longleftrightarrow e^- + {}^2\text{H}, \quad (6)$$
$$\text{(po2h)}: \quad \bar{\nu_e} + p + p \longleftrightarrow e^+ + {}^2\text{H}, \quad (7)$$
$$\text{(el3he)}: \quad \nu_e + {}^3\text{H} \longleftrightarrow e^- + {}^3\text{He}, \quad (8)$$
$$\text{(po3h)}: \quad \bar{\nu_e} + {}^3\text{He} \longleftrightarrow e^+ + {}^3\text{H}. \quad (9)$$

For neutrino absorptions on deuterons of Eqs. (4) and (5), we use the data of vacuum cross section [20]. To account for the medium modification on deuteron mass such as ΔE_{AZ}^{Coul}, ΔE_{AZ}^{Pauli}, and ΔE_{AZ}^{self}, we introduce the shifted neutrino injection energies as $E_\nu^* = E_\nu + m_{^2\text{H}}^* - m_{^2\text{H}}$ with masses of deuteron in medium, $m_{^2\text{H}}^*$, and in vacuum, $m_{^2\text{H}}$. The in-medium mass is evaluated by the same mass model in the EOS. As an example, the neutrino absorption rate of Eq. (4) is expressed as

$$1/\lambda(E_\nu) = n_{^2\text{H}} \int \mathrm{d}p_e \left[\frac{\mathrm{d}\sigma_{\nu^2\text{H}}}{\mathrm{d}p_e}[E_\nu^*]\right](1 - f_e(E_e)], \quad (10)$$

where $n_{^2\text{H}}$ is deuteron number density and f_e denotes the Fermi-Dirac distribution of electrons. The rate of the electron/positron capture on two nucleons forming a deuteron (leftward reactions of Eqs. (4) and (5)) can be evaluated through the detailed balance with the rate of the absorption.

Electron captures on deuterons of Eqs. (6) and (7) can be estimated with the assumption that the matrix elements of electron and positron captures are equivalent to those of neutrino absorptions for Eqs. (4) and (5) as

$$\frac{\mathrm{d}\sigma_{e^2\text{H}}}{\mathrm{d}p_\nu} \sim \frac{1}{2}\frac{\mathrm{d}\sigma_{\nu^2\text{H}}}{\mathrm{d}p_e}, \quad (11)$$

where $\frac{1}{2}$ comes from the difference in spin degrees of freedom between neutrino and electron. This is a reasonable approximation for CCSNe conditions, in which injection energies of leptons are not greatly large and, hence, the energy deposit to the relative motion between two nucleons is negligible. Three bound nucleons, ^{3}H and ^{3}He, interact with neutrinos via breakup or charge exchange, the latter of which is dominant neutrino opacity source. Therefore, we treat only the charge exchange reaction as described in Eqs. (8) and (9). Those rates are calculated as

$$1/\lambda(E_\nu) = n_{^3\mathrm{H}} \left[\frac{G_F^2 V_{ud}^2}{\pi(\hbar c)^4}\right] p_e E_e [1 - f_e(E_e)] B(GT), \tag{12}$$

where $n_{^3\mathrm{H}}$ is triton number density, $B(GT) = 5.87$, and $V_{ud} = 0.967$. Here, in-medium effects on nuclear masses are not taken into account.

4. Full Boltzmann code for CCSNe

All simulations presented in this article are performed by our multi-dimensional (multi-D) neutrino-radiation-hydrodynamic code [21, 22]. We solve Boltzmann equations for neutrino transport. The basic equation of our Boltzmann solver starts with the conservative form of the Boltzmann equation in general relativity:

$$\begin{aligned}
&\frac{1}{\sqrt{-g}} \frac{\partial}{\partial x^\alpha}\bigg|_{q_i} \left[\left(e^\alpha_{(0)} + \sum_{i=1}^{3} \ell_i e^\alpha_i\right)\sqrt{-g} f\right] \\
&- \frac{1}{\nu^2}\frac{\partial}{\partial \nu}(\nu^3 f \omega_{(0)}) + \frac{1}{\sin\bar{\theta}}\frac{\partial}{\partial \bar{\theta}}(\sin\bar{\theta} f \omega_{(\bar{\theta})}) \\
&+ \frac{1}{\sin^2\bar{\theta}}\frac{\partial}{\partial \bar{\phi}}(f \omega_{(\bar{\phi})}) = S_{\mathrm{rad}},
\end{aligned} \tag{13}$$

where g, x^α are the determinant of the metric, coordinates of space-time, respectively, and f is the neutrino distribution function; $e^\alpha_{(\mu)} (\mu = 0, 1, 2, 3)$ denote a set of the tetrad bases for a local orthonormal frame; ℓ_i are directional cosines for the direction of neutrino propagation with respect to $e^\alpha_{(i)}$. The three components of ℓ_i can be written as

$$\begin{aligned}
\ell_{(1)} &= \cos\bar{\theta}, \\
\ell_{(2)} &= \sin\bar{\theta}\cos\bar{\phi}, \\
\ell_{(3)} &= \sin\bar{\theta}\sin\bar{\phi},
\end{aligned} \tag{14}$$

where $\bar{\theta}$ and $\bar{\phi}$ stand for the polar and azimuthal angles. We further define coordinates q_i in momentum space- $q_1 = \nu, q_2 = \bar{\theta}$ and $q_3 = \bar{\phi}$ with ν being the neutrino energy in this local orthonormal frame and also expressed as $\nu \equiv -p_\alpha e^\alpha_{(0)}$ with the four momentum of neutrino, p^α. In this paper neutrinos are assumed to be massless. $\omega_{(0)}, \omega_{(\bar{\theta})}, \omega_{(\bar{\phi})}$ are given as

$$\begin{aligned}
\omega_{(0)} &\equiv \nu^{-2} p^\alpha p_\beta \nabla_\alpha e^\beta_{(0)}, \\
\omega_{(\bar{\theta})} &\equiv \sum_{i=1}^{3} \omega_i \frac{\partial \ell_{(i)}}{\partial \bar{\theta}}, \\
\omega_{(\bar{\phi})} &\equiv \sum_{i=2}^{3} \omega_i \frac{\partial \ell_{(i)}}{\partial \bar{\phi}}, \\
\omega_i &\equiv \nu^{-2} p^\alpha p_\beta \nabla_\alpha e^\beta_{(i)}.
\end{aligned} \tag{15}$$

These ω's can be expressed with the Ricci rotation coefficients. $S_{\rm rad}$ on the right hand side of Eq. (13) originates from the collision term for neutrino-matter interactions.

In the 3+1 formulation of GR, the line element is expressed as

$$ds^2 = (-\alpha^2 + \beta^k \beta_k)dt^2 + 2\beta_i dt dx^i + \gamma_{ij} dx^i dx^j, \tag{16}$$

where α, β^i and γ_{ij} denote the lapse function, shift vector and spatial 3-metric, respectively. In our extended Boltzmann code, the time-like basis $e^{\alpha}_{(0)}$ is chosen so that it should coincide with the unit vector n^{α} normal to the spatial hypersurface with $t = $ const. This choice is a natural extension from our previous SR Boltzmann solver. Then three other spatial tetrad bases are taken so that they should be tangential to the spatial hypersurface. In this paper we assume that the space-time is flat and is foliated with flat spatial hypersurfaces, on which we deploy the polar coordinates $(x^1 = r, x^2 = \theta, x^3 = \phi)$. Then non-vanishing components of the 3 metric are $\gamma_{rr} = 1, \gamma_{\theta\theta} = r^2$ and $\gamma_{\phi\phi} = r^2 \sin\theta^2$. The spatial tetrad bases are chosen so that the $e_{(1)}$ be parallel to the radial coordinate, and $e_{(2)}$ be tangential to the surface spanned by ∂_t and ∂_θ, and $e_{(3)}$ be orthogonal to the other two:

$$\begin{aligned}
e^{\alpha}_{(1)} &= (0, \gamma_{rr}^{-1/2}, 0, 0) \\
e^{\alpha}_{(2)} &= \left(0, -\frac{\gamma_{r\theta}^{-1/2}}{\sqrt{\gamma_{rr}(\gamma_{rr}\gamma_{\theta\theta} - \gamma_{r\theta}^2)}}, \sqrt{\frac{\gamma_{rr}}{\gamma_{rr}\gamma_{\theta\theta} - \gamma_{r\theta}^2}}, 0\right) \\
e^{\alpha}_{(3)} &= \left(0, \frac{\gamma^{r\phi}}{\sqrt{\gamma^{\phi\phi}}}, \frac{\gamma^{\theta\phi}}{\sqrt{\gamma^{\phi\phi}}}, \sqrt{\gamma^{\phi\phi}}\right).
\end{aligned} \tag{17}$$

We employ the central scheme for numerically solving the Newtonian hydrodynamic equations (see also [23]). The direct method of matrix inversion are employed to solve Poisson equation of Newtonian gravity. For all simulations, we employ the same weak interaction rates as those used in [8] except for electron and positron captures on nuclei. In our simulations, we cover the spatial domain of $0 \le r \le 5000$km by spherical polar coordinate. We deploy 384 grid points for radial direction and 128 grid points for lateral direction. In momentum space of neutrinos, we use 20, 10 and 6 grid points for energy ($0 \le \epsilon \le 300$MeV), lateral angular grid ($0° \le \bar{\theta} \le 180°$) and azimuthal angular grid in neutrino momentum space, respectively. In this study, we employ 11.2 solar masses progenitor in [24].

5. Result

The shock wave produced at core bounce expands rather gradually with time, but the prompt convection pushes the shock wave further outward. The increase of shock radius enhances the neutrino heating, which further accelerates the shock expansion. As shown in Figure 1, multi-dimensional fluid instabilities such as neutrino driven convection and standing accretion shock instability (SASI) distorts the post-shock flow and then the morphology of post-shock flows is much different from spherical symmetry. Around $t \sim 150$ms the shock wave goes through rapid expansion in the north side due to kinetic attack by turbulent motions behind the shock wave, and then it reaches in $r \sim 600$km around $t \sim 200$ms (see Figure 2).

Overall dynamics is very similar as LS model in [8] which also shows the sign of shock revival. Our results also share with other CCSNe simulations the qualitative trend that s11.2 progenitor explodes more easily than other more massive progenitors. It is important to emphasize that the inner structure of proto-neutron star (PNS) is different from those of LS model [8]. We also find that weak interactions between neutrinos and light nuclei quantitatively affect both neutrino cooling and heating. More detailed analysis of these effects are currently underway which will be presented in the forthcoming paper.

IOP Conf. Series: Journal of Physics: Conf. Series **1225** (2019) 012003 doi:10.1088/1742-6596/1225/1/012003

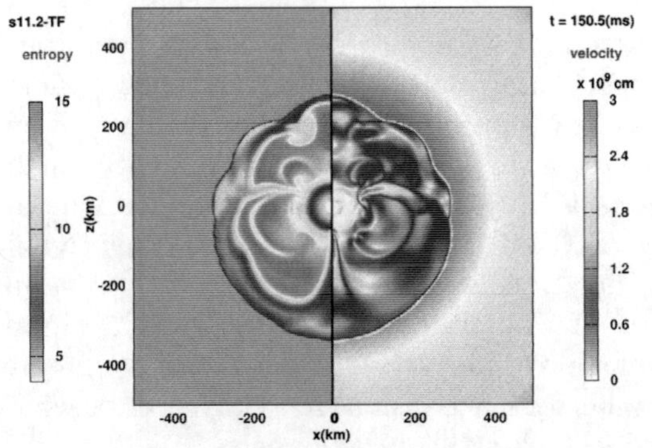

Figure 1. Entropy (left) and velocity (right) map at $t = 150$ms after bounce.

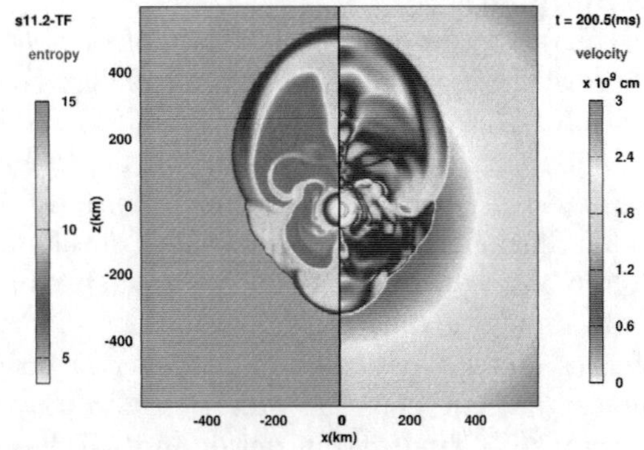

Figure 2. Same as Figure 1 but for $t = 200$ms after bounce.

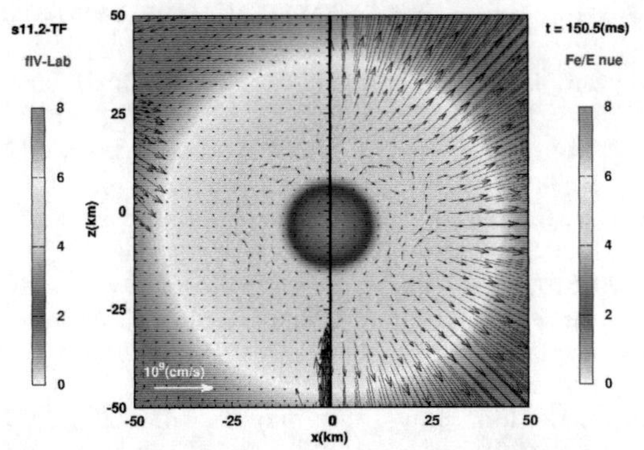

Figure 3. Entropy contour with fluid velocity vector (left) and electron-type neutrino flux vector (right) at $t = 150$ms after bounce.

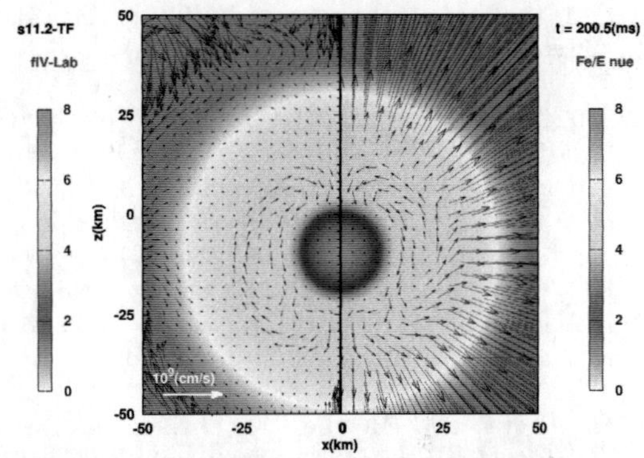

Figure 4. Same as Figure 3 but for $t = 200$ms after bounce. The offset of PNS can be clearly seen.

Although the global shock dynamics is similar as LS model in [8], we obtain a new phenomenon in this simulation, which is a recoil of PNS. The recoil of PNS can be seen by comparing between Figures 3 and 4. As mentioned before, the shock wave expands rapidly in the north side, which gives a linear momentum to PNS towards south side. In our CCSNe code the shift vector is utilized to specify the movement of spatial coordinates so that they could track the PNS motion approximately (see [22] in more details).

Although the detailed analysis of PNS recoil is currently underway, the asymmetric gravitational field gives the driving force of the kick. The asymmetric neutrino pressure in the optically thick region also play an important role to determine the acceleration of PNS, meanwhile other previous simulations took ad-hock prescriptions inside of PNS. Thanks to the moving mesh technique, we self-consistently treat both global asymmetry of shock expansion and a recoil of PNS. It should be noted however, our CCSNe simulations do not guarantee a conservation of total linear momentum. We are currently improve our code in order to reduce a violation of linear momentum, which will be also presented in the forth coming paper.

6. Summary

In this article, we make a report of our recent progress of CCSNe project. We substantially improve the input physics such as nuclear EOS and weak interactions recently and then perform spatially axisymmetric simulations of 11.2 solar masses progenitor. We find that the shock wave expands with $\ell = 1$ asymmetry which triggers the recoil of PNS. The recoil of PNS can be consistently handled by our moving mesh technique in full Boltzmann radiation-hydrodynamic code. The moving mesh technique is developed based on the 3+1 formalism of general relativistic Boltzmann equation, indeed, we make use of shift gauge to trace the motion of PNS. Our result is the first evidence of appearance of PNS recoil in CCSNe simulations with elaborate input physics and sophisticated treatment of neutrino transport. The recoil of PNS would generate the asymmetric neutrino radiation and may affect the shock expansion in CCSNe, all of which are currently analyzed in detail and will be reported in our forthcoming papers.

References

[1] T. Takiwaki, K. Kotake and Y. Suwa, Astrophys. J. **749**, 98 (2012) doi:10.1088/0004-637X/749/2/98 [arXiv:1108.3989 [astro-ph.HE]].
[2] E. J. Lentz *et al.*, Astrophys. J. **807**, no. 2, L31 (2015) doi:10.1088/2041-8205/807/2/L31 [arXiv:1505.05110 [astro-ph.SR]].
[3] T. Kuroda, T. Takiwaki and K. Kotake, Astrophys. J. Suppl. **222**, no. 2, 20 (2016) doi:10.3847/0067-0049/222/2/20 [arXiv:1501.06330 [astro-ph.HE]].
[4] B. Müller, T. Melson, A. Heger and H.-T. Janka, Mon. Not. Roy. Astron. Soc. **472**, no. 1, 491 (2017) doi:10.1093/mnras/stx1962 [arXiv:1705.00620 [astro-ph.SR]].
[5] C. D. Ott, L. F. Roberts, A. da Silva Schneider, J. M. Fedrow, R. Haas and E. Schnetter, Astrophys. J. **855**, no. 1, L3 (2018) doi:10.3847/2041-8213/aaa967 [arXiv:1712.01304 [astro-ph.HE]].
[6] E. O'Connor and S. Couch, Astrophys. J. **865**, no. 2, 81 (2018) [arXiv:1807.07579 [astro-ph.HE]].
[7] D. Vartanyan, A. Burrows, D. Radice, A. Skinner and J. Dolence, arXiv:1809.05106 [astro-ph.HE].
[8] H. Nagakura *et al.*, Astrophys. J. **854**, no. 2, 136 (2018) doi:10.3847/1538-4357/aaac29 [arXiv:1702.01752 [astro-ph.HE]].
[9] M. Oertel, M. Hempel, T. Klähn and S. Typel, Rev. Mod. Phys. **89**, no. 1, 015007 (2017) doi:10.1103/RevModPhys.89.015007 [arXiv:1610.03361 [astro-ph.HE]].
[10] R. B. Wiringa, V. G. J. Stoks and R. Schiavilla, Phys. Rev. C **51**, 38 (1995) doi:10.1103/PhysRevC.51.38 [nucl-th/9408016].
[11] J. Carlson, V. R. Pandharipande and R. B. Wiringa, Nucl. Phys. A **401**, 59 (1983). doi:10.1016/0375-9474(83)90336-6
[12] B. S. Pudliner, V. R. Pandharipande, J. Carlson and R. B. Wiringa, Phys. Rev. Lett. **74**, 4396 (1995) doi:10.1103/PhysRevLett.74.4396 [nucl-th/9502031].
[13] G. Audi, M. Wang, A. H. Wapstra, F. G. Kondev, M. MacCormick and X. Xu, Nucl. Data Sheets **120**, 1 (2014). doi:10.1016/j.nds.2014.06.126
[14] Koura H, Tachibana T, Uno M and Yamada M. 2005. Crossref:. doi:10.1143/PTP.113.305 Prog.Theor.Phys.,113,305-25
[15] K. Langanke and G. Martinez-Pinedo, Nucl. Phys. A **673**, 481 (2000) doi:10.1016/S0375-9474(00)00131-7 [nucl-th/0001018].
[16] K. Langanke *et al.*, Phys. Rev. Lett. **90**, 241102 (2003) doi:10.1103/PhysRevLett.90.241102 [astro-ph/0302459].
[17] T. Oda, M. Hino, K. Muto, M. Takahara and K. Sato, Atom. Data Nucl. Data Tabl. **56**, 231 (1994). doi:10.1006/adnd.1994.1007
[18] G. M. Fuller, W. A. Fowler and M. J. Newman, Astrophys. J. **252**, 715 (1982). doi:10.1086/159597
[19] T. Fischer, G. Martínez-Pinedo, M. Hempel, L. Huther, G. Röpke, S. Typel and A. Lohs, EPJ Web Conf. **109**, 06002 (2016) doi:10.1051/epjconf/201610906002 [arXiv:1512.00193 [astro-ph.HE]].
[20] S. Nakamura, T. Sato, V. P. Gudkov and K. Kubodera, Phys. Rev. C **63**, 034617 (2001) Erratum: [Phys. Rev. C **73**, 049904 (2006)] doi:10.1103/PhysRevC.73.049904, 10.1103/PhysRevC.63.034617 [nucl-th/0009012].
[21] H. Nagakura, K. Sumiyoshi and S. Yamada, Astrophys. J. Suppl. **214**, no. 2, 16 (2014) doi:10.1088/0067-0049/214/2/16 [arXiv:1407.5632 [astro-ph.HE]].
[22] H. Nagakura, W. Iwakami, S. Furusawa, K. Sumiyoshi, S. Yamada, H. Matsufuru and A. Imakura, Astrophys. J. Suppl. **229**, no. 2, 42 (2017) doi:10.3847/1538-4365/aa69ea [arXiv:1605.00666 [astro-ph.HE]].
[23] H. Nagakura, H. Ito, K. Kiuchi and S. Yamada, Astrophys. J. **731**, 80 (2011) doi:10.1088/0004-637X/731/2/80 [arXiv:1009.2326 [astro-ph.HE]].

[24] S. E. Woosley, A. Heger and T. A. Weaver, Rev. Mod. Phys. **74**, 1015 (2002). doi:10.1103/RevModPhys.74.1015

IOP Conf. Series: Journal of Physics: Conf. Series **1225** (2019) 012004 doi:10.1088/1742-6596/1225/1/012004

Modeling subgrid combustion processes in simulations of thermonuclear supernovae

Dean M Townsley[1], **Alan C Calder**[2] **and Broxton J Miles**[3]

[1] Department of Physics and Astronomy, University of Alabama, Tuscaloosa, AL 35487-0324, USA

[2] Department of Physics and Astronomy, Stony Brook University, Stony Brook, NY 11794-3800, USA

[3] Department of Physics, North Carolina State University, Raleigh, NC 27695-8202 USA

E-mail: Dean.M.Townsley@ua.edu

Abstract. Supernovae of type Ia are thought to arise from the thermonuclear incineration of a carbon-oxygen white dwarf stellar remnant. However, the detailed explosion scenario and stellar evolutionary origin scenario – or scenarios – which lead to observed supernovae are still quite uncertain. One of the principal tests of proposed scenarios is comparison with the explosion products inferred, for example, from the spectrum of the supernovae. Making this comparison requires computation of the combustion dynamics and products through simulation of proposed scenarios. Here we discuss two specific proposed explosion scenarios, the deflagration-detonation transition and the helium shell double detonation, With these two examples in mind, we proceed to discuss challenges to computational modeling of the combustion taking place in these explosions. Both subsonically and supersonically propagating reaction fronts are discussed, called deflagrations and detonations respectively. Several major stages of the combustion occur on length and time scales that are many orders of magnitude smaller than those accessible in simulations of the explosion. Models which attempt to capture this sub-grid behavior and the verification of those models is briefly discussed.

1. Introduction

Type Ia supernovae have spectral properties, including a lack of hydrogen or helium and a strong silicon feature, that distinguish them as a well-characterized class. Thermonuclear incineration of a white dwarf star can explain these astrophysical stellar transients, with radioactive decay of nickel providing the power in their optically bright phase. Here we discuss two example scenarios from those proposed and use these to then motivate a brief discussion of the challenges met in simulating these scenarios due to the scales of the combustion fronts involved.

2. Examples of proposed scenarios

While it is fairly clear that Type Ia supernovae are produced by the incineration of a carbon-oxygen rich white dwarf star, there are a variety of suggested scenarios for how this might occur. (See [1, 2] for recent reviews.) Specific scenarios generally have both advantages and shortcomings, and each is at a different level of development in modeling of its detailed predictions. Here we will briefly discuss just two scenarios, with the aim of motivating the need to perform accurate and verifiable simulations that can be used to test these predictions against observations. Many aspects of these scenarios depend on the combustion mode taking place and these will be discussed more fully in section 3. Each combustion mode burns either helium or carbon and oxygen to heavier elements in a localized reaction front. However while a deflagration front will move subsonically, a detonation will move supersonically, and this aspect will cause them to have very different impacts on the overall explosion.

2.1. Deflagration-detonation transition

White dwarfs have a convenient natural ignition mechanism due to the Chandrasekhar mass limit. Near this limiting mass, around 1.4 $M_\odot$, due to the electrons which support the star becoming relativistic, the central density becomes a singular function of the star's mass. Thus if mass is added to a white dwarf star, some nuclear process will eventually take place due to the compression near the center of the star. In the case of a carbon-oxygen rich interior, the process that takes place first is carbon fusion, which releases energy, creating a runaway that eventually incinerates the star. However, there is a basic problem with this scenario. If a Chandrasekhar mass white dwarf is incinerated, it makes almost purely iron-group material, whereas observed type Ia supernovae have half of the mass of their ashes in silicon group material.

The requisite silicon-group material can be created if the combustion occurs slowly enough that the white dwarf can expand to lower densities before all of it burns. The first broadly successful thermonuclear supernova model, the W7 case from [3], did this somewhat artificially with a tunable parameter. The deflagration-detonation transition model [4] proposes that the burning begins slowly in the subsonic deflagration mode, and then transitions to the faster detonation mode. This allows the star to expand enough to give ejected yields similar to those observed in actual objects [5].

However there are many assumptions that must be made for this scenario to work, and several have turned out to be genuinely challenging to make realistic. There are many issues, but we will highlight just a few. It is not easy to construct stellar systems in which the white dwarf will increase in mass in the necessary way. Also, the distribution of ignition points of the deflagration stage needs to be fairly symmetric, but recent simulations of this process find that it should not be so [6, 7]. Additionally, while 1-dimensional models provided a good match to the variety of explosions observed, recent simulations of multi-dimensional models indicate that the relation between brightness and decline rate is opposite that observed [8, 9]. Thus while something like the deflagration-detonation transition model has been favored for some time, and remains a leading scenario, there are reasons to investigate others as well.

2.2. Helium shell double detonation

As noted above, Chandreskhar mass white dwarfs, if incinerated, will make more iron-group and less silicon-group elements than observed in type Ia supernovae. Lower mass white dwarfs, when incinerated quickly by, for example, a detonation, will make a balance of element more similar to that observed [10]. However, a viable ignition mechanism for sub-Chandrasekhar mass white dwarfs is more elusive and not as natural as in the Chandrasekhar-mass case. One possible ignition scenario is called the helium shell double detonation. In this scenario, burning starts in an accreted helium layer, which only needs to be a few hundredths of the star's mass to ignite. After a detonation sweeps through the helium shell, a shock is left propagating inward through the carbon-oxygen core, which eventually focuses enough to ignite detonation of the rest of the star. When first proposed, this scenario was thought unlikely because a shell thick enough to host a detonation would make a titanium and nickel-rich outer layer that is not observed in supernovae [11].

More recently, as the deflagration-detonation transition scenario has run into challenges, the double detonation scenario has been revisited. It was found that if the helium layer burning turned out to be less complete than previously predicted, this would make a viable scenario [12].

However, it appeared that very strong enrichment of the helium layer by underlying carbon would be needed to make a layer that would host a detonation while not producing too much titanium and chromium. Since that work it has been found that by including more complete reactions in the simulation and a critical nitrogen isotope in the accreted material, this may be a workable scenario [13]. It is also promising that simulations of the early stages of the merger process of two white dwarfs show that ignition of the helium layer may be a generic feature of white dwarf mergers [14].

3. Modeling small-scale reaction structure in burning

Challenges in accurately simulating thermonuclear supernovae arise largely from the wide range of scales that the combustion and explosion processes cover. To simulate the explosion, the whole star, 10^8 cm, must be simulated. However, combustion processes occur on scales as small as a micron. Here we will discuss briefly how these challenges are met for the two major modes of combustion active in thermonuclear supernovae: deflagration and detonation.

3.1. Deflagration

We use the term deflagration to indicate reaction fronts that propagate due to thermal diffusion. In thermonuclear supernovae these fronts typically burn carbon and oxygen to silicon or iron group elements. Their speed is determined by the balance of heat diffusion and reaction and has been computed in both steady state and hydrodynamic calculations [15, 16]. In quiet, i.e. laminar, flow, this flame has a 1-dimensional structure that proceeds from unburned to burned across a multi-stage reaction front. This is typically thought of as a flame surface when the reaction stages are short compared to the lateral extent of the flow geometry.

In the presence of shear flow the flame surface will become wrinkled as indicated in figure 1. Turbulent shear is the type of shearing flow of primary interest at subgrid scales in thermonuclear supernovae. A distinctive feature of turbulence is that shearing occurs on all scales in a mostly random way, though the shearing strength depends on scale in a predictable way. Figure 2 compares the strength of the shearing field due to Kolmogorov turbulence to the properties of the flame approximately midway through a thermonuclear supernova. Several important scales are labeled. The grid size at which typical simulations are performed is a few 10^5 cm. The width of the carbon burning stage of the laminar flame is many orders of magnitude smaller than this, of order 10^{-2} cm. Each of these scales is indicated with a vertical line.

The spectrum of turbulent shear on many scales will generally act to wrinkle the flame surface. This action, however, does depend on a competition between the flame propagation and the eddy turnover. Considering the flow shown in Figure 1, it is clear that if the flame surface propagates faster than the eddy overturns, the eddy will not effectively wrinkle the flame. Additionally, wrinkled flames tend to self-smooth by burning out small-scale structure. As a result of this action, there is a scale below which the flame surface will not be effectively wrinkled, termed the Gibson scale, and indicated by l_G in Figure 2. This is the scale below which the flame speed is faster than the typical shear velocity of a turbulent eddy.

Capturing and modeling this structure of subgrid turbulence flame interaction is done with a model, which inherently has some free parameters that must be chosen [17, 18]. The typical strategy is to treat some coarsened structure at the grid scale, which represents the coarse propagation of the reaction front. The speed at which this coarsened structure propagates and other aspects of its behavior are then controlled with a model. This model will then include various approximations and features intended to capture the physics of the subgrid turbulence-flame interaction. The overall effect is much like accelerating the flame propagation, leading to a turbulent flame speed $s_t = \Xi s_\ell$, where s_ℓ is the laminar flame speed and Ξ is a wrinkling factor enhancing the propagation.

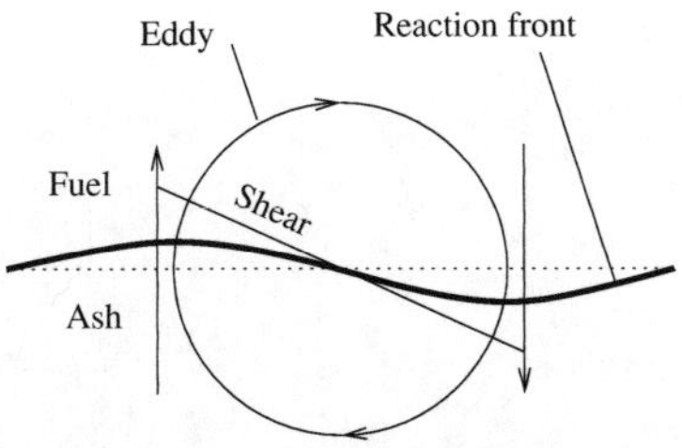

Figure 1. Action of an eddy in the flow to shear and wrinkle the reaction front surface.

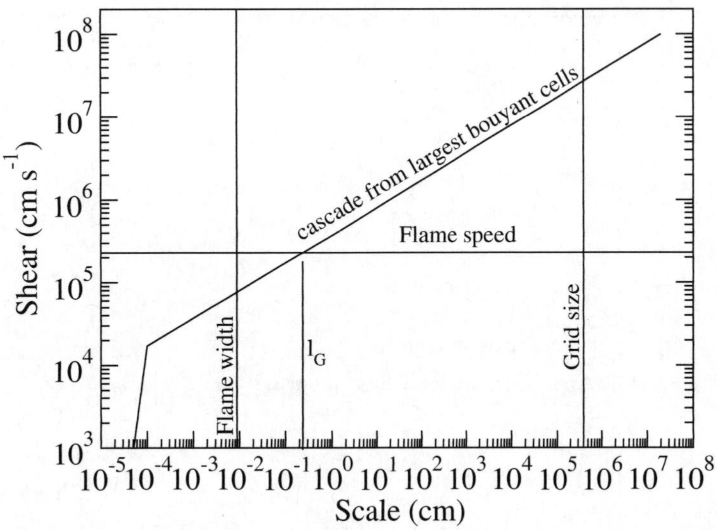

Figure 2. Comparison of shear strength in turbulent cascade to velocity and length scales of the flame. (adapted from [17])

3.2. Detonation

Detonations are reaction fronts that are propagated through fuel by a shock that is self-sustained by the energy released in burning. In thermonuclear supernovae, detonations also present the challenge of unresolved detailed reaction kinetics in simulations that compute full-star explosions. Figure 3 shows the basic thermodynamic and composition structure of a detonation propagating in a carbon-oxygen mixture. See [19] for more details on how this steady-state structure is computed. The various thermodynamic properties – temperature and density – and the mass fraction abundance of the most abundant species are shown as a function of distance behind the propagating shock. This figure shows the detonation structure for a moderate density expected in the exploding white dwarf star, 10^7 g cm^{-3}.

As mentioned previously, the typical grid scale for a full-star simulation is a few 10^5 cm. This means that at a wide range of densities, the length scale for consumption of the fuel, but carbon and oxygen, is unresolved. Exactly which stages are unresolved and how complete the burning can be depends strongly on the density of the fuel. Also the overall size of the star, around 10^8 cm, is important for determining when the burning stops. At densities above approximately that shown in Figure 3, 10^7 g cm^{-3}, there is enough time and space for the burning to produce iron-group material, as shown near the large scales on the right side of Figure 3. At lower densities, more stages become resolved and the burning will be truncated with the production of silicon-group material.

One of the constraints imposed by the overall size of the star is that the detonation front will typically be curved instead of planar [20]. This arises just because the detonation must originate in some localized place and then move radially away from that spot. The resulting propagation will, in general, not align with the spherical structure of the star due to the initiation point being away from the star's center. While this curvature is macroscopic, that is, resolved in the simulation, the resulting weakening of the shock and post-shock flow results in slower burning even on microscopic scales. This leads to early truncation of burning and the need to treat these unresolved scales with some model [21].

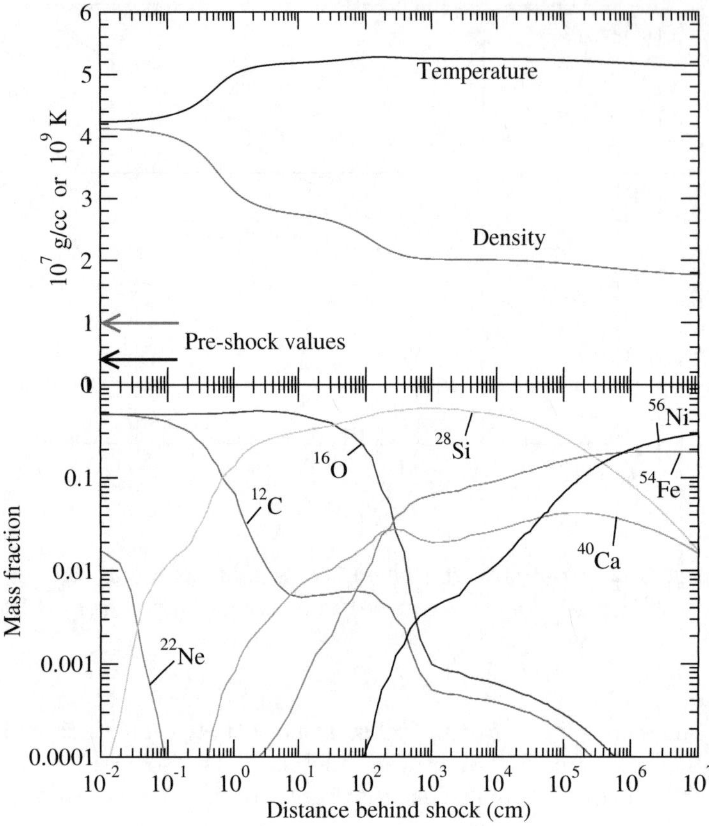

Figure 3. Thermodynamic and composition structure behind the shockfront in a detonation in a carbon-oxygen mixture at a density of 10^7 g cm^{-3}.

3.3. Method and verification

The small-scale physics governing combustion in thermonuclear supernovae must be accounted for in two distinct ways. First, explosion simulations, which are coarse in resolution by necessity of capturing the whole star, must use some model for the combustion. These models have been discussed briefly above. For deflagrations, techniques include flame thickening [22, 23, 19] or front-tracking [24]. For detonations, generally some form of reaction limiting is used [25, 19, 10]. In either case, frequently a reduced nuclear reaction network is used in the simulation as well.

The second component of the calculation begins by tracing the history of fluid elements in the simulated explosion. This produces density and temperature histories, $\rho(t)$ and $T(t)$ respectively, each of which we call a track. In post-processing, a complete nuclear reaction network, usually with 200 or more species, is used to compute, for each track, the ashes of the overall combustion process. These yields can then be compared to observations, either via the spectrum of the supernova, or via abundances of material in the solar system.

Our recent work has focused on using ancillary information from the simulation, beyond the T, ρ history, to make the post-processed abundances more physically realistic and verifiable. This is necessitated by the fact that, as shown in previous sections, many combustion stages are unresolved, and therefore are not represented accurately in the track produced by the simulation. In order to compensate for this, we use the physics of the particular combustion mode to reconstruct these unresolved combustion stages. In the case of the deflagration, the pressure near the reaction front is used to inform a self-heating calculation that mimics how a laminar flame behaves [19]. For detonations, we use information about the curvature and density

IOP Conf. Series: Journal of Physics: Conf. Series **1225** (2019) 012004 doi:10.1088/1742-6596/1225/1/012004

gradients to reconstruct the unresolved stages based on steady-state models [21]. Verification now becomes more achievable, as the detailed model of the unresolved combustion stages can be closely compared to the outcome of the simulation. Also, the uncertainties are controlled by the degree to which approximations like the steady state and constant pressure assumptions are satisfied, which can be mostly quantified.

4. Summary and conclusion

Simulations of proposed scenarios for thermonuclear supernovae must accurately and verifiably capture the physics of combustion processes that occur at scales smaller than the grid scale. Here we have briefly discussed the combustion modes involved, techniques used to model them, and some example scenarios in which these combustion modes appear. Modelling the unresolved combustion is addressed in two steps. Hydrodynamic simulations of the full explosion use a model which is intended to capture the energy release dynamics of the burning. The thermal history determined by these simulations are then used along with more detailed reaction physics and steady-state models for unresolved processes to compute final yields from the explosion. Comparison of these yields to those inferred from observed supernovae will then allow identification of which scenarios might occur in nature, how often, and how they might be distinguished.

Acknowledgments

DMT and BJM received support from NASA through the Astrophysics Theory Program (NNX17AG28G). BJM was also supported in part by the United States Department of Energy under an Early Career Award (DOE grant no. SC111026) and by the Research Corporation for Science Advancement under a Cottrell Scholar Award to Carla Frölich. ACC received support from the US Department of energy under grant DE-FG02-87ER40317.

References

[1] Seitenzahl I R and Townsley D M 2017 *Nucleosynthesis in Thermonuclear Supernovae* (Springer International Publishing AG) p 1955 in *Handbook of Supernovae* eds. Alsabti, A. W. and Murdin, P.
[2] Röpke F K and Sim S A 2018 *Space Science Reviews* **214** 72 (*Preprint* 1805.07268)
[3] Nomoto K, Thielemann F K and Yokoi K 1984 *Astrophys J* **286** 644–658
[4] Khokhlov A M 1991 *Astron Astrophys* **245** 114–128
[5] Hoeflich P and Khokhlov A 1996 *Astrophys J* **457** 500 (*Preprint* astro-ph/9602025)
[6] Zingale M, Nonaka A, Almgren A S, Bell J B, Malone C M and Woosley S E 2011 *Astrophys J* **740** 8
[7] Nonaka A, Aspden A J, Zingale M, Almgren A S, Bell J B and Woosley S E 2012 *Astrophys J* **745** 73 (*Preprint* 1111.3086)
[8] Seitenzahl I R, Ciaraldi-Schoolmann F, Röpke F K, Fink M, Hillebrandt W, Kromer M, Pakmor R, Ruiter A J, Sim S A and Taubenberger S 2013 *MNRAS* **429** 1156–1172 (*Preprint* 1211.3015)
[9] Miles B J, van Rossum D R, Townsley D M, Timmes F X, Jackson A P, Calder A C and Brown E F 2016 *Astrophys J* **824** 59 (*Preprint* 1508.05961)
[10] Shen K J, Kasen D, Miles B J and Townsley D M 2018 *Astrophys J* **854** 52 (*Preprint* 1706.01898)
[11] Livne E and Arnett D 1995 *Astrophys J* **452** 62
[12] Kromer M, Sim S A, Fink M, Röpke F K, Seitenzahl I R and Hillebrandt W 2010 *Astrophys J* **719** 1067–1082 (*Preprint* 1006.4489)
[13] Shen K J and Moore K 2014 *Astrophys J* **797** 46 (*Preprint* 1409.3568)
[14] Guillochon J, Dan M, Ramirez-Ruiz E and Rosswog S 2010 *Astrophys J Let* **709** L64–L69 (*Preprint* 0911.0416)
[15] Timmes F X and Woosley S E 1992 *Astrophys J* **396** 649–667
[16] Chamulak D A, Brown E F and Timmes F X 2007 *Astrophys J Let* **655** L93–L96 (*Preprint* astro-ph/0612507)
[17] Jackson A P, Townsley D M and Calder A C 2014 *Astrophys J* **784** 174 (*Preprint* 1402.4527)
[18] Schmidt W, Niemeyer J C, Hillebrandt W and Röpke F K 2006 *Astron Astrophys* **450** 283–294 (*Preprint* astro-ph/0601500)

IOP Conf. Series: Journal of Physics: Conf. Series **1225** (2019) 012004 doi:10.1088/1742-6596/1225/1/012004

[19] Townsley D M, Miles B J, Timmes F X, Calder A C and Brown E F 2016 *Astrophys J Suppl* **225** 3 (*Preprint* 1605.04878)
[20] Dunkley S D, Sharpe G J and Falle S A E G 2013 *MNRAS* **431** 3429–3443
[21] Miles B J, Townsley D M, Shen K J, Timmes F X and Moore K 2018 *ArXiv e-prints* (*Preprint* 1806.07820)
[22] Calder A C, Townsley D M, Seitenzahl I R, Peng F, Messer O E B, Vladimirova N, Brown E F, Truran J W and Lamb D Q 2007 *Astrophys J* **656** 313–332 (*Preprint* astro-ph/0611009)
[23] Townsley D M, Jackson A P, Calder A C, Chamulak D A, Brown E F and Timmes F X 2009 *Astrophys J* **701** 1582–1604 (*Preprint* 0906.4384)
[24] Röpke F K, Niemeyer J C and Hillebrandt W 2003 *Astrophys J* **588** 952–961 (*Preprint* astro-ph/0211202)
[25] Kushnir D, Katz B, Dong S, Livne E and Fernández R 2013 *Astrophys J Let* **778** L37 (*Preprint* 1303.1180)

IOP Conf. Series: Journal of Physics: Conf. Series **1225** (2019) 012005 doi:10.1088/1742-6596/1225/1/012005

Toward Resolved Simulations of Burning Fronts in Thermonuclear X-ray Bursts

M. Zingale[1], K. Eiden[1], Y. Cavecchi[2,3], A. Harpole[1], J. B. Bell[4], M. Chang[1], I. Hawke[3], M. P. Katz[5], C. M. Malone[6], A. J. Nonaka[4], D. E. Willcox[4], and W. Zhang[4]

[1]Department of Physics and Astronomy, Stony Brook University, Stony Brook, NY 11794-3800 USA
[2]Department of Astrophysical Sciences, Princeton University, Peyton Hall, Princeton, NJ 08544, USA
[3]Mathematical Sciences and STAG Research Centre, University of Southampton, SO17 1BJ, UK
[4]Center for Computational Sciences and Engineering, Lawrence Berkeley National Lab, Berkeley, CA 94720 USA
[5]NVIDIA Corporation, 2788 San Tomas Expressway, Santa Clara, CA, 95050 USA
[6]Los Alamos National Laboratory, Los Alamos, NM, 87545 USA

E-mail: `michael.zingale@stonybrook.edu`

Abstract. We discuss the challenges of modeling X-ray bursts in multi-dimensions, review the different calculations done to date, and discuss our new set of ongoing simulations. We also describe algorithmic improvements that may help in the future to offset some of the expense of these simulations, and describe what may be possible with exascale computing.

1. Introduction

X-ray bursts (XRBs) are fascinating astrophysical explosions. A neutron star accretes fuel from its binary companion, building up just a thin layer ($\sim$ 5–10 m) before the immense gravitational acceleration at the neutron star surface compresses and heats the fuel to the point of thermonuclear runaway. A brief flash of X-rays follows, the accreted layer diffuses the heat produced from reactions, and the process repeats. We can observe multiple bursts from a single source, and the information their lightcurves encodes tells us about the underlying neutron star, and ultimately the nuclear equation of state. A challenging aspect of interpreting the lightcurves is understanding the radiation transport through the hot ash layers. In particular, the composition can affect the effective temperature, which in turn affects the radius we measure for the neutron star [1]. Convection in the burning layer can even bring ashes up to the photosphere which alters what we see [2]. Accurate measurements of neutron star radii from XRBs could greatly constrain the nuclear equation of state [3, 4]. Finally, brightness oscillations during the rising phase of the bursts and other features seen in lightcurves are interpreted as arising from the spreading of a hotspot across the neutron star—demonstrating that the burst begins in a localized region [5, 6]. See [7] for a recent review of XRBs.

Simulations of XRBs can help us to understand the observations, but many technical challenges remain. The primary difficulties are simply the range of length and timescales

IOP Conf. Series: Journal of Physics: Conf. Series **1225** (2019) 012005 doi:10.1088/1742-6596/1225/1/012005

involved. At the largest lengthscale, the neutron star radius is 10 km and the Rossby lengthscale, where rotation and lateral pressure gradients balance, is $\sim$ 1 km [8]. At the smallest scales, a conductive He flame has a thickness of 10s of centimeters [9] and the pressure scale height of the fuel layer is on the order of meters. To capture the rise timescale of the XRB, we would need to model seconds, but the flame itself is subsonic, so long timescale evolution is difficult.

Present and past simulations of XRBs employ a variety of approximations to get past these difficulties. The variety of approximations used enables a complementary picture of XRBs to be built up, and has greatly advanced our understanding of these systems. The breadth of simulations to date has been impressive. We summarize the different approaches below.

One-dimensional simulations, e.g. [10, 11], assume spherical symmetry and model the burst through the vertical column depth of fuel. These simulations can capture the luminosities and burst recurrence times well. They can utilize very large nuclear reaction networks and have been used to understand the nucleosynthesis and rp-process [12–14], as well as to understand how rate uncertainties can affect the outcomes of the bursts. Many successive bursts can be modeled and these simulations can show us the evolution of the ash layer with time. The approximation of 1D however means that we cannot learn about lateral variations across the neutron star, including flame spreading.

Global shallow water simulations [8] capture the large scale spreading of the burning front by using a very simple vertical structure. These simulations were instrumental in showing that the Coriolis force plays an important role in confining a spreading region and that the flame spreading can explain brightness oscillations during the rise of the burst lightcurve.

Our group and others have modeled the convection preceding the runaway using low Mach number hydrodynamic methods for pure helium bursts [15, 16] and mixed hydrogen/helium bursts [17] in 2D, as well as full 3D simulations of mixed bursts [18], and followed the development of turbulence. These simulations can give an understanding of the pre-burning front regime, including the strength and character of the turbulence, but cannot model large scale lateral variations because of the implicit assumptions built into the low Mach number model [19].

While detonations are computationally easier to model than flames (we largely get the speed through the jump conditions in the Riemann problem without resolving the structure), detonations require extreme conditions not found in XRBs [20, 21]. This means that calculations that want to capture the nucleosynthesis yields from spreading burning fronts in XRBs need to model a deflagration.

Flame spreading was first modeled in a series of calculations employing an algorithm from atmospheric science, where vertical hydrostatic equilibrium is enforced and wide-aspect ratio zones give a horizontal CFL number that is large enough to obtain long timescale evolution [22]. These calculations display key features of the propagation mechanism, namely a balance between hydrodynamics and the pure flame physics. The flame proceeds mainly via conduction, thus being a deflagration. The flame front, however, is not vertical but inclined, the angle with the horizontal being very small: $\theta \sim H/R_{\rm Ro} \sim 10^{-3}$. Here H is the scale height and $R_{\rm Ro}$ is the Rossby radius. It is the hydrodynamical balance between the Coriolis force, gravity and pressure gradients that determines the Rossby radius and therefore the inclination angle of the flame front. The extended surface of the flame front is what leads to propagation speeds of order 10^5 km/s, in good agreement with observations. Due to its dependence on the Coriolis force, the flame speed changes across the surface of the NS (being faster at the equator), leading to potentially detectable effects if the rise of the burst is observed with enough resolution [23]. We note that there may be preexisting turbulence ahead of the flame from simmering occurring there, and no calculations have considered the effects of the flame interacting with this turbulence, as we characterized in [18]. Flame-turbulence interactions have long been seen as a means to accelerate a flame in Type Ia supernovae, so future studies should look at turbulence in the layer. Capturing turbulence, of course, requires high resolution, so the ideas described here will be needed.

The NS systems exhibiting XRBs are expected to have magnetic fields in the range 10^7–10^{10} G. Inclusion of a vertical magnetic field proved to have a strong influence on the nature and the speed of flame propagation [24], due to the interaction between the Coriolis force and the magnetic tension, which further acts towards determining the inclination angle of the flame front.

A longstanding goal is to perform simulations where we resolve the flame structure, to allow us to accurately capture the nucleosynthesis, and watch it move across the neutron star surface. The results of these simulations will enable strict comparison to observations made with the new generation of X-ray telescopes such as *eXTP* and *STROBE-X* [25–27] which will offer high collecting area and time resolution. We show some in-progress calculations that are a step toward this, discuss their remaining approximations, and also discuss what new algorithmic techniques might be needed to realize this goal with the advent of exascale computing.

2. Resolved Flame Studies

We have begun a set of flame spreading calculations using the compressible hydrodynamics code Castro [28], part of the AMReX astrophysics suite [29]. Castro uses adaptive mesh refinement, supports a general equation of state and reaction networks, has a conservative gravity formulation [30], and is optimized to run on current supercomputers. All of our code base is open source and available on GitHub[1], and all of the solvers, problem setup, initial models, and analysis scripts for each of our science problems are distributed with the codes.

We setup a problem by creating a pair of 1D hydrostatic XRB models (hot and cool) representing the post- and pre-flame states and apply these laterally across the domain to setup a hot region at the left end of our domain in equilibrium with the cool model on the right. Thermal diffusion and nuclear reactions quickly produce a laterally propagating flame that begins to move through our domain. We resolve the diffusion scale, using a resolution of 10 cm, and use adaptive mesh refinement (two jumps of 4×) to keep a buffer of low density material at low resolution above the surface to allow for expansion. To make the lengthscales tractable, we rotate the neutron star at 2000 Hz—much faster than expected. This helps confine the spreading region to a more reasonable horizontal domain. Nevertheless, our simulation uses 12288 zones laterally, on the finest grid. We use the stellar conductivities from [9] and a 13-isotopic alpha-chain reaction network. To make the timescales easier, we boost the flame speed by a factor of 10 by scaling both the conductivity and energy generation rate by the same factor. This is an approximation we hope to relax in the near future.

Figure 1 shows the flame structure for an initial 2D run at 0.007 s. Vertical profiles at a few radial distances from the origin are shown for the same quantities in Figure 2. The burning layer is only about 10 m high. Underneath is ^{56}Ni, representative of the underlying neutron star, and above it is a very low density (10^{-4} g cm^{-3}) buffer. Only a portion of the vertical extent is shown. The top panel shows the temperature structure. The initial perturbation reached to 2×10^4 cm, so at this point, the extent of the burnt region has nearly doubled. The flame structure is quite wide, since there are multiple burning stages beyond just helium to carbon, and this shows up in the composition. This is seen more clearly in the next panel which shows the mean molecular weight of the nuclei ($\bar{A}$). We see that the further behind the head of the flame we are, the heavier the ash nuclei. We also see the head of the flame appears lifted off of the base of the burning layer—unburned material underlies the spreading flame. This is reminiscent of the flame structure seen in the earlier simulations of [22]. The third panel shows the energy generation rate. It is strongest just above the neutron star surface, behind the flame. We again see that the head of the flame is detached from the base of the atmosphere, with the

[1] `http://github.com/amrex-astro/`

IOP Conf. Series: Journal of Physics: Conf. Series **1225** (2019) 012005 doi:10.1088/1742-6596/1225/1/012005

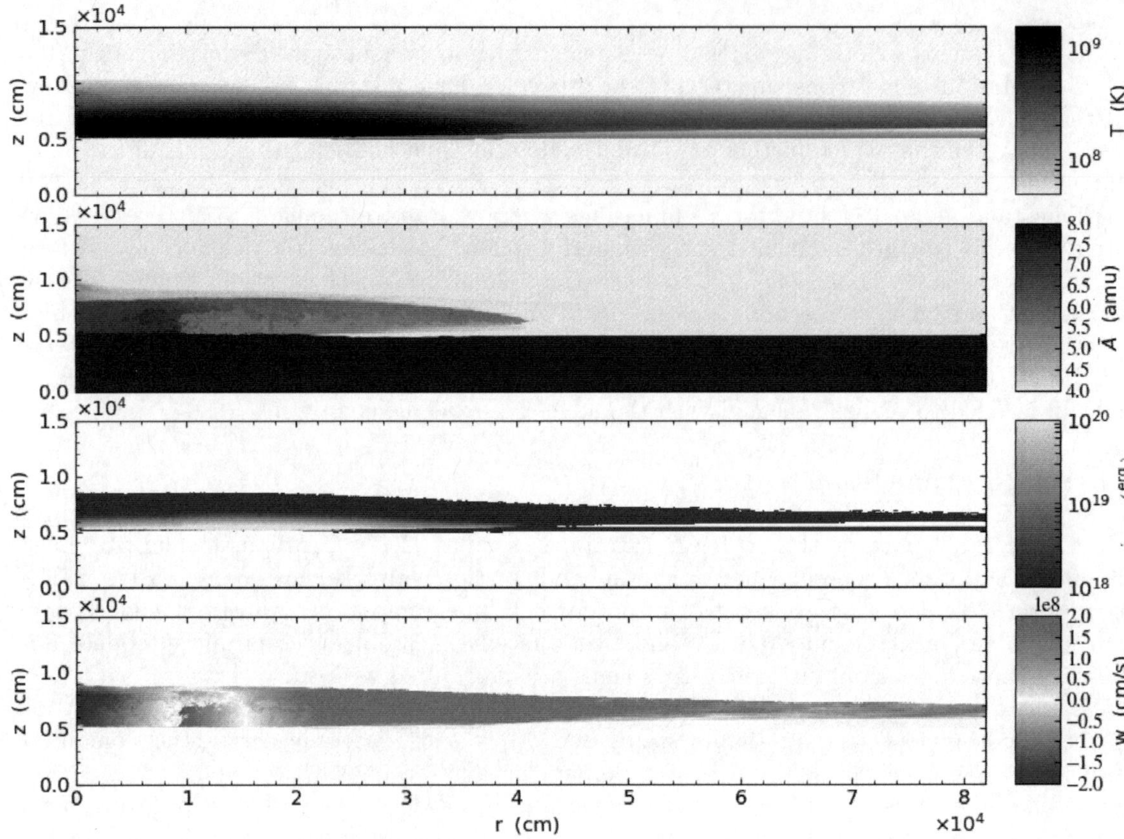

Figure 1. Helium flame spreading across the surface of a neutron star. Shown are temperature (top), the mean molecular weight of the ash (second), the nuclear energy generation rate (third), and the velocity out of the simulation plane (bottom). The out-of-plane velocity contours indicate where the Coriolis force has created a hurricane-like wind structure which confines the flame, leading to an inclined front. It is at such an interface that the temperature and burning rate are highest.

strongest burning at the head of the flame slightly higher up in the atmosphere. Finally, the bottom panel shows the velocity through the plane of the simulation—this is induced by the Coriolis force as the burning fluid expands and begins to spread laterally. A tight hurricane structure has setup as a result of the flame spreading. This is the geostrophic balance discussed in [8]. These calculations are ongoing and will be the subject of a detailed study in the near future.

3. Large Scale Simulations and Algorithmic Developments

The 2D calculations shown in the previous section require about 10k node-hours of computational time to get to about 10 ms (running on the Edison machine at NERSC). We expect qualitatively different results in 3D calculations: for example, the shear from the hurricane structure acting along the flame front may cause instabilities that can affect the flame propagation. With our current simulation methodology, a 3D version of this calculation is estimated to require 10M node-hours or more—this is about the size of an annual allocation. This simulation covers less than 1% of the neutron star surface, so a fully resolved calculation of burning over the entire neutron star is out of reach: a naive scaling calculation suggests that 15 years of Moore's Law

IOP Conf. Series: Journal of Physics: Conf. Series **1225** (2019) 012005 doi:10.1088/1742-6596/1225/1/012005

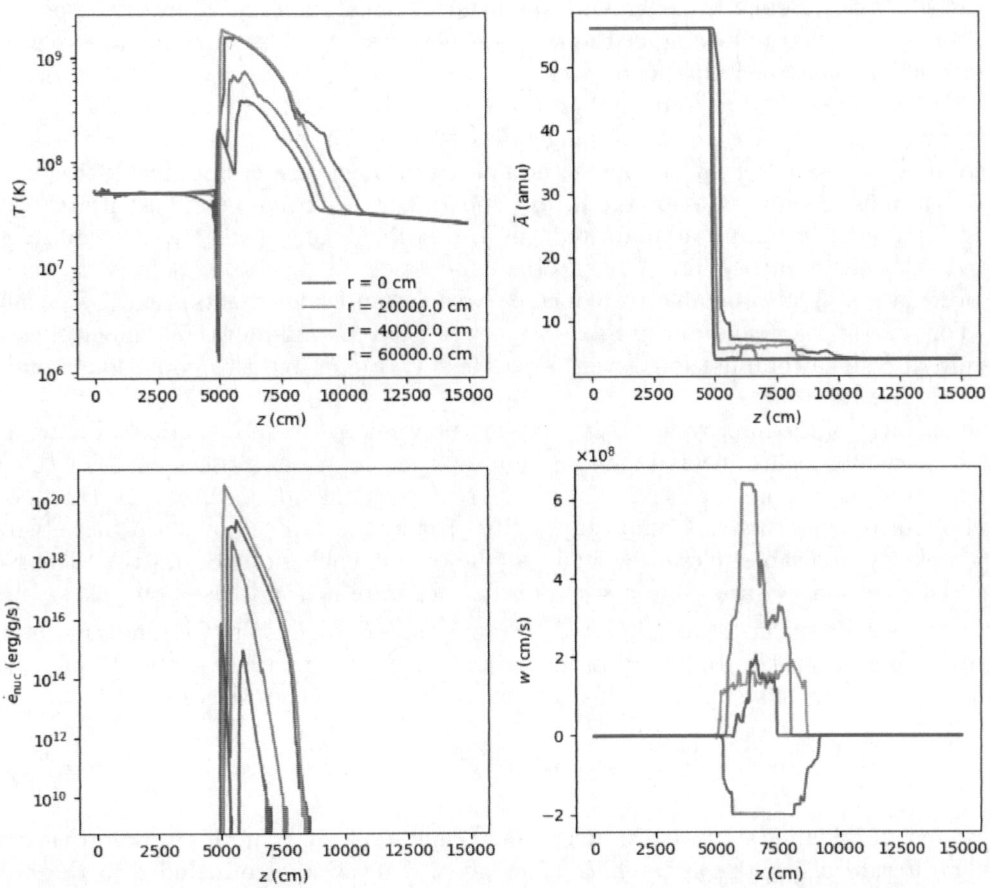

Figure 2. Vertical profiles at several radial distances from the vertical axis through the plots shown in Figure 1. These profiles sample the region behind and ahead of the flame.

improvements are required.

Improvements are needed to push the scale of simulations to the point where we can model XRBs at the full star scale with resolved nuclear physics. As the computational cost is roughly a product of the number of zones, the number of zone updates, and the cost per zone update, we need to look for improvements in each of these. The number of zones depends on the physical model employed, the meshing strategy, and the accuracy required. The number of zone updates depends on the smallest zone size and the largest characteristic speed in the model. The computational cost per zone update is dominated by the time spent in the reactions and equation of state. We discuss a few possible improvements that can come both from implementing new algorithms and from exploiting new hardware.

Machine architectures are evolving, with heterogeneous architectures becoming more common. Our recent development in Castro has focused on moving the hydrodynamics to GPUs [29]. The XRB problem is an ideal case for this, since hydrodynamics with constant gravity (for plane-parallel domains) ports in a straightforward manner to GPUs. The current GPU version of the Castro hydrodynamics solver is about 10× faster using GPUs on a node than the entire node of CPU cores (comparing 4 NVIDIA Volta GPUs to a dual-socket IBM Power9 CPU using 40 cores). We see similar speed ups with reaction networks, with these GPU ports under active development. This means a GPU calculation could lower the above costs by a factor of 10. GPU offloading of the reaction networks will also allow us to explore larger networks and more detailed nuclear physics.

Another way we can reduce the number of zones for given accuracy is to use a more accurate hydrodynamic method. We are developing fully fourth-order (in space and time) coupling of hydrodynamics and reactions using the method of spectral deferred corrections (SDC) in Castro, instead of the more commonly employed Strang splitting [31]. This provides us with two benefits. First, the improved coupling actually reduces the stiffness of the reactions, requiring fewer righthand side calls and therefore reducing the expense of the reactions. Second, by moving to fourth order, we may also be able to reduce our resolution requirements needed for converged flames. For smooth wave solutions we may expect to need around half the number of zones per dimension, and since computational work scales like $(\Delta x)^4$ in 3D, this could lower costs by a factor of around 10.

As an example of the improvements in coupling between hydro and reactions, Figure 3 shows the mass fraction of helium behind a detonation over the course of 2 timesteps. For this test, a 19 isotope reaction network was used. The points represent individual calls to the righthand side function of our reaction network, and the density of points indicates how hard the integrator is working. We use the same tolerances for both the Strang-split method and the SDC method. For the SDC method, we are using a second-order accurate method based on [32]. We predict a time-centered advection term, $[\boldsymbol{\mathcal{A}}(\boldsymbol{\mathcal{U}})]_i^{n+1/2}$ using standard unsplit Godunov methods, but explicitly include a reactive source term in the tracing of the interface states, $\mathbf{R}(\boldsymbol{\mathcal{U}}_i)$. We can then solve the reactive system, using this advective term as a piecewise-constant-in-time source

$$\frac{d\boldsymbol{\mathcal{U}}_i}{dt} = -[\boldsymbol{\mathcal{A}}(\boldsymbol{\mathcal{U}})]_i^{n+1/2} + \mathbf{R}(\boldsymbol{\mathcal{U}}_i) \tag{1}$$

This system can be integrated with standard ODE methods. To achieve second-order accuracy, we need to iterate, using the updated $\boldsymbol{\mathcal{U}}$ to create a reactive source included in the predictor for the advective update, and then reintegrate the system. The result of this iteration is that the advection sees the effects of reactions over the timestep and the reactions see the effects of advection. For the zone tracked in Figure 3, each SDC iteration called the righthand side function of the network 2 times less than the Strang case. This savings is dependent on the network and thermodynamic state, but our tests thus far show that the SDC methods need fewer network calls to integrate a timestep. We also clearly see how the Strang system evolves in a discontinuous fashion. We are continuing to develop this new integration strategy, with full fourth-order in space and time reacting hydrodynamics methods to follow shortly. The XRB simulations are our target application. We note that there has been some development in the community on implicit time-integration methods for compressible flow, but for the XRB problem, the wide range of Mach number seen in the domain (exceeding unity near the top of the atmosphere, ahead of the flame) removes much of the advantages of implicit timestepping, so we focus on explicit methods.

For the ODE integrations shown we use the variable-order, variable-step integrator VODE [33] with its implementation of the backward-differentiation formulas. We note that for both Strang splitting and SDC, VODE must start the integration at first order and correspondingly small timesteps since no higher-order information about the solution is available (our reaction network right-hand sides provide only the first time derivative of the solution). Based on testing with a 13-isotope approximate alpha-chain network, VODE takes many small timesteps (e.g. ~ 200) before it can stably increase to a high order representation of the solution, but after it does so, it can often complete the integration in only a handful of steps (e.g. ~ 5–6). Observing that during a Castro timestep, the first Strang step is a direct continuation of the integration in the second Strang step of the previous Castro timestep, we explored saving the VODE state following

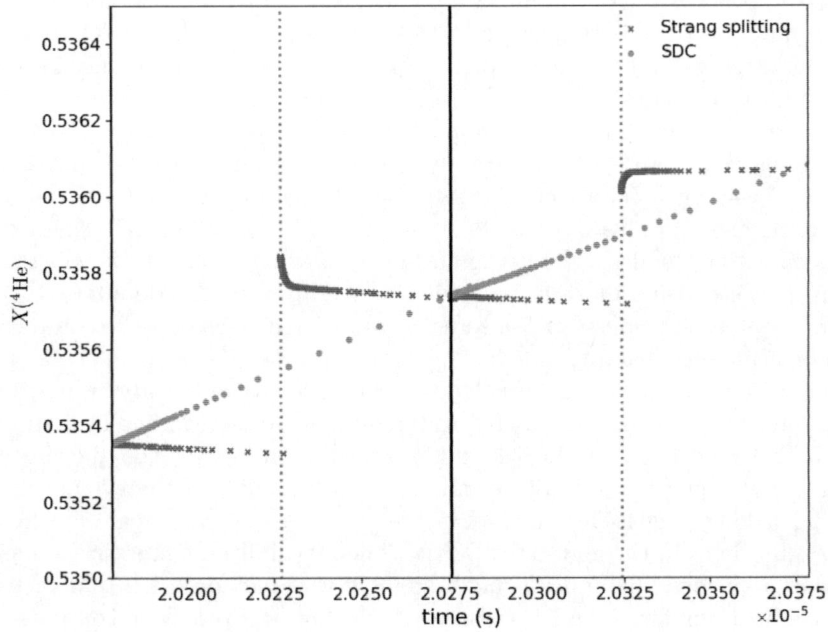

Figure 3. Helium mass fraction as a function of time over two timesteps for a detonation test problem with a 19-isotope reaction network. The points represent individual calls to the righthand side function of our reaction network (see Eq. 1) and the density of points indicates how hard the integrator is working. For the Strang case, we see that the first half of the reactions brings us off of the smooth solution (represented by the SDC curve), then the advection overcorrects, resulting in a discontinuous jump, and finally the second half of the reactions works to bring the state back to the solution curve.

the second Strang step and restarting VODE during the first Strang step. This approach allowed us to start the first Strang step with a high order representation of the solution and complete it in very few VODE timesteps ($\lesssim 10$ steps), speeding up the Castro timestep advance by 50% for our detonation test problem using the 13-isotope network. However, this came at the memory cost of storing $7N_{eq} + 90$ scalar variables on our grid, where for our networks $N_{eq} = N_{spec} + 2$ (in addition to the number of species N_{spec} we also integrate temperature and specific energy). For a large-scale science problem with a large network, this could quickly become prohibitive, so we are exploring ways to mitigate either the memory requirements for restarting the integration or the small timestep start up costs for the integration method.

Other ways to make the simulation less expensive can include a subgrid model. Flame models are commonly used for Type Ia supernovae. There however, the flame is thin compared to the scale height, so a simple model can represent the unresolved flame on the grid. For the XRB, the flame thickness is a non-trivial fraction of the pressure scale height and further, the laminar flame properties will change strongly over the height of the burning layer (see [9]). This makes traditional flame models a poor fit. Methods that use even something as simple as subzone burning (e.g. [34]) may help capture the energetics with lower resolution. This can be implemented using the AMR machinery. This may help a little, but will still be insufficient for full star models.

If we continue to resolve the flame structure, then we need to be smarter (and more aggressive) about our grid refinement strategy. Currently, we put the entire fuel layer on the finest grid, but for the material ahead of the flame, we could keep it at lower resolution until the flame reaches it. The main obstacle is that we need to keep the fuel ahead of the flame in hydrostatic equilibrium, and keeping it refined accomplishes this well. Using well-balanced schemes [35] can help us retain hydrostatic equilibrium (HSE) in the lower refined regions, reducing the number of zones needed to maintain fidelity, but we would still need a way to regrid the layer ahead of the flame to higher resolution to accurately capture the burning dynamics. To accomplish this, we envision building a refinement strategy that does interpolation of the data onto the newly refined grids enforcing hydrostatic equilibrium. This can allow us to keep only a narrow fully refined region that "slides" along the surface of the neutron star keeping pace with the advancing flame. When using mesh refinement the overall costs are dominated by the number of zones at the finest resolution, so such a scheme would greatly reduce the computational costs. This is a longer term development.

All of the above developments still involve fully compressible hydrodynamics, but multiscale methods [36] where we couple different hydrodynamics solvers together in a single simulation may ultimately be needed to model the full star while resolving the burning front. Such techniques rely on it being possible to decompose the dynamics of the system into two scales, e.g. a short and a long lengthscale, and that the processes at each scale are at most weakly coupled with each other. In the context of XRBs, one possibility for multiscale modeling would be to model the large scale dynamics using a shallow water model (like that of [8]) to capture the effects of the Coriolis force, and use a low Mach number or a fully compressible model in the vicinity of the burning front to capture the turbulent burning processes. Given that the computational costs of the shallow water model would be negligible compared to those of the low Mach/compressible model, the overall cost of this multiscale model would be determined by the size of the low Mach/compressible region only. This approach would therefore allow us to capture dynamics across the full star without sacrificing resolution around the flame. As with the more aggressive gridding strategy above, the hybrid approach reduces the number of zones with the largest computational cost through mesh refinement, and also reduces the computational costs where possible by employing cheaper physical models in regions where this is possible. An example of coupling compressible and low Mach number models implemented using the AMReX framework, but in a terrestrial context, is [37]. An initial implementation, [38], in an astrophysical context, coupling shallow water to compressible models in both Newtonian and relativistic gravity, indicates that considerable computational efficiencies are possible, but that close attention to the coupling between the models is needed.

4. Summary

XRBs are multiscale, multiphysics problems that are challenging to model. Nevertheless, significant progress has been made in multidimensional models of these events, through a set of complementary algorithmic approaches. Future advances in computer architectures and algorithms will allow for even more realistic models of these events, and help us connect to observations, ultimately telling us about the neutron star itself.

Acknowledgments

The work at Stony Brook was supported by DOE/Office of Nuclear Physics grant DE-FG02-87ER40317 and contract 7418390 with Lawrence Berkeley National Laboratory as part of the Exascale Compute Project ExaStar collaboration. The work at LBNL was supported by the DOE Office of Advanced Scientific Computing Research under Contract No, DE-AC02-05CH11231. YC is supported by the European Union Horizon 2020 research and innovation programme under the Marie Sklodowska-Curie Global Fellowship grant agreement No 703916. An award of computer time was provided by the Innovative and Novel Computational Impact on Theory and Experiment (INCITE) program. This research used resources of the Oak Ridge

Leadership Computing Facility at the Oak Ridge National Laboratory, which is supported by the Office of Science of the U.S. Department of Energy under Contract No. DE-AC05-00OR22725. This research used resources of the National Energy Research Scientific Computing Center, which is supported by the Office of Science of the U.S. Department of Energy under Contract No. DE-AC02-05CH11231. Visualizations were done using yt [39]. This research has made use of NASA's Astrophysics Data System Bibliographic Services.

References

[1] Suleimanov V, Poutanen J and Werner K 2011 *Astronomy and Astrophysics* **527** A139 (*Preprint* 1009.6147)

[2] Kajava J J E, Nättilä J, Poutanen J, Cumming A, Suleimanov V and Kuulkers E 2017 *Monthly Notices of the Royal Astronomical Society* **464** L6–L10 (*Preprint* 1608.06801)

[3] Steiner A W, Lattimer J M and Brown E F 2010 *Astrophysical Journal* **722** 33–54 (*Preprint* 1005.0811)

[4] Özel F, Baym G and Güver T 2010 *Physical Review D* **82** 101301 (*Preprint* 1002.3153)

[5] Bhattacharyya S and Strohmayer T E 2006 **636** L121–L124 (*Preprint* arXiv:astro-ph/0509369)

[6] Bhattacharyya S and Strohmayer T E 2007 **666** L85–L88

[7] Galloway D K and Keek L 2017 *ArXiv e-prints* (*Preprint* 1712.06227)

[8] Spitkovsky A, Levin Y and Ushomirsky G 2002 *Astrophysical Journal* **566** 1018–1038

[9] Timmes F X 2000 *Astrophysical Journal* **528** 913–945 source code obtained from http://cococubed.asu.edu/code_pages/kap.shtml

[10] Woosley S E, Heger A, Cumming A, Hoffman R D, Pruet J, Rauscher T, Fisker J L, Schatz H, Brown B A and Wiescher M 2004 *Astrophysical Journal Supplement* **151** 75–102

[11] Fisker J L, Schatz H and Thielemann F K 2008 *Astrophysical Journal Supplement* **174** 261–276

[12] Schatz H, Bildsten L, Cumming A and Wiescher M 1999 *Astrophysical Journal* **524** 1014–1029 (*Preprint* arXiv:astro-ph/9905274)

[13] Schatz H, Aprahamian A, Barnard V, Bildsten L, Cumming A, Ouellette M, Rauscher T, Thielemann F and Wiescher M 2001 *Physical Review Letters* **86** 3471–3474 (*Preprint* arXiv:astro-ph/0102418)

[14] Schatz H 2006 *International Symposium on Nuclear Astrophysics - Nuclei in the Cosmos* p 2.1

[15] Lin D J, Bayliss A and Taam R E 2006 *Astrophysical Journal* **653** 545–557

[16] Malone C M, Nonaka A, Almgren A S, Bell J B and Zingale M 2011 *Astrophysical Journal* **728** 118 (*Preprint* 1012.0609)

[17] Malone C M, Zingale M, Nonaka A, Almgren A S and Bell J B 2014 *Astrophysical Journal* **788** 115

[18] Zingale M, Malone C M, Nonaka A, Almgren A S and Bell J B 2015 *Astrophysical Journal* **807** 60 (*Preprint* 1410.5796)

[19] Almgren A S, Bell J B, Rendleman C A and Zingale M 2006 *Astrophysical Journal* **637** 922–936 paper I

[20] Zingale M, Timmes F X, Fryxell B, Lamb D Q, Olson K, Calder A C, Dursi L J, Ricker P, Rosner R, MacNeice P and Tufo H M 2001 *Astrophysical Journal Supplement* **133** 195–220

[21] Harpole A and Hawke I 2018 *ArXiv e-prints* (*Preprint* 1806.07301)

[22] Cavecchi Y, Watts A L, Braithwaite J and Levin Y 2013 *Monthly Notices of the Royal Astronomical Society* **434** 3526–3541 (*Preprint* 1212.2872)

[23] Cavecchi Y, Watts A L, Levin Y and Braithwaite J 2015 *Monthly Notices of the Royal Astronomical Society* **448** 445–455 (*Preprint* 1411.2284)

[24] Cavecchi Y, Levin Y, Watts A L and Braithwaite J 2016 *Monthly Notices of the Royal Astronomical Society* **459** 1259–1275 (*Preprint* 1509.02497)

[25] Wilson-Hodge C A, Ray P S, Gendreau K, Chakrabarty D, Feroci M, Maccarone T, Arzoumanian Z, Remillard R A, Wood K, Griffith C and STROBE-X Collaboration 2017 *American Astronomical Society Meeting Abstracts* (*American Astronomical Society Meeting Abstracts* vol 229) p 309.04

[26] Zhang S N *et al.* 2016 *Space Telescopes and Instrumentation 2016: Ultraviolet to Gamma Ray* (*Proc. SPIE* vol 9905) p 99051Q (*Preprint* 1607.08823)

[27] in 't Zand J J M *et al.* 2018 *Science China Physics, Mechanics & Astronomy* **62** 29506 ISSN 1869-1927 URL https://doi.org/10.1007/s11433-017-9186-1

[28] Almgren A S, Beckner V E, Bell J B, Day M S, Howell L H, Joggerst C C, Lijewski M J, Nonaka A, Singer M and Zingale M 2010 *Astrophysical Journal* **715** 1221–1238 (*Preprint* 1005.0114)

[29] Zingale M, Almgren A S, Barrios Sazo M G, Beckner V E, Bell J B, Friesen B, Jacobs A M, Katz M P, Malone C M, Nonaka A J, Willcox D E and Zhang W 2017 *ArXiv e-prints* Accepted to Proceedings of AstroNum 2017 (*Preprint* 1711.06203)

[30] Katz M P, Zingale M, Calder A C, Swesty F D, Almgren A S and Zhang W 2016 *Astrophysical Journal* **819** 94 (*Preprint* 1512.06099)

[31] Strang G 1968 *SIAM J. Numerical Analysis* **5** 506–517

[32] Nonaka A, Bell J B, Day M S, Gilet C, Almgren A S and Minion M L 2012 *Combustion Theory and Modelling* **16** 1053–1088 (*Preprint* https://doi.org/10.1080/13647830.2012.701019) URL https://doi.org/10.1080/13647830.2012.701019

[33] Brown P N, Byrne G D and Hindmarsh A C 1989 *SIAM J. Sci. Stat. Comput.* **10** 1038–1051

[34] Wang W, Shu C W, Yee H and Sj??green B 2012 *Journal of Computational Physics* **231** 190 – 214 ISSN 0021-9991 URL http://www.sciencedirect.com/science/article/pii/S0021999111005250

[35] Käppeli R and Mishra S 2016 *Astronomy and Astrophysics* **587** A94

[36] Weinan E 2011 *Principles of multiscale modeling* (Cambridge University Press)

[37] Motheau E, Duarte M, Almgren A and Bell J B 2018 *Journal of Computational Physics* **372** 1027–1047 ISSN 00219991 URL https://linkinghub.elsevier.com/retrieve/pii/S0021999118300469

[38] Harpole A 2018 *Multiscale modelling of neutron star oceans* Ph.D. thesis University of Southampton URL https://eprints.soton.ac.uk/id/eprint/422175

[39] Turk M J, Smith B D, Oishi J S, Skory S, Skillman S W, Abel T and Norman M L 2011 *Astrophysical Journal Supplement* **192** 9 (*Preprint* 1011.3514)

IOP Conf. Series: Journal of Physics: Conf. Series **1225** (2019) 012006 doi:10.1088/1742-6596/1225/1/012006

The structure of magnetic turbulence in the heliosheath region observed by *Voyager 2* at 106 AU

Federico Fraternale
Dipartimento di Scienza Applicata e Tecnologia, Politecnico di Torino, Torino, 10131, Italy

Nikolai V Pogorelov
Department of Space Science, University of Alabama in Huntsville, Huntsville, AL 35805, USA

John D. Richardson
Kavli Institute for Astrophysics and Space Research, Massachusetts Institute of Technology, Cambridge MA, 02139, USA

Daniela Tordella
Dipartimento di Scienza Applicata e Tecnologia, Politecnico di Torino, Torino, 10131, Italy

E-mail: `federico.fraternale@polito.it`

Abstract. It is currently believed that the turbulent fluctuations pervade the outermost heliosphere. Turbulence, magnetic reconnection, and their link may be responsible for magnetic energy conversion in these regions. The governing mechanisms of such anisotropic and compressible magnetic turbulence in the inner heliosheath (IHS) and in the local interstellar medium (LISM) still lack a thorough description. The present literature mainly concerns large scales which are not representative of the inertial-cascade dynamics of turbulence. Moreover, lack of broadband spectral analysis makes the IHS dynamics critically understudied. Our recent study [1] shows that 48 s magnetic-field data from the *Voyager* mission are appropriate for a spectral analysis over a frequency range of six decades, from 5×10^{-8} Hz to 10^{-2} Hz. Here, focusing on the *Voyager 2* observation interval from 2013.824 to 2016.0, we describe the structure of turbulence in a sector zone of the IHS. A spectral break around 7×10^{-7} Hz (magnetic structures with size $\ell \approx 1.3$ Astronomical Units) separates the energy-injection regime from the inertial-cascade regime of turbulence. A second scale is observed around 6×10^{-5} Hz ($\ell \approx 0.017$ AU) and corresponds to a peak of compressibility and intermittency of fluctuations.

1. Introduction

The *Voyagers* (V1, V2) are the only operating spacecraft providing us with *in situ* data from the outermost part of heliosphere. The inner heliosheath (IHS) is the region of space between the termination shock (TS) and the heliopause (HP). The HP is a tangential discontinuity that separates the solar wind (SW) from the local interstellar medium (LISM).

Major scientific questions are related to the physical processes responsible for the conversion of magnetic energy and SW heating, acceleration and transport of energetic particles [2, 3, 4, 5, 6], existence and topology of the sector - swept by the heliospheric current sheet, HCS - and unipolar regions of the IHS [7, 8], and eventually the coupling between the interstellar and heliospheric magnetic fields [9]. For details, we refer the reader to three comprehensive reviews of the heliosheath plasma and related physical processes [10, 11, 12].

These topics are tightly linked to the turbulent nature of the IHS/LISM plasma and magnetic fields [13, 14, 15, 16, 17, 1], and the potential presence of magnetic reconnection [18, 7, 19, 20, 9]. Heliospheric numerical simulations are shedding light on the three-dimensional heliospheric structure (see, e.g. [21, 22, 23, 9, 24]). Notably, taking into account the solar-cycle variations made it possible to reproduce many observed features of the IHS and LISM bulk plasma flow and magnetic field. Moreover, it has been shown that the transition to chaos is possible, and the turbulence may be responsible for the SW heating and the observed average values of the heliospheric magnetic field [22, 20]. However, resolving the inertial range of turbulent fluctuations (say, $\ell \lesssim 1.5$ AU) numerically is still unfeasible except for a very small computational domain, which makes the spectral analysis of *Voyagers*' data over a broad range of spatial and temporal scales highly desirable to improve and constrain the models. Unfortunately, the intrinsically space- and time-local, one-dimensional nature of spacecraft measurements, lack of plasma data at V1, presence of data gaps, and high level of noise in general, make such analysis nontrivial.

The plasma and magnetic fields observed by V2 during 2015 have been discussed in details in [25], where the presence of two merged interaction regions (MIR) and a global merged interaction region (GMIR) was shown. Previously, in [26], it was shown that near 2013.824 at 103 AU V2 entered the IHS sector region, which was likely due to the increasing latitudinal extent of the HCS related to the near-maximum solar activity during solar cycle 24.

This study extends our recent work [1], where a spectral analysis of V1 and V2 magnetic field data was performed for several IHS and LISM periods (in both unipolar and sector zones). This work was carried out for a range of scales unprecedented in the literature (spacecraft-frame frequencies $10^{-8} < f < 10^{-2}$ Hz, with corresponding spatial scales in the range $5 \times 10^{-5} < \ell < 65$ AU). Following the same line of research, here we focus on the most recent V2 data publicly available. In particular, we analyze the V2 magnetic field measurements from 2013.824 (day-of-year 300) to 2016.0. We investigate the spectral properties of magnetic field fluctuations in the *energy-injection* and *inertial-cascade* ranges of turbulence, with focus on the variance anisotropy, the presence of compressible modes, and high-order multi-scale statistics. We believe that these analyses will be of importance for the improvement of existing theoretical and numerical IHS models.

In section 2 we present details of the data set used for this study and the methodology adopted for multi-scale analyses. In section 3 results are shown: 3.1 discusses the energy-injection regime of fluctuations and 3.2 is focused on the inertial-cascade of turbulence. Conclusions follow in section 4.

2. Data

In the time interval from 2013.824 to 2016.0, V2 was traveling at an heliocentric distance of 106.5±3.4 AU, latitude of $-30.5°$ and longitude $-217.5°$ in HelioGraphic Inertial (HGI) coordinates. Since the sector-boundary crossing in 2013.824, V2 has been traveling inside the sector region of the IHS [27]. This study considers magnetic field *in situ* data provided by the V2 MAG experiment [28]. The field's magnitude and angles are shown in Figure 1. To explore a broad range of scales, we used data at the highest sampling rate publicly available, i.e., the 48 s resolution (processed data can be downloaded from the NASA's Space Physics Data Facility `https://spdf.gsfc.nasa.gov/`). Data are provided in the HGI RTN coordinate reference system, but for the purpose of this study it was convenient to adopt a reference frame having one axis aligned with the average field $\mathbf{B_0}$ (the $\|$ axis). The $\perp_1$ and the $\perp_2$ axes form the plane orthogonal to $\mathbf{B_0}$, with $B_{\perp 1}$ belonging the T-N plane. Given that in the IHS $\mathbf{B_0}$ is nearly aligned with the T direction, it follows that $\perp_1$ is approximately aligned with N and $\perp_2$ with R.

Computing the power spectral density (PSD) is challenging due to the sparsity of the *Voyager* data set. In the interval considered in this study, 72% of 48 s data points are missing. Typically, the data gaps of 8–16 hours in the IHS are largely due to ground tracking issues and occur once a day. The average frequency of the largest gap is $f_{\rm gap} = 2.4 \times 10^{-5}$ Hz. The PSD is computed on the basis a comparative analysis of four spectral estimation methods (Compressed Sensing CS, linear interpolation CI, optimization of model spectra OP, Fourier transform of gap-free subsets SS). These techniques have been successfully adopted already in our previous papers on the SW turbulence at 5 AU [29, 30, 31], and in the recent study [1] regarding IHS and LISM turbulence. CS and SS has also been used in [32] to analyze turbulence in the Earth's magnetosphere, in proximity of the magnetopause. Scripts and detailed descriptions of all methods are provided in the Supplemental Material of [31] and in Chap. 4 of [33]. Some tests specific to IHS data can be found in Appendix A of [1].

The accuracy of *Voyager* observations is affected by the level of noise. This includes the magnetometer's sensibility (0.006 nT), and various sources of noise such as the telemetry system, the interference with other instruments, and the calibration process. The resulting 1σ error of magnetic field components is estimated to set around 0.03 nT [34]. The actual level and distribution of the noise are unknown and may differ from the above estimate. If a white-noise model with the amplitude of 0.03 nT is considered, the PSD is constant and equal to 0.029 $\rm nT^2s$. This threshold is represented in all figures with a gray band. It can be noticed, however, that the actual level of noise may be lower for the period considered here, since a serious spectral flattening is not observed until $f \approx 10^{-3}$ Hz. In fact, this frequency intercepts the spectra at $P \approx 10^{-3}$ $\rm nT^2s$, which would correspond to a white noise of 0.0056 nT amplitude (see Figure 2(a)). Consequently, as discussed in §3, the present spectral analysis may be affected by the noise in the frequency range $4 \times 10^{-4} \lesssim f \lesssim 10^{-2}$ Hz, which likely includes the transitional MHD-to-kinetic regime of magnetic turbulence.

Higher-order statistics of magnetic field increments (the structure functions, in the time domain) are computed from both 48 s data and 1824 s averages, which helps to estimate the effect of noise. The p-th structure function of the j-th magnetic field component is defined as $S_{p,j}(\tau) = \langle |\Delta B_j|^p \rangle$, where $\Delta B_j(t;\tau) = B_j(t) - B_j(t+\tau)$ is the magnetic field increment of the j-th component for the time lag τ, and angle brackets denote the time-average for the period analyzed (as usual, this assumes ergodicity of the underlying physical processes). We computed $S_{p,j}$ from available data points as

$$S_{p,j}(\tau_k) = \frac{1}{N(\tau_k)} \sum_{i=1}^{N(\tau_k)} |B_j(t_i) - B_j(t_i + \tau_k)|^p, \quad (1)$$
$$\tau_k = k \cdot \Delta t_s \quad k = 1, \ldots, n,$$

where $\Delta t_s = 48$ s (or 1824 s), and $n = 401600$ (or 10200) is the total number of points of the data set. $N(\tau)$ is the number of the available increments $\Delta B(t_i, \tau)$. Note that for contiguous data sets $N(\tau)$ is necessarily a linearly-decreasing function of the time lag, due to the limitedness of the data sequence. In the case of *Voyager* data sets in the IHS, however, due to data gaps $N(\tau)$ is also oscillating, with minima at multiples of $\tau_{\rm gap} = 1/f_{\rm gap} \approx 11.5$ hours, that corresponds to the periodicity of large gaps. Points with smaller statistical significance (when $N(\tau) < 0.25 \max[N(\tau')]$, $\tau' \in [\tau - 48h,\ \tau + 48h])$) have been disregarded in the computation of power-law exponents (Table 2) and are shown in gray in Figure 3(a). The figure also reports the fluctuation intensity ratio $\mathcal{I}(\tau) = \langle |\mathbf{\Delta B}|/B_\tau \rangle$ between the the magnitude of the increments defined above and the average magnitude between the instant t and $t + \tau$, $B_\tau(t;\tau)$.

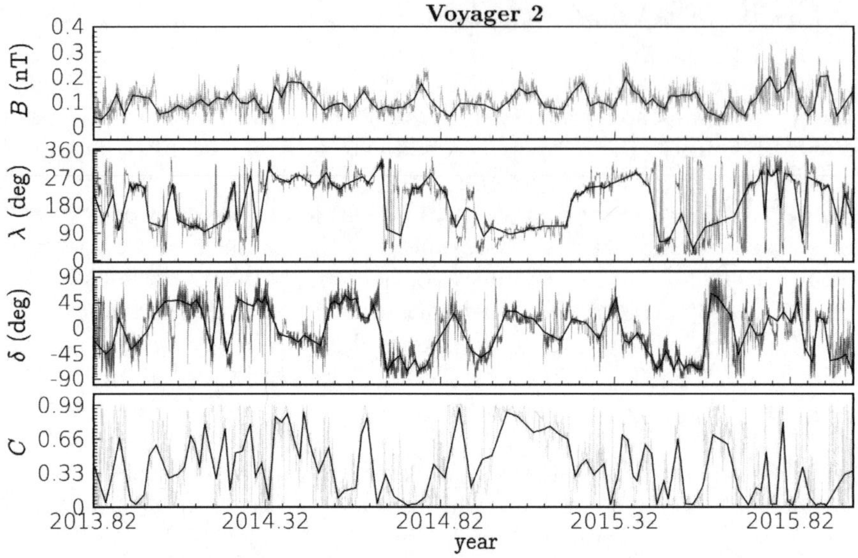

Figure 1. *Voyager 2* data considered in the present study, observation period 2013.824–2016.0. Top to bottom: the magnetic-field magnitude $B = |\mathbf{B}|$; the azimuthal angle $\lambda = \tan^{-1}(B_\mathrm{T}/B_\mathrm{R})$; the elevation angle $\delta = \sin^{-1}(B_\mathrm{N}/B)$; and the fraction of fluctuating magnetic energy in the B-parallel direction, computed as $C = \left\langle [\mathbf{B_0} \cdot \boldsymbol{\delta}\mathbf{B}/(B_0 \delta B)]^2 \right\rangle$, with $\delta\mathbf{B} = \mathbf{B} - \mathbf{B}_0$. Data points with $|B_j| < 0.03$ nT have been disregarded in the computation of λ and δ. (High-resolution data is taken from the COHO website `https://cohoweb.gsfc.nasa.gov/coho/`).

3. Results and discussion

Figure 2(a) shows the power spectral density (PSD, or P) of the magnetic field components. The red, green and blue curves stand for $\delta B_{\parallel}$, $\delta B_{\perp 1}$, $\delta B_{\perp 2}$, respectively, while the total magnetic energy $E_m(f) = P[B_{\parallel}] + P[B_{\perp 1}] + P[B_{\perp 2}]$ is represented in black. Figure 2(b) shows the spectral variance anisotropy and a proxy for spectral compressibility. The former is defined as $A_j(f) = P[B_j]/E_m$, while the latter is $C(f) = P[|\mathbf{B}|]/E_m$, the ratio between the spectrum of the magnetic field magnitude and the trace. It is considered here as a proxy for the density fluctuations, as they are typically strongly correlated with the fluctuations of magnetic field magnitude [35]. Note that most of notations and definitions that we use in the present work are the same as in [1].

Inevitably, single-spacecraft measurements cannot provide the omni-directional spectrum but only the reduced one (1D), containing the contribution of all vector wavenumbers. Since the spacecraft speed during the analyzed period is about 0.1 of the bulk wind speed ($V_\mathrm{SW} \approx 150$ km/s) and about 0.3 of the Alfvén speed ($V_\mathrm{A} \approx 53$ km/s), we used the Taylor's hypothesis to convert the spacecraft-frame (SC) frequencies to wavenumbers [36]. The average magnetic field being nearly orthogonal to the wind direction, such wavenumbers can be interpreted as perpendicular to $\mathbf{B_0}$, $k_\perp \approx 2\pi f_\mathrm{SC}/V_\mathrm{SW}$. This information can be used to compare the present results with theoretical findings on the spectral scaling laws of anisotropic turbulence [37], and to estimate the order of magnitude of magnetic structures in the direction of the wind (the average azimuthal and elevation angles of the thermal plasma flow are $\lambda_v = 43°$ and $\delta_v = -25°$, respectively). However, it should be reminded that the application of the Taylor's hypothesis in the IHS is more critical than in the supersonic SW upstream the termination shock, and that it may not be applicable in the high-frequency range of the spectrum, especially if the dispersive waves play an important role.

Table 1. Average quantities, in the V2 observation period 2013.824–2016.0.

	Parameter	Value	
$d_{\rm SC}$	Spacecraft-Sun radial distance	106.5±3.44	AU
$V_{\rm SW}$	Average solar wind speed	150.2±27.3	km/s
B_0	Average magnetic field	0.03	nT
B	Magnetic field average strength	0.11	nT
$n_{\rm p}$	Thermal protons density	(1.95±0.8) $\times 10^{-3}$	cm $^{-3}$
$T_{\rm p}$	Thermal protons temperature	(5.29±2.5)$\times 10^4$	K
$V_{\rm A}$	Alfvén speed	52.8	km/s
$r_{\rm cp}$	Thermal protons Larmor radius	2880	km
$r_{\rm i}$	Thermal protons inertial radius	5150	km
$r_{\rm cp\ 1keV}$	1-keV protons Larmor radius	43000	km
$f_{\rm cp\ PL}$	Thermal protons gyrofrequency (plasma frame)	1.63 $\times 10^{-3}$	Hz
$f_{\rm cp\ SC}$	Thermal protons Larmor frequency (SC frame)	2.6 $\times 10^{-2}$	Hz
$f_{\rm ip\ SC}$	Thermal protons inertial frequency (SC frame)	1.4 $\times 10^{-2}$	Hz
$f_{\rm cp\ 1keV\ SC}$	1-keV protons Larmor frequency (SC frame)	1.7 $\times 10^{-3}$	Hz
$f_{\rm e\ SC}$	1-eddy-turnover frequency (SC frame)	3.9 $\times 10^{-7}$	Hz

Table 1 offers the average quantities, typical length scales, and frequencies. Table 2 summarizes the results of the fluctuation analysis. The left panel of Figure 3 shows the structure functions of magnetic field increments for parallel fluctuations ($S_{p,\|}$, red curves) and perpendicular fluctuations ($S_{p,\perp} = [S_{p,\perp 1} + S_{p,\perp 2}]/2$, blue curves). In the insert of panel (a), the fluctuation intensity ratio $\mathcal{I}$ is shown. The right panel shows the scale-dependent kurtosis of magnetic increments for the three field components, $K_j = S_{4,j}/S_{2,j}^2$, a measure of intermittency.

The range of scales considered in this study, $f \in [10^{-8}, 10^{-2}]$ Hz, allows us to identify the large-scale, MHD, energy-injection and inertial-cascade regimes of turbulence. In principle, the transition to the kinetic regime should also be observed. In fact, the gyrofrequency of thermal protons, $f_{\rm cp}$, is around 1.6 mHz in the plasma reference frame, and 0.03 Hz if converted to the spacecraft-frame frequency through the Larmor radius $r_{\rm cp} \approx 3000$ km. Ion inertial-scale structures ($r_i \approx 5000$ km) may be detected at a frequency of 0.01 Hz as well. However, as discussed in §2, noise in the data limits the investigation to scales larger than 5×10^{-4} Hz, at least at the current stage of the research.

Note that about 95% of thermal energy of ions in the heliosheath belongs to the population of pickup-ions (PUI) [38, 39, 40]. Thus, PUIs are expected to have a relevant mediation effect on the turbulence, such as that documented in [41, 42, 43]. The gyroradius of a 1 keV pickup proton is about 40000 km, which may be detected in the V2 spectrum at frequencies around 1.5 mHz.

3.1. Energy-injection regime (EI)

A large-scale energy-injection regime is identified at spacecraft-frame frequencies less than $f_{\rm b1} \approx 7 \times 10^{-7}$ Hz, where a spectral break takes place, as shown in Figure 2. This frequency corresponds to a spatial scale around $\ell_{\rm b1} \approx 1.3$ AU. In this regime, the power spectral density decays slowly, with a spectral index $\alpha_{\rm EI} \approx -1.26$. We have shown that the extension of the EI range can vary significantly in different heliosheath regions, in particular the threshold frequency is larger during unipolar periods [1]. We also note that the EI energy decay rate observed here is faster than that observed in earlier periods. The break at $f_{\rm b1}$ is indeed difficult to recognize by looking at the spectrum only. However, it is clearly distinguishable from the power-law variation of the structure functions and the kurtosis shown in Figure 3 ($\tau_{\rm b1} = 1/f_{\rm b1}$).

In the SW upstream of the TS, this EI range with the frequency scaling of $\sim 1/f$ is found

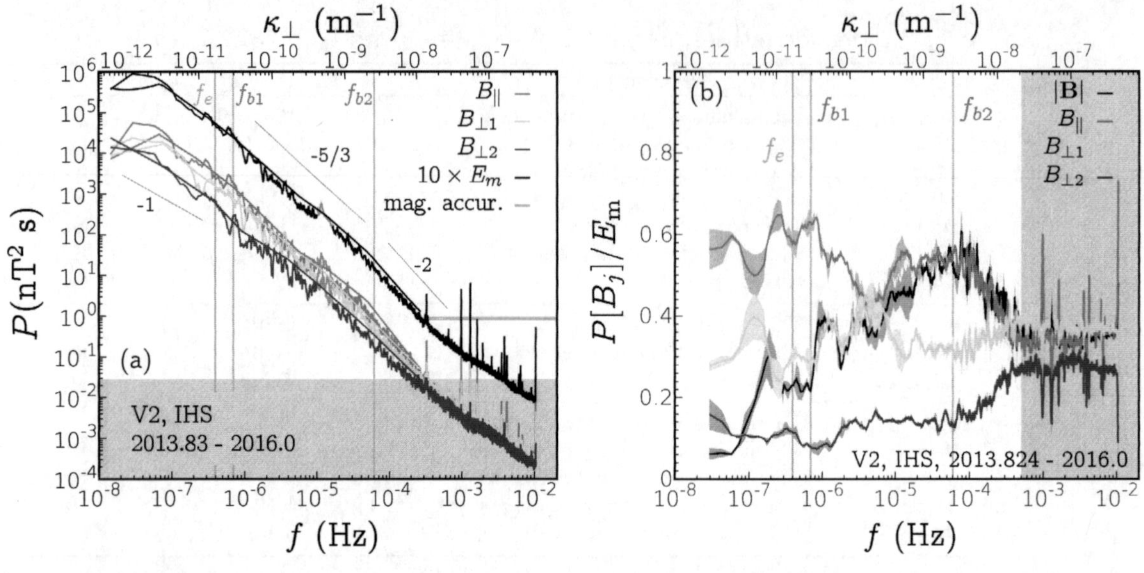

Figure 2. Power spectral density of magnetic field fluctuations, anisotropy and compressibility. (a) PSD, the three components are represented in red (||), green ($\perp_1$), and blue ($\perp_2$). The trace is shown in black, $E_m(f) = P[B_{\|}] + P[B_{\perp 1}] + P[B_{\perp 2}]$. The smooth lines show the result of the OP method, the rugged curves are build from the CI and CS methods, as described in Appendix A of [1]. The gray band and the horizontal line represent the instrumental uncertainty threshold, modeled as a white noise with 0.03 nT amplitude per each field component. As a reference, the three spectral slopes of -1, -5/3 and -2 are shown in the EI, IC1 and IC2 regimes, respectively. Some sharp spikes in the PSD are instrument-related, as they are harmonics of the sampling time. The bump at 1.5×10^{-5} Hz is an artifact due to the data gaps and it is thus shown in grey. (b): spectral compressibility ($C(f) = P[|\mathbf{B}|]/E_m$, black curve) and spectral anisotropy for each j-th field component ($A_j(f) = P[B_j]/E_m$, colored curves, as in the left panel). The difference between the three spectral estimation methods is also shown.

to be related to the Alfvénicity of fluctuations and, in particular, it is more extended in fast-wind streams [45]. In fact, large-scale Alfvénic waves in this regime did not experience yet a sufficient nonlinear interaction to produce a turbulent cascade and form a reservoir of energy for turbulence at smaller scales. The Sun rotation provides some forcing to the system. It acts at $f_{\rm sun} \approx 4.5\times10^{-7}$ Hz. This determines the nominal width of magnetic sectors, which is around 2 AU in the IHS. Moreover, the causality condition implies that fluctuations with spacecraft-frame frequencies below a specific threshold $f_{\rm e}$ cannot be considered as "true" fluctuations, since they are either waves of wavelengths larger than the distance between the spacecraft and their source, or equivalently, vortexes that did not experience yet one eddy-turnover (the typical nonlinear time scale). As a reasonable approximation for this scale we use $f_{\rm e} \approx \pi V_{\rm SW}^2/[V_{\rm A}(r_{\rm SC} - r_{\rm source})]$, where $r_{\rm SC}$ is the spacecraft location and $r_{\rm source}$ the location of the fluctuation's source. We consider the termination shock as a source location ($r_{\rm source} \approx 84$ AU), which yields the value of $f_{\rm e} \approx 4\times10^{-7}$ Hz reported in the figures. If the Sun is considered as the source point, ($r_{\rm source} = 0$), $f_{\rm e} \approx 10^{-8}$ Hz. The former choice better agrees with the observed location of the large-scale spectral break. Notice also that the IHS width should be considered as the outer scale of the system (see the sub-panel of Figure 3a). *Voyager 1* crossed the HP at 121.5 AU, 27 AU away from the point where it crossed the TS.

The EI regime is also identified from the structure functions of magnetic field increments as shown in Figure 3(a) for both parallel (with respect to $\mathbf{B}_0$) and perpendicular fluctuations. In fact, the spectral break fairly well corresponds to a change in the behavior of the structure functions, which follow flatter trends - not typical of fully-developed turbulence - for time lags $\tau > \tau_{b1}$. The EI range is not intermittent, as the kurtosis is close to Gaussian values (see the left panel of Figure 3). At $\tau \gtrsim 10^7$ s ($\ell \sim 10$ AU), a peak and further flattening of S_p is observed. We note that these scales are close to the outer scale of the system, but the statistic is insufficient to derive conclusions.

It should be noted that compressibility is also small ($C \approx 0.2$), as shown by the black curve in the right panel of Figure 2. In this regard, we emphasize that the high energy of $\delta B_{\|}$ (red curve in Figure 2) at these scales is not related to the presence of compressible modes, but rather to changes in the magnetic field polarity, clearly visible from the time history of the azimuthal angle in the second panel of Figure 1. In the limit of incompressible fluctuations (constant $|\mathbf{B}|$), it has been recently shown for near-Earth SW that the Kolmogorov's turbulent cascade is saturated and cannot subsist at scales, where $\mathcal{I}(\tau)$ exceeds the unity, resulting in a $\sim 1/f$ spectral power law in the EI range [44]. Here, this relationship is observed in the IHS for the first time, as shown in Figure 3(a) (black curve in the sub-panel).

3.2. Inertial-cascade regime (IC)

The range of fluctuations with frequency between f_{b1} and 10^{-3} Hz can be referred to as the magnetohydrodynamic inertial-cascade regime of magnetic turbulence. Within IC, we highlight a second typical scale of magnetic fluctuations. In fact, it is seen that a spectral steepening takes place at $f_{b2} \approx 6 \times 10^{-5}$ Hz ($\ell_{b2} \approx 2.5 \times 10^6$ km). This scale splits the IC range in two subranges which will be named IC1 and IC2, respectively. For $f_{b1} < f < f_{b2}$, we observe a defined power-law energy decay with a spectral index $\alpha_{IC1} \approx -1.5$ for E_m. This may be consistent with an anisotropic Iroshnikov–Kraichnan scaling. However, here the role of $\delta B_{\|}$ is important, as they contribute substantially to the fluctuation of the magnetic field's magnitude, as shown in Figure 2(b). The fraction of fluctuating energy due to compressible fluctuations increases from 0.2 to 0.6 in IC1, reaching the maximum at f_{b2}. The inertial cascade regime is featured by a power-law behavior of the structure functions. In particular, in the IC1 range S_p show defined power laws with exponents typical of MHD turbulence (Figure 3). The exponents $\zeta_{p,\|}$, $\zeta_{p,\perp}$ are computed by linear regression in the log-log plane, excluding points with lower statistical significance due to data gaps (shown in gray in panel (a)), and are reported in Table 2. The comparison with existing theoretical models is better done computing the exponents relative to S_3. This was done via the *extended self-similarity* principle (ESS) by which $S_p \sim S_3^{\zeta^{\mathrm{ESS}}}$ [46]. It is found that S_p as a function of S_3 exhibits a significantly broader scaling range than $S_p(\tau)$, extending beyond the inertial range. This makes it possible to perform an accurate computation of the scaling exponents. The kurtosis shown in Figure 3(b) displays an increase from larger to smaller scales in the IC1 range, which fits the power laws $K_{\|} \sim \tau^{-0.25}$, $K_{\perp_1} \sim \tau^{-0.14}$, $K_{\perp_2} \sim \tau^{-0.19}$, where the fit of 48 s and 1824 s data gives the uncertainty of ± 0.02 for the exponent. The maximum intermittency is achieved around $\tau_{b2} = 1/f_{b2}$. In the IC2 range, the intermittency reduces and Gaussian values of K are found for $\tau \lesssim 2000$ s. This may be an artifact due to noise in data. However, the peak of activity at τ_{b2} seems physical and deserves further investigation in future studies. Indeed, it is worth noticing that the power spectrum is steeper on such scales, especially for the $\delta B_{\|}$ component, with a spectral index $\alpha_{IC2} \approx 2.2$ (Figure 2).

4. Conclusions

This study reports on the *Voyager 2* observations of magnetic turbulence in the inner heliosheath from 2013.824 to 2016.0, when V2 was at 106.5±3.4 AU from the Sun. It is believed that during this period the spacecraft was inside the sector region of the inner heliosheath. The present paper shows follow-up results of our recent work [1], where the spectral properties of magnetic

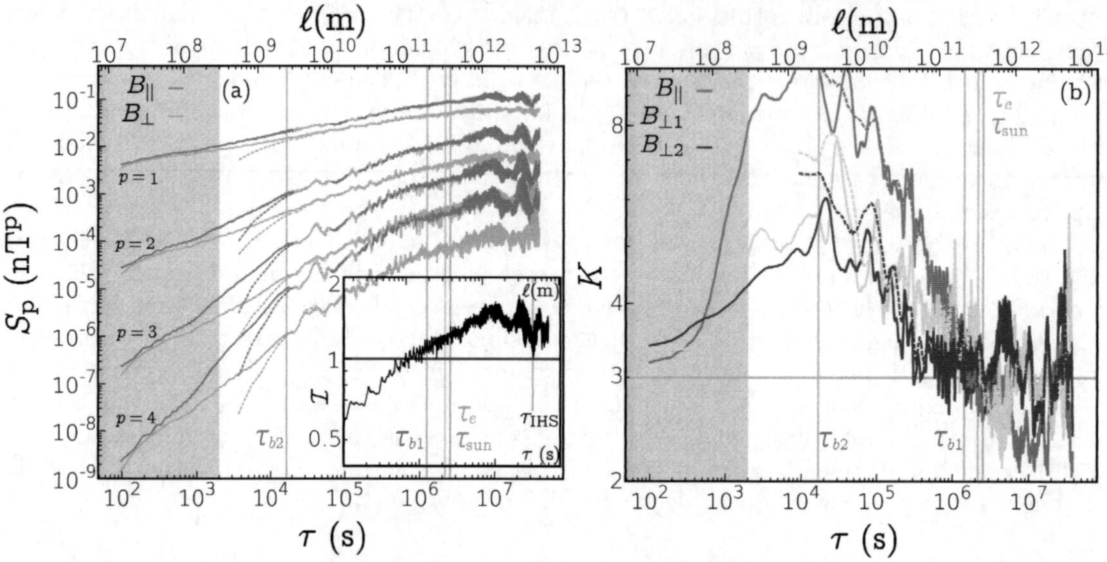

Figure 3. (a) Structure functions of magnetic field increments, S_p, for $p = 1, 2, 3, 4$. Red curves represent the structure functions of parallel fluctuations $S_{p,\|}$, while light-blue curves represent the average between $S_{p,\perp 1}$ and $S_{p,\perp 2}$ (dashed lines represent results from 1824 s data). Oscillations are due to the data gaps in the time sequence. When the counter $N(\tau)$ is less than 1/4 of the maximum value, the color switches to gray. The vertical lines indicate the two spectral breaks ($\tau_{\rm b1}$ and $\tau_{\rm b2}$, respectively), the 1-eddy-turnover time scale ($\tau_{\rm e}$), and the Sun rotation period ($\tau_{\rm sun}$). The panel insert shows $\mathcal{I}(\tau) = \langle|\mathbf{\Delta B}|/B_\tau\rangle$. Note that at the EI-to-IC spectral break $\mathcal{I}(\tau_{\rm b1}) \approx 1$, the upper limit for the turbulent cascade expected for incompressible fluctuations [44]. Here, we also indicate $\ell_{\rm IHS} \approx V_{\rm SW}/\tau_{\rm IHS} \approx 30$ AU, that represents the IHS's width in the wind direction, under the Taylor's approximation. (b) Scale-dependent kurtosis of magnetic increments for all the field components. It is observed that both S_p and K fit power laws in the inertial-cascade range between the spectral breaks ($K \sim \tau^{[-0.25,-0.14]}$). The energy-injection range is not intermittent.

field fluctuations have been shown for a collection of several unipolar and sector IHS regions, and for LISM intervals, for a spectral bandwidth over six decades.

We identify two scales that may be characteristic to the turbulence in this region. The first scale is located at $f_{\rm b1} \approx 7 \times 10^{-7}$ Hz (spacecraft-frame frequency), which corresponds to structures of size $\ell_{\rm b1} \approx 1.3$ AU in the solar wind direction - under the Taylor's approximation. This scale seems discriminating the energy-injection range of magnetic field fluctuations from a second regime which can be interpreted as the inertial-cascade range of turbulence.

The energy-injection regime regime consists of incompressible and non-intermittent fluctuations, following a power law for the reduced power spectra with spectral index $\alpha_{\rm EI} \approx -1.25$. At these scales, the magnitude of magnetic field increments is between one and two times the average magnetic field's strength.

The inertial-cascade regime displays a spectral index $\alpha_{\rm IC1} \approx -1.5$ and power-law growing intermittency. Here, we observe a second scale located at $f_{\rm b2} \approx 6 \times 10^{-5}$ Hz ($\ell_{\rm b2} \approx 0.017$ AU), where a faster cascade is observed until 5×10^{-4} Hz, especially in the B-parallel fluctuations ($\alpha_{\rm IC2} \approx -2$). Concurrently, here the maximum compressibility and intermittency are observed. Higher frequencies, up to 0.01 Hz, should include the transition from the magnetohydrodynamic regime to the kinetic regime. However, this range could be affected by noise in the data and

Table 2. Magnetic-field fluctuation properties from time-domain analysis (first six lines) and spectral analysis. Note that in the definitions below $\delta B_j(t) = B_j(t) - B_{j,0}$, $B_0 = |\mathbf{B_0}|$ is the magnitude of the average magnetic field and $B = \langle|\mathbf{B}|\rangle$ is the average magnetic field's magnitude.

	Parameter	Value	
$\boldsymbol{E_m}$	Average magnetic energy	0.013	nT^2
$\boldsymbol{C}$	Average compressibility	0.41	
$\boldsymbol{I_{\parallel}}$	Average $\parallel$ intensity $\langle\delta B_{\parallel}/B_0\rangle$, $\langle\delta B_{\parallel}/B\rangle$	1.97, 0.66	
$\boldsymbol{I_{\perp 1}}$	Average $\perp_1$ intensity $\langle\delta B_{\perp_1}/B_0\rangle$, $\langle\delta B_{\perp_1}/B\rangle$	1.41, 0.47	
$\boldsymbol{I_{\perp 2}}$	Average $\perp_2$ intensity $\langle\delta B_{\perp_2}/B_0\rangle$, $\langle\delta B_{\perp_2}/B\rangle$	0.86, 0.29	
$\boldsymbol{I}$	Average magnitude intensity $\langle|\delta\mathbf{B}|/B_0\rangle$, $\langle|\delta\mathbf{B}|/B\rangle$	2.92, 0.99	
$\boldsymbol{f_{b1}}$	EI-to-IC spectral break	7×10^{-7}	Hz
$\boldsymbol{f_{b2}}$	IC1-to-IC2 spectral break	6×10^{-5}	Hz
$\boldsymbol{\alpha_{EI}}$	Spectral index of E_m in the EI range	-1.26±0.07	
$\boldsymbol{\alpha_{IC1}}$	Spectral index of E_m in the IC1 range	-1.49±0.02	
$\boldsymbol{\alpha_{IC2}}$	Spectral index of E_m in the IC2 range	-1.87±0.07	
$\boldsymbol{\zeta_{1,\parallel}}$ ($\boldsymbol{\zeta^{ESS}_{1,\parallel}}$)	1st SF-exponent of $\delta B_{\parallel}$	0.35 (0.45)	
$\boldsymbol{\zeta_{2,\parallel}}$ ($\boldsymbol{\zeta^{ESS}_{2,\parallel}}$)	2nd SF-exponent of $\delta B_{\parallel}$	0.62 (0.78)	
$\boldsymbol{\zeta_{3,\parallel}}$ ($\boldsymbol{\zeta^{ESS}_{3,\parallel}}$)	3rd SF-exponent of $\delta B_{\parallel}$	0.80 (1)	
$\boldsymbol{\zeta_{4,\parallel}}$ ($\boldsymbol{\zeta^{ESS}_{4,\parallel}}$)	4th SF-exponent of $\delta B_{\parallel}$	0.95 (1.16)	
$\boldsymbol{\zeta_{1,\perp}}$ ($\boldsymbol{\zeta^{ESS}_{1,\perp}}$)	1st SF-exponent of $\delta B_{\perp}$	0.27 (0.39)	
$\boldsymbol{\zeta_{2,\perp}}$ ($\boldsymbol{\zeta^{ESS}_{2,\perp}}$)	2nd SF-exponent of $\delta B_{\perp}$	0.50 (0.73)	
$\boldsymbol{\zeta_{3,\perp}}$ ($\boldsymbol{\zeta^{ESS}_{3,\perp}}$)	3rd SF-exponent of $\delta B_{\perp}$	0.71 (1)	
$\boldsymbol{\zeta_{4,\perp}}$ ($\boldsymbol{\zeta^{ESS}_{4,\perp}}$)	4th SF-exponent of $\delta B_{\perp}$	0.89 (1.22)	

was not considered in the present discussion. The influence of energetic particle populations as well as the potential presence of turbulent magnetic reconnection should be considered in future research in order to shed light into the nature of these observations.

Acknowledgments

F.F. acknowledge support from the postdoctoral grant FOIFLUT 37/17/F/AR-B. N.P. was supported, in part, by NASA grants NNX14AJ53G, NNX16AG83G, 80NSSC18K1649NS, and 80NSSC19K0260, and NSF PRAC award OAC-1811176. J.D.R. was supported under NASA contract 959203 from JPL to MIT. Computational resources for spectral and statistical analysis were provided by HPC@POLITO (`http://www.hpc.polito.it`).

References

[1] Fraternale F, Pogorelov N V, Richardson J D and Tordella D 2019 *Astrophys. J.* **872** 40
[2] McComas D J and Schwadron N A 2006 *Geophys. Res. Lett.* **33** L04102
[3] Fisk L A and Gloeckler G 2009 *Adv. Space Res.* **43** 1471–1478
[4] Heerikhuisen J, Pogorelov N V, Zank G P, Crew G B, Frisch P C, Funsten H O, Janzen P H, McComas D J, Reisenfeld D B and Schwadron N A 2010 *Astrophys. J. Lett.* **708** L126–L130
[5] Pogorelov N V, Bedford M C, Kryukov I A and Zank G P 2016 *J. Phys. Conf. Series* **767** 012020
[6] Zhao L L, Adhikari L, Zank G P, Hu Q and Feng X S 2017 *Astrophys. J.* **849** 88
[7] Hill M E, Decker R B, Brown L E, Drake J F, Hamilton D C, Krimigis S M and Opher M 2014 *Astrophys. J.* **781** 94
[8] Richardson J D and Decker R B 2015 *J. Phys. Conf. Ser.* **577** 012021
[9] Pogorelov N V, Heerikhuisen J, Roytershteyn V, Burlaga L F, Gurnett D A and Kurth W S 2017 *Astrophys. J.* **845** 9
[10] Richardson J D and Burlaga L F 2013 *Space Sci. Rev.* **176** 217–235
[11] Zank G P 2015 *Annu. Rev. Astron. Astrophys.* **53** 449

[12] Pogorelov N V, Fichtner H, Czechowski A, Lazarian A, Lembege B, Roux J A I, Potgieter M S, Scherer K, Stone E C, Strauss R D, Wiengarten T, Wurz P, Zank G P and Zhang M 2017 *Space Sci. Rev.* **212** 193–248
[13] Burlaga L F, Ness N F and Acuna M H 2006 *Astrophys. J.* **642** 584
[14] Burlaga L F, Ness N F and Acuna M H 2006 *J. Geophys. Res.-Space Physics* **111** A09112
[15] Burlaga L F and Ness N F 2009 *Astrophys. J.* **703** 311
[16] Burlaga L F, Florinski V and Ness N F 2015 *Astrophys. J. Lett.* **804** L31
[17] Burlaga L F, Florinski V and Ness N F 2018 *Astrophys. J.* **854** 20
[18] Drake J F, Opher M, Swisdak M and Chamoun J N 2010 *Astrophys. J.* **709** 963–974
[19] Drake J F, Swisdak M, Opher M and Richardson J D 2017 *Astrophys. J.* **837** 159
[20] Pogorelov N V, Borovikov S N, Bedford M C, Heerikhuisen J, Kim T K, Kryukov I A and Zank G P 2013 Modeling solar wind flow with the multi-scale fluid-kinetic simulation suite *Numerical modeling of space plasma flows astronum-2012* vol 474 ed Pogorelov N V, Audit E and Zank G P pp 165–171 7th Annual International Conference on Numerical Modeling of Space Plasma Flows, Big Isl, HI, JUN 25-29, 2012
[21] Pogorelov N V, Borovikov S N, Zank G P, Burlaga L F, Decker R A and Stone E C 2012 *Astrophys. J. Lett.* **750** L4
[22] Pogorelov N V, Suess S T, Borovikov S N, Ebert R W, McComas D J and Zank G P 2013 *Astrophys. J.* **772** 2
[23] Pogorelov N V, Borovikov S N, Heerikhuisen J and Zhang M 2015 *Astrophys. J. Lett.* **812** L6
[24] Kim T K, Pogorelov N V and Burlaga L F 2017 *Astrophys. J. Lett.* **843** L32
[25] Burlaga L F, Ness N F and Richardson J D 2018 *Astrophys. J.* **861** 9
[26] Burlaga L F, Ness N F and Richardson J D 2017 *Asptrophys. J.* **841** 47
[27] Richardson J D and the Voyager Team 2016 Voyager observations in the outer heliosphere and interstellar medium AIP Conf. Proc. vol 1720, *SOLAR WIND 14: Proceedings of the Fourteenth International Solar Wind Conference*, ed. L. Wang et al. (Melville, NY: AIP), 080001
Solar Wind 14 (*AIP Conf. Proc.* vol 1720)
[28] Behannon K W, Acuna M H, Burlaga L F, Lepping R P, Ness N F and Neubauer F M 1977 *Space Sci. Rev.* **21** 235–257
[29] Fraternale F, Gallana L, Iovieno M, Opher M, Richardson J D and Tordella D 2016 *Physica Scripta* **91** 394–401
[30] Iovieno M, Gallana L, Fraternale F, Richardson J D, Opher M and Tordella D 2016 *European Journal of Mechanics B/Fluids* **55** 394–401
[31] Gallana L, Fraternale F, Iovieno M, Fosson S M, Magli E, Opher M, Richardson J D and Tordella D 2016 *J. Geophys. Res. - Space Physics* **121** 3905–3919
[32] Sorriso-Valvo L, Catapano F, Retin A, Le Contel O, Perrone D, Roberts O W, Coburn J T, Panebianco V, Valentini F, Perri S, Greco A, Malara F, Carbone F, Veltri P, Pezzi O, Fraternale F, Di Mare F, Marino F, Giles B, Moore T E, T R C, Torbert R B, Burch J L and Khotyaintsev Y V 2019 *Phys. Rev. Lett.* **122** 035102
[33] Fraternale F 2017 *Internal waves in fluid flows. Possible coexistence with turbulence* Ph.D. thesis Politecnico di Torino
[34] Berdichevsky D B 2009 Voyager Mission, Detailed processing of weak magnetic fields; II - Update on the cleaning of Voyager magnetic field density B with MAGCALs (Washington, DC: NASA)
[35] Roberts D A, Klein L W, Goldstein M L and Matthaeus W H 1987 *J. Geophys. Res.* **92** 11021–11040
[36] Taylor G I 1938 *Proc. R. Soc. Lond. A* **164** 476–490
[37] Zhou Y, Matthaeus W H and Dmitruk P 2004 *Rev. Mod. Phys.* **76** 1015–1035
[38] Malama Y G, Izmodenov V V and Chalov S V 2006 *Astron. Astrophys.* **445** 693–701
[39] Decker R B, Krimigis S M, Roelof E C, Hill M E, Armstrong T P, Gloeckler G, Hamilton D C and Lanzerotti L J 2008 *Nature* **454** 67–70
[40] Zank G P, Heerikhuisen J, Pogorelov N V, Burrows R and McComas D 2010 *Astrophys. J.* **708** 1092–1106
[41] Smith C W, Hamilton K, Vasquez B J and Leamon R J 2006 *Astroph. J. Lett.* **645** L85–L88
[42] Aggarwal P, Taylor D K, Smith C W, Joyce C J, Fisher M K, Isenberg P A, Vasquez B J, Schwadron N A, Cannon B E and Richardson J D 2016 *Astrophy. J.***822** 94
[43] Zank P G, 2016 *Geosci. Lett.***3** 22 **822** 94
[44] Matteini L, Stansby D, Horbury T S, Chen C H K 2018 *Astrophys. J. Lett.* **869** L32
[45] Roberts D A 2010 *J. Geophys. Res.* **115** A12101
[46] Benzi R, Ciliberto S, Tripiccione R, Baudet C, Massaioli F and Succi S 1993 *Phys. Rev. E.* **48** R29–R32

IOP Conf. Series: Journal of Physics: Conf. Series **1225** (2019) 012007 doi:10.1088/1742-6596/1225/1/012007

Coupled MHD-Focused Transport Simulations for Modeling Solar Particle Events

Jon A. Linker[1], Ronald M. Caplan[1], Nathan Schwadron[2], Matthew Gorby[2], Cooper Downs[1], Tibor Torok[1], Roberto Lionello[1], and Janvier Wijaya[1]

[1]Predictive Science Inc., San Diego, CA 92121
[2]University of New Hampshire, Durham, NH 03824

E-mail: linkerj@predsci.com, caplanr@predsci.com, nschwadron@guero.sr.unh.edu, Matthew.Gorby@unh.edu, cdowns@predsci.com, tibor@predsci.com, lionel@predsci.com, janvier.wijaya@gmail.com

Abstract. We describe the initial version of the Solar Particle Event (SPE) Threat Assessment Tool or STAT. STAT relies on elements of Corona-Heliosphere (CORHEL) and the Earth-Moon-Mars Radiation Environment Module (EMMREM), and allows users to investigate coronal mass ejection (CME) driven SPEs using coupled magnetohydrodynamic (MHD) and focused transport solutions. At the present time STAT focuses on modeling solar energetic particle (SEP) acceleration in and transport from the low corona, where the highest energy SEP events are generated. We illustrate STAT's capabilities with a model of the July 14, 2000 "Bastille Day" event, including innovative diagnostics for understanding the three-dimensional distribution of particle fluxes and their relation to the structure of the underlying CME driver. A preliminary comparison with NOAA GOES measurements is shown.

1. Introduction

Solar Particle Events (SPEs) represent a significant hazard for humans and technological infrastructure [1]. SPEs can harm aircraft avionics, communication and navigation systems. They also represent a possible risk to the health of airline crews and passengers on polar flights. In space, SPEs can be hazardous for crews of Low Earth Orbit spacecraft and the International Space Station, especially when engaged in extravehicular activity. They may also imperil crews of future manned lunar or interplanetary missions.

Understanding the genesis and temporal development of SPEs spans vast regions of space and very different physical regimes. The largest SPEs are typically associated with X-class flares and very fast (> 1000 km/s) coronal mass ejections (CMEs). These events usually originate in solar active regions that contain strong, highly energized magnetic fields. The violent release of this stored magnetic energy drives the eruption. The solar energetic particles (SEPs) accelerated close to the Sun can travel near the speed of light, and arrive at Earth minutes to hours after the eruptive event, well ahead of the CME. SEP generation continues as the CME propagates outward.

The exact role of flares and CMEs in particle acceleration is presently under debate; however, the compressions and shock waves associated with CMEs have long been recognized as a key driver of SEP acceleration. The compressional properties in turn depend in part on

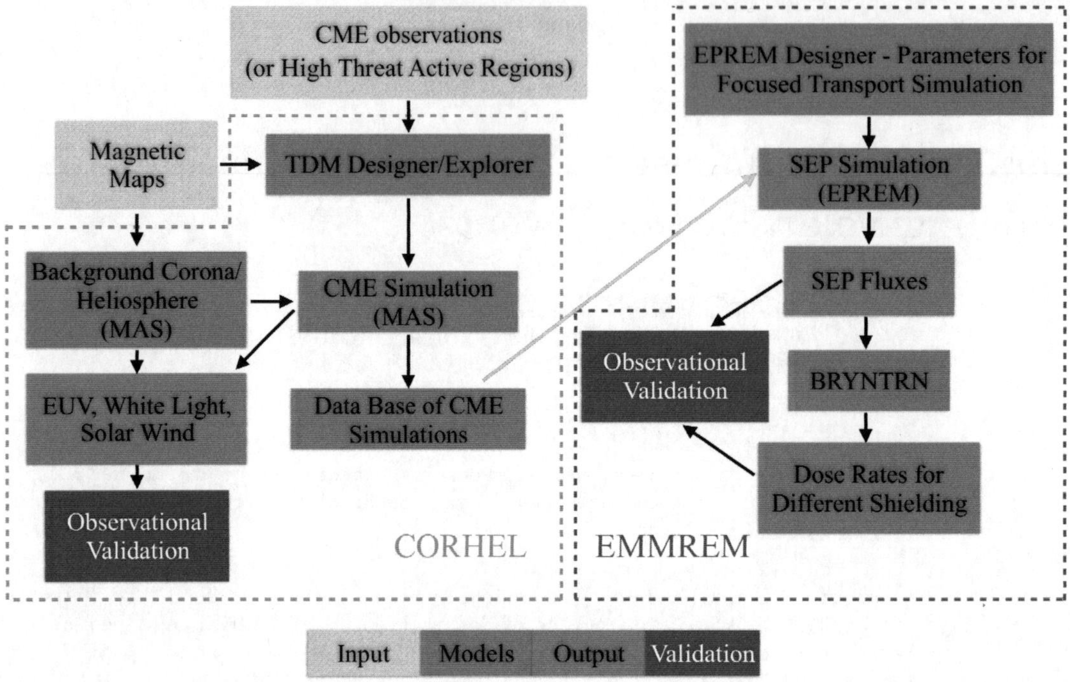

Figure 1. Flow chart depicting the STAT framework. The components derived from CORHEL and EMMREM are shown.

the characteristics of the plasma through which the CME propagates. While much can be learned from theories/models that use simplified treatments of the local acceleration environment and focus on specific aspects of the physics, the complexity of SEP observations have driven increasingly more sophisticated attempts to represent the variation in the coronal/heliospheric environment more realistically. One approach has been to use the magnetic field and plasma properties of CMEs simulated with magnetohydrodynamic (MHD) models to calculate particle acceleration and transport on selected field lines [2]. Considerable work has been done on the acceleration and transport of SEPs at shocks beyond 20 Rs [3, 4, 5] including multi-dimensional approaches [6, 7]. Reference [8] has used the Enlil MHD model [9] to study field line connections for multiple events in the inner heliosphere.

Modeling SEP acceleration and transport from the lower corona presents some of the greatest challenges, as this is where the highest energy particles originate, and the local plasma environment has the greatest variability. To investigate this complex but critical problem, we couple MHD simulations of CMEs in the low corona with 3D solutions of the focused transport equation for SEPs. Our simulations employ CORHEL (Corona-Heliosphere, [10]) and EMMREM (Earth-Moon-Mars Radiation Environment Module, [11]), and have been used to model SEPs generated by a CME initiated in a generic coronal configuration [12], which has led to insights into the possible origin of some of the observed properties of SEPs [13]. Recently, further development of this capability has culminated in the creation of STAT (SPE Threat Assessment Tool) to model real events. STAT allows users to run the Energetic Particle Radiation Environment Module (EPREM, a component of EMMREM) for precomputed Magnetohydrodynamic Algorithm outside a Sphere (MAS, a component of CORHEL) simulations of real CME events to simulate SEP events and provide diagnostics that can be compared with observations. It has recently been delivered to the Community

Coordinated Modeling Center (CCMC) at NASA Goddard Space Flight Center (GSFC) for eventual runs on demand by the community.

Figure 1 summarizes STAT. The present version allows users to perform coupled CME-SEP simulations for previously computed MHD simulations of CME events. The MHD simulations have been performed with MAS/CORHEL and are stored in a database. These are full thermodynamic MHD simulations of CMEs including the solar wind, which is the level of description required for realistic modeling of SEP events. STAT has been delivered with three CME event simulations and more will be added to the database as they become available. The user selects the parameters for a coupled CME-SEP simulation with EPREM designer; this includes choosing the CME event simulation and parameters for the solution of the focused transport equation. At the completion of the run, automated visualizations and diagnostics (described further in section 4) are provided. TDm Designer/Explorer allows users to create stable, pre-event flux-rope magnetic configurations using modified Titov-Demoulín flux ropes [14]; these are computed with magnetogram-derived boundary conditions for user-selected time periods. At the present time, CME simulations with TDm designer use simplified physics (plasma pressure = 0, no energy equation). In the future, it will allow CCMC users to perform thermodynamic CMEs simulations and add these to the database.

In this paper, we describe the elements of this new modeling capability and show an example for a specific event: the July 14, 2000 ("Bastille Day") flare/CME. In the following sections we describe the MHD modeling (section 2) and SEP modeling (section 3) capabilities, and the development of innovative diagnostic techniques to facilitate understanding of the results (section 4). We summarize in section 5.

2. MHD Modeling

2.1. The Background Corona

The properties of compressional regions such as shocks that drive SEP acceleration depend critically on the properties of the local plasma environment. This is especially true in the coronal environment close to the Sun, where the plasma and magnetic properties can vary by orders of magnitude, which leads to significant variation in the local Alfvén (V_A) and sound speeds (C_S). A realistic model of the background solar corona is therefore an essential component of any effort to model the acceleration of SEPs in this region. Simulations of the solar corona for specific time periods using the MAS model have a long history of development and application [15, 16, 17, 18, 10, 19, 20, 21]; MAS is written in Fortran and achieves parallelism using the Message Passing Interface (MPI). It has been shown to scale efficiently to thousands of processors. The primary input data to MAS/CORHEL are synoptic maps of the photospheric magnetic field from observatories such as the National Solar Observatories (NSO) SOLIS and GONG, the Michelson Doppler Imager (MDI) aboard the the Solar and Heliospheric Observatory (SOHO), and the Helioseismic and Magnetic Imager (HMI) aboard the Solar Dynamics Observatory (SDO). The model can also ingest maps from surface flux transport models, e.g., [22] and [23]. In the "thermodynamic" MHD model, a realistic energy equation that accounts for anisotropic thermal conduction, radiative losses, and coronal heating is included. This allows the plasma density and temperature to be computed with sufficient accuracy to simulate EUV and X-ray emission observed from space. This accuracy is vital for obtaining realistic V_A and C_S, but greatly increases the computational requirements. Reference [24] describes the equations solved; 66 million grid points were used in the calculation described in this paper. A semi-implicit algorithm is used to allow for time-steps larger than the fast magneto-sonic wave limit [25, 26]. This feature is crucial for efficient computation of the detailed structure of solar active regions (where V_A is very large) within a global model. Recently, a wave-turbulence-driven (WTD) methodology [27, 28] has been applied to coronal heating in the model; this was used to predict the structure of the solar corona prior to the August 21, 2017 total solar eclipse [29].

An important aspect of the modeling approach for CORHEL is to divide the coronal and heliospheric domains, and to allow for different codes and approximations in each domain. Coronal solutions are computed from the solar surface to beyond the magnetosonic critical points (usually 20-30R_S) and are used to provide the inner boundary condition for heliospheric solutions [30].

2.2. Modeling CMEs

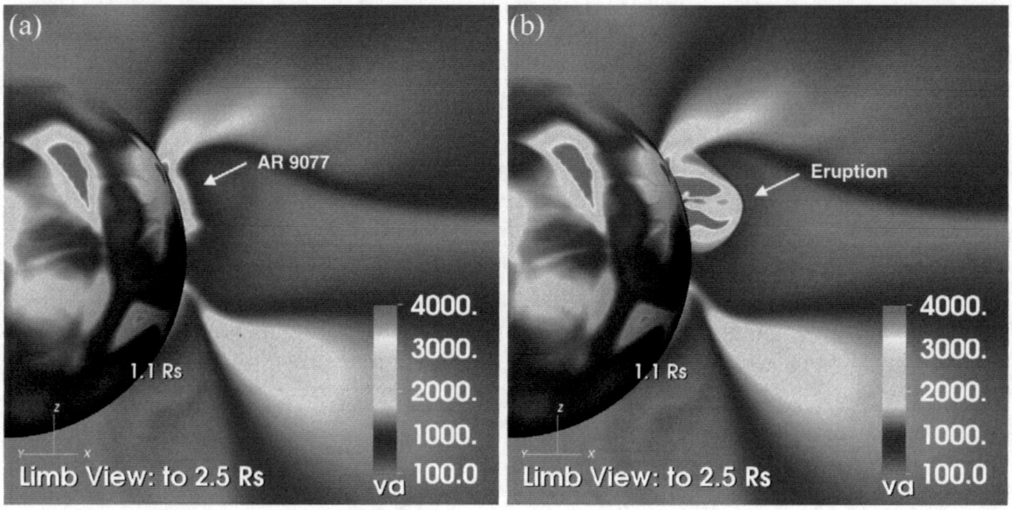

Figure 2. Three-dimensional (3D) view of the Alfvén speed (V_A) in the simulated corona (km/s) for the Bastille Day event, contoured in color on a sphere at 1.1 R_S and in a meridional plane cutting through the center of the active region. (a) V_A just prior to eruption. (b) 2.4 minutes after the start of the eruption.

The largest SEP events start in the low corona; understanding and predicting the particle fluxes of these events requires modeling the initiation and propagation of CMEs in a realistic low coronal environment. To robustly simulate CMEs, we employ an extension to the Titov-Demoulín ([31], hereafter referred to as TD) model, a useful analytical model of the stable pre-event structure of a CME that has found wide application [32, 33, 34]. The flux rope can be destabilized by introducing a perturbation such as the frequently observed canceling flows near the polarity inversion line (PIL, the line separating positive and negative polarities). In the original TD model, the entire magnetic field (both the flux rope field and the overlying arcade that stabilizes it) is described by the equilibrium. To model real/observed CMEs in the corona, we include the background field based on observed photospheric magnetic fields. To produce an equilibrium configuration that satisfies this, the improved model accounts for the pre-existing coronal magnetic field as well as the magnetic field created by the flux rope current itself. Specifically, the flux rope parameters are chosen to account for the stabilizing tension of this pre-existing field. We refer to this as the TDm or modified TD model. With this technique, we introduce a flux rope that is close to, but beneath, the threshold for eruption [30, 14].

To simulate CMEs in global coronal models with TDm, we first separately develop a pre-event flux-rope configuration using a zero-beta MHD model (solution of the momentum equation and Faraday's law with plasma pressure = 0, an option in MAS). These solutions can be computed rapidly (a few minutes) on supercomputers. This relaxed configuration is then inserted into the global coronal model (a full thermodynamic MHD solution with solar wind) and relaxed further, prior to initiating the eruption. We take care to preserve the original B_{r0} (the lower boundary condition, derived from magnetogams) in developing the flux rope configuration, as this a key observational input to the model; the procedure is described by [35] and [24].

The July 14, 2000 CME event (often referred to as the Bastille Day event) commenced with an X5.7 flare in NOAA active region 9077. The CME had a propagation speed of $\approx 1700\,\mathrm{km\,s^{-1}}$ in the SOHO/LASCO coronagraph field of view [36]. The flare was followed by an intense radiation storm that resulted in one of the 16 ground level events of cycle 23 [37]. To model this event, we used a chain of seven overlapping TDm flux ropes, which together form a single rope, to account for the highly elongated and curved PIL in the source region of the eruption. After relaxing this configuration, we introduced converging flows that cancel flux along the PIL to trigger the eruption. Our results, including the interplanetary propagation of the CME, are described by [35] and [24]; here we note some aspects most relevant to modeling SEPs.

Figure 2 shows the variable structure of V_A in the simulated corona. The "bubble" of large V_A protruding upward just after the eruption starts (labeled "eruption" in Figure 2b) is the erupting CME carrying strong magnetic fields upward into the corona. The figure shows the very different local plasma properties the CME can encounter as it evolves. Figure 3(a) shows magnetic field vectors low in the corona projected on a plane cutting through the active region, tracing out the approximate location of the CME flux rope soon after the eruption has started. Superimposed on these are color contours of $\nabla \cdot \mathbf{V}$ (where $\mathbf{V}$ is velocity), with large, negative values in red (these are the regions of strongest compression). The surface magnetic field is contoured in grey on the solar surface (barely visible in this view). Strong compression can be seen in front of the CME, on the flanks, and beneath the flux rope where strong plasma flows associated with reconnection are jetting outward. Figure 3(b) shows the magnitude of $\mathbf{V}$ divided by the maximum fast mode speed $v_f = \sqrt{(V_A^2 + C_S^2)}$.

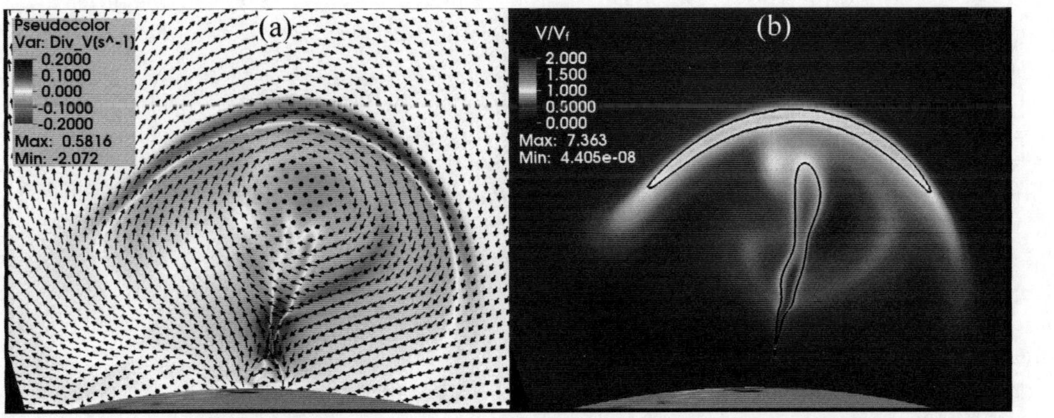

Figure 3. Plasma properties of the simulated CME in a plane cutting through the active region, shortly after eruption. (a) $\nabla \cdot \mathbf{V}$ (compression of the plasma) as color contours superimposed on the magnetic field vectors projected on the plane. (b) Local plasma speed divided by the fast most speed v_f (mach number) in the same plane as (a).

This CME simulation is in the STAT database and can be used as the basis for solutions of the focused transport equation to model particle acceleration and transport, as described in the next section.

3. Focused Transport Modeling of Energetic Particles

To model particle acceleration and transport in the corona and inner heliosphere, we couple time-dependent MHD solutions (like those described in the previous section) to EPREM, which has been designed to couple with MHD models [38]. It is written in C and uses MPI for parallelism. EPREM computes energetic particle distributions along a 3D Lagrangian grid of "nodes" that propagate out with the solar wind in the evolving magnetic fields of the inner heliosphere. The details of the method of solution are described by [11]; here we provide a brief overview. EPREM accounts for the time-dependent transport of pickup ions, suprathermal, and energetic particles along and across magnetic field lines using the evolving 3D field and flow topology provided by the MHD simulation. It produces time histories of the particle velocity distribution functions as a function of pitch angle and energy. Particle transport and energy change parallel to the interplanetary magnetic field is treated in the EPREM model with the focused transport equation [39, 40], using the formalism outlined by [41]:

$$\left(1 - \frac{\mathbf{V}\cdot\mathbf{e_b}v\mu}{c^2}\right)\frac{df}{dt} \tag{1}$$

$$+v\mu\mathbf{e_b}\cdot\nabla f \tag{2}$$

$$+\frac{(1-\mu^2)}{2}\left[-v\mathbf{e_b}\nabla\ln B - \frac{2}{v}\mathbf{e_b}\cdot\frac{d\mathbf{V}}{dt} + \mu\frac{d\ln(n^2/B^3)}{dt}\right]\frac{\partial f}{\partial\mu} \tag{3}$$

$$+\left[-\frac{\mu\mathbf{e_b}}{v}\cdot\frac{d\mathbf{V}}{dt} + \mu^2\frac{d\ln(n/B)}{dt} + \frac{(1-\mu^2)}{2}\frac{d\ln(B)}{dt}\right]\frac{\partial f}{\partial\ln p} \tag{4}$$

$$= \frac{\partial}{\partial\mu}\left(\frac{D_{\mu\mu}}{2}\frac{\partial f}{\partial\mu}\right) \tag{5}$$

where convection (1), particle streaming (2), adiabatic focusing (3), cooling (4), and pitch angle scattering (5) are accounted for. Here d/dt is the convective derivative, $\mathbf{V}$ is the solar wind flow velocity, $\mathbf{e_b}$ is the unit vector along the magnetic field, μ is the cosine of the pitch angle, n is the plasma density, B is the magnetic field strength, p is the particle momentum, and v is the particle speed.

The advantage of this Lagrangian formulation is that most of the transport coefficients are obtained simply by differencing the bulk plasma quantities (e.g., density, field strength, and plasma velocity) at each node moving with the solar wind flow. In this formulation, the grid is made up of nodes that propagate out with the solar wind flow; in the rotating frame of the Sun the nodes follow magnetic field lines. The connected lists of node lines are referred to as streams. This formulation enables highly efficient computation of energetic particle distribution functions at each node. It is further described by [11]; an example of a node mesh is shown in Figure 4 of that paper. EPREM also solves for cross-field diffusion and particle drift using a separate convection-diffusion equation, as described by [11]. For the results shown in this paper, only parallel transport was solved for. [12] describes results that did include these terms. EPREM has been used to solve for pickup ion distributions [11, 42], energetic particle distributions [43, 44], energies from keV up to relativistic GeV energies [45], and has been used to investigate SEP acceleration low in the corona using MHD solutions [46, 12, 13].

In practice, the time-dependent MHD data is supplied to EPREM as a sequence of files that are read in as the EPREM calculation proceeds, and the MHD values are used to compute the terms in the above equation. The initial background MHD solution is used to propagate the nodes of the Lagrangian grid throughout the corona and heliosphere. As the simulation proceeds, nodes are carried out of the domain (beyond 1 AU) and are dropped from the calculation. New nodes are added to the calculation at the lower boundary and propagate into the domain. Closed field regions in the MHD solution (e.g. within helmet streamers) have zero or very weak flow, and are therefore not seeded with nodes. In principle, MHD solutions for both the coronal and heliospheric domains can be included in the EPREM calculation. At the present time we only employ the MHD coronal domain within EPREM, and fill the heliosphere using a simple spiral magnetic field created with a radially constant solar wind speed. This restricts us to modeling the first few hours of an SEP event; once the CME leaves the coronal portion of the domain, possible SEP acceleration from the CME propagation in the heliosphere is not modeled. Particle acceleration in the model occurs primarily through the plasma and magnetic field compressional terms in (3); therefore we expect regions in the MHD simulation undergoing strong compression to provide strong particle acceleration.

As part of developing STAT, we greatly optimized the EPREM computations and made the code more suitable for runs-on-demand by implementing a number of structural changes. These included changing static arrays to dynamic allocation, using temporary stride-1 arrays, re-ordering loops, in-lining functions called within heavily-nested loops, and pre-computing expensive floating point operations outside of loops when possible. We also found that the computation would slow dramatically in cases where the nodes "bunch-up" (such as in the slow flows after a CME launch, or in close proximity to the CME shock), requiring a large number of sub-cycles (for time-step stability). In these cases the nodes were generally much closer than the grid spacing of the MHD values, making these extensive calculations superfluous. We modified the algorithm to ignore nodes within a minimum distance (generally much less than the MHD grid size). This modification greatly reduced the computation time while preserving the same physical solution. As a result of optimization, low resolution runs (96 streams) that required substantial computation (10-20 hours of wall clock time on hundreds of cores) can be completed in less than an hour on a single 24-core node. Typical resolution runs (such as shown in the next section) use 384 streams and 2000 nodes per stream and require several hours on ~100 cores. This is a small fraction of the computational requirements for detailed thermodynamic MHD simulations of CMEs.

3.1. Key Parameters

Within the compressional acceleration paradigm for SEP acceleration and transport, three important parameters that influence the solutions are the parallel mean free path for scattering ($\lambda_{\|}$), the seed population of suprathermal particles (J_{seed}), and diffusion perpendicular to the magnetic field ($\kappa_{\perp}$). These parameters are poorly constrained by observations, so at the present time we use relatively simple approximations. Indeed, one of the purposes of creating STAT is to allow for exploration of the effects of these parameters. We assume $\lambda_{\|} = \lambda_0 (r/r_{1AU})^2 (R_g/R_{g0})^{1/3}$, where R_g is particle rigidity, r is heliocentric distance, and $R_{g0} =$ 1 GV, and $\lambda_0 = 1$ AU. This results in a mean free path that varies substantially over the domain. Other scalings are available, and in principle a physical equation based on the propagation of MHD turbulence could be solved to describe its behavior [47].

For the seed population we follow [46, 48] and assume a differential flux that scales as

$$J_{\text{seed}}(T, r) = J_0 \left(\frac{r_1}{r}\right)^2 \left(\frac{T}{T_0}\right)^{-\gamma_s} \exp\left(-T/T_0\right) \tag{6}$$

where T is kinetic energy, $T_0 = 1$ MeV, $\gamma_s = 1.96$ and J_0 is given in protons cm^{-2} s^{-1} sr^{-1} MeV^{-1} (chosen to reflect the activity level of the time period being modeled). For the results shown in this paper, we investigated the parallel transport processes only, by neglecting particle drifts and assuming $\kappa_{\perp} = 0$. These processes will be explored in future work; [12] describes an initial simulation that included these effects.

4. Diagnostics for Coupled CME-SEP models

The ability to globally simulate a particle event provides the opportunity to directly test whether the CME/shock paradigm can explain its major features. To fully leverage this simulation capability, we need to obtain a global picture of the acceleration in relation to the CME properties. This is not easy, because the simulated particle data reside on the (Lagrangian frame) nodes that are scattered within the domain. We have developed two innovations (automated field tracing and interpolated particle fluxes) to allow automated visualizations and diagnostics to be created at the completion of EPREM calculations.

4.1. Visualizing CME Flux Ropes

A key aspect of understanding and diagnosing simulations of CMEs is obtaining visual diagnostics of the CME evolution and propagation, especially of the magnetic field evolution. In a complex 3D simulation based on realistic magnetic fields, tracing field lines from static locations usually yields an unsatisfactory picture of the evolution. We have found that advecting tracing points with the flow can help to capture key features of the time-dependent behavior. In this approach, we assign footpoint positions to fluid elements, and use the flow field of the simulation to advect these fluid elements, which we then use as the location from which to trace magnetic field lines. This approach provides powerful diagnostics for the CME flux rope, and can allow visualizations to be generated automatically, i.e., without the need for painstaking selection of representative field lines. We have implemented these tracer points into our MAS MHD simulations, so that visualizations of the field lines in the erupting CME can be generated as part of a post-processing pipeline. To combine this capability with the results from EPREM requires a method for visualizing the particle fluxes, as described in the next section.

4.2. Visualization of Particle Fluxes

A challenge for rapidly understanding the results from EPREM simulations is that it is difficult to identify where, in 3D space, the high energy particles are being accelerated and transported, without scanning through many different plots of stream data. The fluxes are not on a standard mesh, but are distributed throughout the volume via the stream locations. We require a method where we can rapidly visualize the location of energetic particle fluxes and relate these to the properties of the simulated CME (such as the location of the erupting flux rope). In order to gain a better understanding of how the particle flux evolves in time and space, we developed a new way to visualize the particle data from EPREM.

EPREM outputs particle fluxes for a specific radius for each EPREM stream, for each energy bin. We define a non-uniform scale vector in r and gather the flux at each of these radii. We integrate over energy to obtain integrated flux values (log10) for the three NOAA GOES energy channels ($F_{\text{int}} > 10, 50$, and 100 MeV). We distribute the integrated fluxes onto a structured spherical grid (as described below). For each time step of the MHD simulation, the resulting 3D logical grid of flux values defined on the grid is written out so that the fluxes can be visualized together with MHD variables.

Creating a structured 3D grid of data points from a non-structured set of points requires an interpolation technique. We experimented with different techniques, including radial basis function interpolation (RBF) [49] and the "brute-force" Shepard [50] weighted average. We found that RBF could generate spurious maxima of flux values in sparsely sampled regions. The Shepard technique gives acceptable results without artifacts, but was time consuming even with parallel computing (~25 minutes on a 24-core node to compute a single GOES flux bin). To increase the efficiency, we developed a new interpolation method. Starting with the same seed points used in the Shepard interpolation, we fill the domain with values using a region-growing methodology. This region-growing algorithm is iterative. For each iteration, each point in the 3D grid is checked. If the point is not already filled with a value, it takes an average of 18 adjacent points (if there is at least one point with a value) and sets the current point to that averaged value (in a temporary array). At the end of the iteration, the main 3D grid is set to the temporary array. The iterations continue until all points are filled. The results are very

IOP Conf. Series: Journal of Physics: Conf. Series **1225** (2019) 012007 doi:10.1088/1742-6596/1225/1/012007

similar to the Shepard interpolation; the major advantage of the new region-growing method is that it takes only 1.1 minutes to compute all three GOES energy bins on 24 cores.

4.3. Visualization and Diagnostics from the Bastille Day Simulation

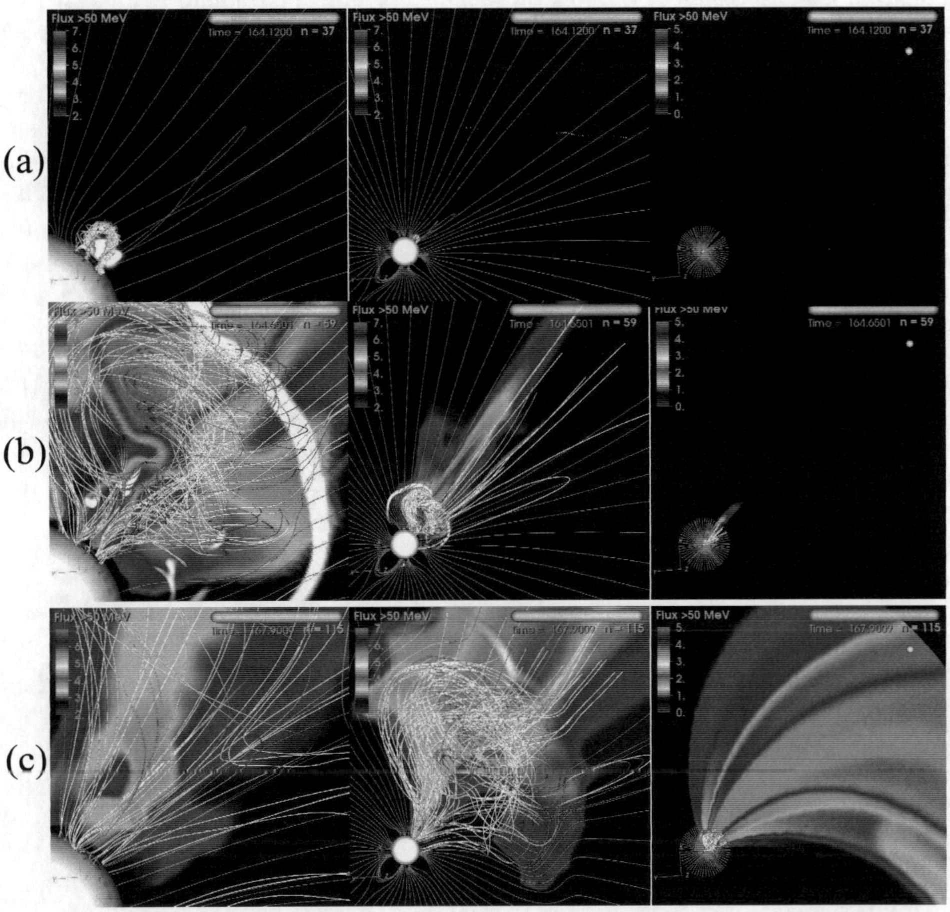

Figure 4. Combined visualization of 3D magnetic field lines, $-\nabla \cdot \mathbf{V}$ (compression), and proton flux for energies above 50 MeV (the GOES 2nd channel). The $>$ 50 MeV flux is shown as color contours, and $-\nabla \cdot \mathbf{V}$ above a threshold value is contoured in white in the heliographic plane. The figures in the leftmost column display the low coronal portion (out to about $4R_S$) of the simulation, the middle frames display the corona to about $17R_S$, and the rightmost frames show the domain to 1 AU (the green sphere on the righthand edge of these images shows the Earth's position. (a) Early in the eruption phase. (b) Time = 1.8 hours in the simulation, about 14 minutes after the start of the eruption. (c) $\sim$1.5 hours after the start of the eruption.

We illustrate the visualization techniques with results for an EPREM calculation coupled with the Bastille Day MHD simulation. Animations of the images are part of the standard report that is provided after a STAT run is completed.

Using tracer points for footpoints (section 4.1) and the interpolated particle fluxes (section 4.2), we create combined visualizations of the particle fluxes with graphical representations of the flux rope field lines; an example is shown in Figure 4. Each frame shows a combined visualization of 3D magnetic field lines, $-\nabla \cdot \mathbf{V}$ (compression), and proton flux for energies above 50 MeV (the second GOES energy channel). The > 50 MeV flux (created from the interpolated particle fluxes as described in section 4.2) is shown as color contours, and $-\nabla \cdot \mathbf{V}$ above a threshold value is contoured in white in the heliographic plane. Each column of frames shows different portions of the domain, with the rightmost frames showing the simulation out to 1 AU. The MHD simulation goes up to 20 R_S, and from $20R_S$ to 1 AU the domain is filled with a simple spiral field as described in section 3. At the start of the eruption (a), the CME flux rope (white field lines) is visible. Magnetic field lines in the background corona are shown in magenta. About 14 minutes after the start of the eruption (shown in (b)), the flux rope has propagated upward. A driven shock (the white, ribbonlike contour) is visible in the corona, strong particle acceleration is in progress, and particles are streaming out of the corona. One and a half hours later, shown in (c), significant fluxes are spread widely at 1 AU. The generation of the highest particle fluxes are associated with the front of the CME low in the corona, but significant particle acceleration occurs at the flanks as well; acceleration of particles from these different regions may be responsible for energy spectra with broken power law distributions [13].

Figure 5 shows the integrated particle fluxes > 10 MeV (the first GOES energy channel) in the heliographic plane for the Bastille Day simulation at three different times. The black diamond shows the Earth's location. The "circles" show the locations of selected EPREM nodes (where values are actually calculated) closest to this plane, and gives a measure of how well different regions are sampled to produce the images created from the interpolated fluxes. In the simulated event, SEPs rapidly reach 1 AU; the Earth's location received significant fluxes but higher fluxes propagated eastward of the Earth.

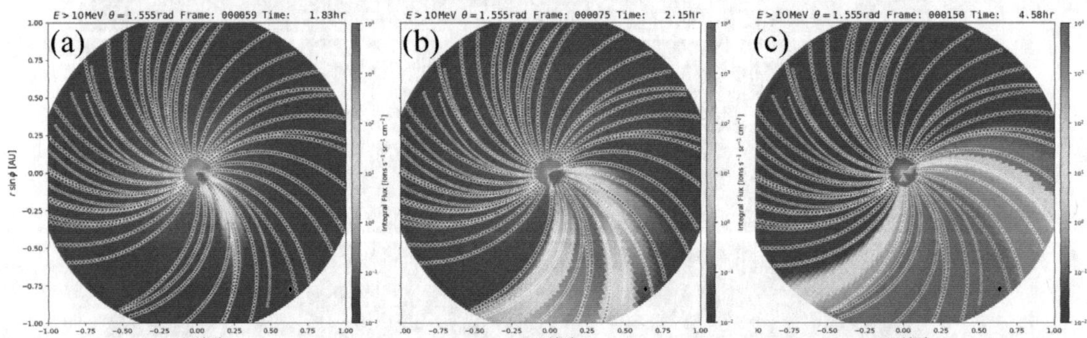

Figure 5. 2D visualization of the simulated particle flux in the first GOES energy bin ($F_{\text{int}} > 10$MeV) in the heliographic ($r{-}\phi$) plane. (a) Time = 1.8 hours in the simulation, about 14 minutes after the start of the eruption. (b) Time = 2.15 hours, about 1/2 hour after eruption. (c) ~3 hours after eruption.

Figure 6 shows the particle fluxes in the first GOES energy bin at time = 4.58 hours (about 3 hours after the start of the eruption, the same time as Figure 5(c)) on a spherical shell at 1 AU. The shell has been stretched out as a sin(latitude)-heliographic longitude map. Strong, but highly variable, particle fluxes are seen near Earth with significant spread in longitude and latitude.

To test the simulation results against observations, we provide plots and data that can be compared with GOES measurements. To understand the particle sources, it is essential to be able relate these results at a specific location to the overall 3D structure of the corona and heliosphere. To facilitate this identification, STAT provides a clickable interface for images like Figure 6 that allows the user to browse particle fluxes at different locations. The user can select

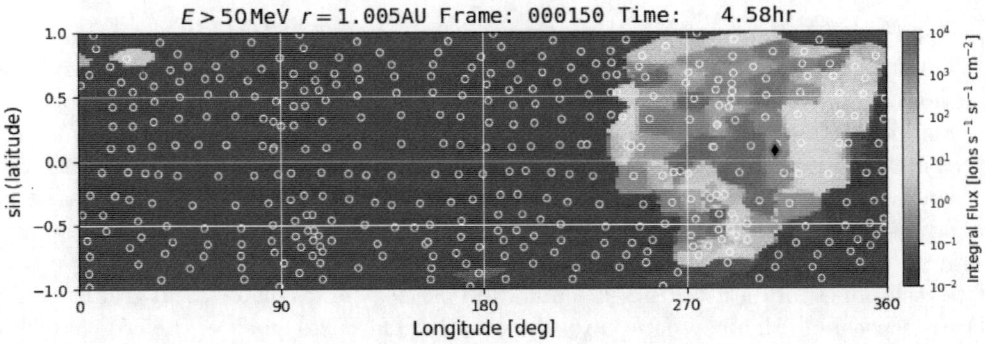

Figure 6. Simulated particle flux in the first GOES energy bin ($F_{\rm int} > 10$MeV) at 1 AU, as a function of sin(latitude) and heliographic longitude. The location of selected EPREM nodes for the particle calculation are shown as circles. Earth's location is the black diamond.

the node locations and view the temporal evolution for the GOES energy bins, as shown in Figure 7. This figure shows the simulated fluxes at three of the streams close to Earth in the simulation. Stream 154 is the circle closest to Earth (partially obscured by the black diamond) in Figure 6; streams 153 and 155 are the adjacent circles. Also shown in Figure 7 is the actual GOES data for this event. The CME propagates out of the coronal domain of the simulation at just past 5 hours, so a comparison with GOES observations is meaningful only up to this time. The behavior of the fluxes show similarity to the GOES observations early in the event, however, the figure illustrates the importance of investigating the variability in the solution, as the three locations give different values. This variability might be lessened if $\kappa_\perp$ and particle drifts were included in the solution.

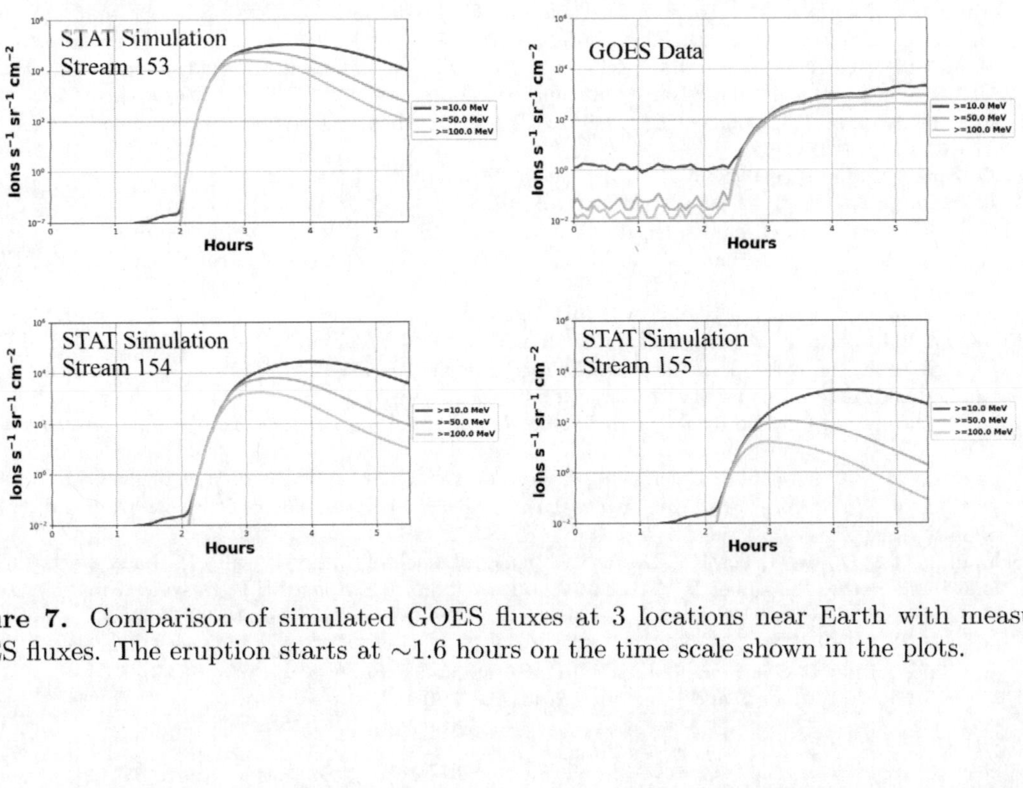

Figure 7. Comparison of simulated GOES fluxes at 3 locations near Earth with measured GOES fluxes. The eruption starts at ~1.6 hours on the time scale shown in the plots.

5. Summary

We have described STAT, a tool for investigating CME-driven SEP acceleration and transport in real events using solutions of the focused transport equation coupled with MHD models. A simulation of the July 14, 2000 CME is used to illustrate the combined model. Full thermodynamic MHD solutions (that account for coronal heating and energy transport) are necessary obtaining realistic Alfven and sound speeds in the simulated corona. In the simulated event, the erupting CME produces strong compressions and shocks very low in the corona. These compressions drive significant particle acceleration in the focused transport simulations, and energetic particle fluxes stream outward and rapidly arrive at Earth orbit. We described innovative techniques for distributing the simulated particle fluxes on a logical grid to understand their 3D structure and visualize them in relation to MHD features, such as the erupting flux rope. Our results show that even for a complex event like Bastille Day, this modeling approach can produce particle fluxes with characteristics reasonably similar to interplanetary measurements, indicating that this is a promising direction for tackling the difficult problem of SPEs.

We consider the results we have presented here to be preliminary, as we are still exploring how the variation of parameters can affect the solutions. We plan to publish a more detailed study of this event when this exploration is completed. Future planned improvements to this modeling include increasing the number of streams in the EPREM calculations, and extending EPREM to include both the coronal and heliospheric domains from the MAS simulations. The latter extension will allow STAT to follow the entire development of an SPE that can occur over days as the CME propagates in interplanetary space.

Acknowledgments

This research was supported by AFOSR, by NASA via the LWS TR&T, HSR, and STTR programs, and by NSF. Computational resources were provided by NSF's XSEDE and NASA's NAS.

References

[1] Schrijver C J, Kauristie K, Aylward A D, Denardini C M, Gibson S E, Glover A, Gopalswamy N, Grande M, Hapgood M, Heynderickx D, Jakowski N, Kalegaev V V, Lapenta G, Linker J A, Liu S, Mandrini C H, Mann I R, Nagatsuma T, Nandy D, Obara T, Paul O'Brien T, Onsager T, Opgenoorth H J, Terkildsen M, Valladares C E and Vilmer N 2015 *Advances in Space Research* **55** 2745–2807 (*Preprint* 1503.06135)

[2] Sokolov I V, Roussev I I, Gombosi T I, Lee M A, Kóta J, Forbes T G, Manchester W B and Sakai J I 2004 ApJ **616** L171–L174

[3] Li G, Zank G P and Rice W K M 2003 *J. Geophys. Res.* **108** 10–1

[4] Li G and Zank G P 2005 *Geophys. Res. Lett.* **32** 2101–+

[5] Verkhoglyadova O P, Li G, Zank G P, Hu Q and Mewaldt R A 2009 ApJ **693** 894–900

[6] Hu J, Li G, Ao X, Zank G P and Verkhoglyadova O 2017 *Journal of Geophysical Research (Space Physics)* **122** 10

[7] Zhang M, Qin G and Rassoul H 2009 ApJ **692** 109–132

[8] Luhmann J G, Mays M L, Odstrcil D, Li Y, Bain H, Lee C O, Galvin A B, Mewaldt R A, Cohen C M S, Leske R A, Larson D and Futaana Y 2017 *Space Weather* **15** 934–954

[9] Odstrcil D 2003 *Advances in Space Research* **32** 497–506

[10] Riley P, Linker J A, Lionello R and Mikic Z 2012 *Journal of Atmospheric and Solar-Terrestrial Physics* **83** 1–10

[11] Schwadron N A, Townsend L, Kozarev K, Dayeh M A, Cucinotta F, Desai M, Golightly M, Hassler D, Hatcher R, Kim M Y, Posner A, PourArsalan M, Spence H E and Squier R K 2010 *Space Weather* **80** S00E02

[12] Schwadron N A, Gorby M, Török T, Downs C, Linker J, Lionello R, Mikić Z, Riley P, Giacalone J, Chandran B, Germaschewski K, Isenberg P A, Lee M A, Lugaz N, Smith S, Spence H E, Desai M, Kasper J, Kozarev K, Korreck K, Stevens M, Cooper J and MacNeice P 2014 *Space Weather* **12** 323–328

[13] Schwadron N A, Lee M A, Gorby M, Lugaz N, Spence H E, Desai M, Török T, Downs C, Linker J, Lionello R, Mikić Z, Riley P, Giacalone J, Jokipii J R, Kota J and Kozarev K 2015 ApJ **810** 97

[14] Titov V S, Török T, Mikic Z and Linker J A 2014 ApJ **790** 163

[15] Mikić Z and Linker J A 1996 The large-scale structure of the solar corona and inner heliosphere *International Solar Wind 8* vol 382 ed Winterhalter D e a (AIP Conf. Proceedings) p 104
[16] Linker J A, Mikić Z, Bisecker D A, Forsyth R J, Gibson S E, Lazarus A J, Lecinski A, Riley P, Szabo A and Thompson B J 1999 J. Geophys. Res. **104** 9809
[17] Mikić Z, Linker J A, Schnack D D, Lionello R and Tarditi A 1999 *Phys. Plasmas* **6** 2217
[18] Lionello R, Linker J A and Mikić Z 2009 ApJ **690** 902–912
[19] Downs C, Linker J A, Mikić Z, Riley P, Schrijver C J and Saint-Hilaire P 2013 *Science* **340** 1196–1199
[20] Titov V S, Mikić Z, Török T, Linker J A and Panasenco O 2017 ApJ **845** 141 (*Preprint* 1707.07773)
[21] Linker J A, Caplan R M, Downs C, Riley P, Mikic Z, Lionello R, Henney C J, Arge C N, Liu Y, Derosa M L, Yeates A and Owens M J 2017 ApJ **848** 70 (*Preprint* 1708.02342)
[22] Schrijver C J and DeRosa M L 2003 *Sol. Phys.* **212** 165–200
[23] Arge C N, Henney C J, Koller J, Compeau C R, Young S, MacKenzie D, Fay A and Harvey J W 2010 *Twelfth International Solar Wind Conference* **1216** 343–346
[24] Török T, Downs C, Linker J A, Lionello R, Titov V S, Mikić Z, Riley P, Caplan R M and Wijaya J 2018 ApJ **856** 75 (*Preprint* 1801.05903)
[25] Lionello R, Mikić Z and Linker J A 1999 *Journal of Computational Physics* **152** 346–358
[26] Caplan R M, Mikić Z, Linker J A and Lionello R 2017 Advancing parabolic operators in thermodynamic MHD models: Explicit super time-stepping versus implicit schemes with Krylov solvers *Journal of Physics Conference Series* (*Journal of Physics Conference Series* vol 837) p 012016 (*Preprint* 1610.01265)
[27] Lionello R, Velli M, Downs C, Linker J A and Mikić Z 2014 ApJ **796** 111 (*Preprint* 1410.1789)
[28] Downs C, Lionello R, Mikić Z, Linker J A and Velli M 2016 *"Closed-Field Coronal Heating Driven by Wave Turbulence," ArXiv e-print 1610.02113, to appear in ApJ* (*Preprint* 1610.02113)
[29] Mikić Z, Downs C, Linker J A, Caplan R M, Mackay D H, Upton L A, Riley P, Lionello R, Török T, Titov V S, Wijaya J, Druckmüller M, Pasachoff J M and Carlos W 2018 *Nature Astronomy* **2** 913–921
[30] Lionello R, Downs C, Linker J A, Török T, Riley P and Mikić Z 2013 ApJ **777** 76
[31] Titov V S and Démoulin P 1999 A&A **351** 707–720
[32] Roussev I I, Forbes T G, Gombosi T I, Sokolov I V, DeZeeuw D L and Birn J 2003 *Ap. J. Lett.* **588** L45–L48
[33] Török T and Kliem B 2005 ApJ **630** L97–L100 (*Preprint* arXiv:astro-ph/0507662)
[34] Török T and Kliem B 2007 *Astronomische Nachrichten* **328** 743–+ (*Preprint* 0705.2100)
[35] Linker J, Torok T, Downs C, Lionello R, Titov V, Caplan R M, Mikić Z and Riley P 2016 MHD simulation of the Bastille day event *American Institute of Physics Conference Series* (*American Institute of Physics Conference Series* vol 1720) p 020002
[36] Andrews M D 2001 Sol. Phys. **204** 179–196
[37] Bieber J W, Dröge W, Evenson P A, Pyle R, Ruffolo D, Pinsook U, Tooprakai P, Rujiwarodom M, Khumlumlert T and Krucker S 2002 ApJ **567** 622–631
[38] Kozarev K, Schwadron N A, Dayeh M A, Townsend L W, Desai M I and PourArsalan M 2010 *Space Weather* **8** S00E08
[39] Ruffolo D 1995 *Ap. J.* **442** 861–874 (*Preprint* arXiv:astro-ph/9408056)
[40] Skilling J 1971 ApJ **170** 265
[41] Kóta J, Manchester W B, Jokipii J R, de Zeeuw D L and Gombosi T I 2005 Simulation of SEP Acceleration and Transport at CME-driven Shocks *The Physics of Collisionless Shocks: 4th Annual IGPP International Astrophysics Conference* (*American Institute of Physics Conference Series* vol 781) ed Li G, Zank G P and Russell C T pp 201–206
[42] Chen J H, Schwadron N A, Möbius E and Gorby M 2015 *Journal of Geophysical Research (Space Physics)* **120** 9269–9280
[43] Dayeh M A, Desai M I, Kozarev K, Schwadron N A, Townsend L W, PourArsalan M, Zeitlin C and Hatcher R B 2010 *Space Weather* **8** S00E07
[44] PourArsalan M, Townsend L W, Schwadron N A, Kozarev K, Dayeh M A and Desai M I 2010 *Space Weather* **8** S00E05
[45] Cucinotta F A, Hu S, Schwadron N A, Kozarev K, Townsend L W and Kim M H Y 2010 *Space Weather* **8** S00E09
[46] Kozarev K A, Evans R M, Schwadron N A, Dayeh M A, Opher M, Korreck K E and van der Holst B 2013 ApJ **778** 43 (*Preprint* 1406.2377)
[47] Sokolov I V, Roussev I I, Skender M, Gombosi T I and Usmanov A V 2009 ApJ **696** 261–267
[48] Dayeh M A, Desai M I, Dwyer J R, Rassoul H K, Mason G M and Mazur J E 2009 ApJ **693** 1588–1600
[49] Broomhead D and Lowe D 1988 *Complex Systems* **2** 321–355
[50] Shepard D 1968 A two-dimensional interpolation function for irregularly-spaced data *Proceedings of the 1968 23rd ACM National Conference* ACM '68 (New York, NY, USA: ACM) pp 517–524 URL http://doi.acm.org/10.1145/800186.810616

IOP Conf. Series: Journal of Physics: Conf. Series **1225** (2019) 012008 doi:10.1088/1742-6596/1225/1/012008

A Dynamical Model of the Heliosphere with the Adaptive Mesh Refinement

Tomoaki Matsumoto[1,2,3], Daikou Shiota[4,5], Ryuho Kataoka[6,7], Hiroko Miyahara[8], Shoko Miyake[9],

[1] Faculty of Sustainability Studies, Hosei University, Fujimi, Chiyoda-ku, Tokyo 102-8160, Japan
[2] Department of Astrophysical Sciences, Princeton University, 4 Ivy Lane, Princeton, NJ 08544, USA
[3] RIKEN Center for Computational Science (R-CCS), 7-1-26 Minatojima-minami-machi, Chuo-ku, Kobe, Hyogo 650-0047, Japan
[4] National Institute of Information and Communications Technology (NICT), 4-2-1, Nukui-Kitamachi, Koganei, Tokyo 184-8795, Japan
[5] Institute for Space-Earth Environmental Research (ISEE), Nagoya University Furo-cho, Chikusa-ku, Nagoya, 464-8601, Japan
[6] National Institute of Polar Research, 10-3, Midori-cho, Tachikawa-city, Tokyo 190-8518, Japan
[7] Department of Polar Science, School of Multidisciplinary Sciences, Sokendai 10-3 Midori-cho, Tachikawa-city, Tokyo 190-8518, Japan
[8] Humanities and Sciences/Museum Careers, Musashino Art University, 1-736 Ogawa-cho, Kodaira-city, Tokyo 187-8505, Japan
[9] National Institute of Technology, Ibaraki College, Nakane, Hitachinaka-shi, Ibaraki 312-8508 Japan

E-mail: matsu@hosei.ac.jp

Abstract. A three-dimensional time-dependent heliospheric model is developed with the adaptive mesh refinement (AMR) technique. The AMR code, SFUMATO, is adopted, which has been originally developed for star-formation problems. The solar wind model is based on the space weather forecast system, SUSANOO, and it is imposed on the inner boundary at $r = 25\,R_\odot$. The solar wind model is derived from magnetogram synoptic maps, which are observed daily. Using the potential field source surface models, the WS formula, and a few empirical laws, the MHD parameters at the inner boundary are reproduced. The heliospheric model here extends up to 14 au, and the AMR grid resolves the heliospheric current sheets and co-rotating interaction regions with refined resolutions. The AMR grid brings considerable improvement in the fine features in the outer region of $r \gtrsim 2\,\mathrm{au}$ compared to the concentric nested grid, of which resolution has the relationship $\Delta x \propto r$, similar to spherical coordinates. The refinement criteria are based mainly on the distance from the Sun, the anti-parallel toroidal magnetic fields, and the density of the solar wind.

1. Introduction

Simulating a change in the heliospheric environment is important for space weather. The magnetic field structure and the speed of the solar wind affect the modulation of galactic cosmic rays (GCRs).

In order to calculate the modulation of GCRs, the simple Parker spiral models have been utilized [1, 2]. This approach is advantageous in that the heliospheric environment is described by several parameters, e.g., the solar wind speed, the tilt angle, and the polarity, and it is easy to perform numerical experiments according to observations of solar activity and cycle. The spiral should, however, have more complicated structures in dynamical and time-dependent phenomena. Several works [3, 4] have used magneto-hydrodynamics (MHD) models to reproduce the modulation of GCRs, while steady-state solutions have been adopted.

Since the heliospheric environment is affected by solar activity, we have been developing a framework for simulating the heliosphere by using MHD simulations. The GCRs are transported efficiently in the heliospheric current sheet (HCS) [5]. The gradient-curvature drift of the GCRs allows them to propagate along the HCS efficiently. The HCS should be reproduced with a fine resolution in the model. In the previous work [2], we solved the stochastic differential equations (SDEs) rather than the individual motion of the GCR proton. The timestep is restricted in the region close to the HCS so that the HCS is resolved at a high spatial resolution, e.g., $\Delta x = 0.1$ au at $r = 2$ au. The required resolution is larger at the points farther from the Sun. We therefore utilized the adaptive mesh refinement (AMR) technique for improving the local resolution. In this paper, we present the current status of the model development.

This paper is organized as follows. In Section 2, the models and methods are shown, and the prescription of the solar wind model and the refinement criteria are described. The results of the simulations are presented in Section 3, and they are summarized and discussed in Section 4.

2. Model and Method

2.1. Basic Equations

We assume the ideal MHD with a point source of gravitational force. The governing equations are given by,

$$\frac{\partial \rho}{\partial t} + \nabla \cdot (\rho \boldsymbol{v}) = 0, \tag{1}$$

$$\frac{\partial}{\partial t}(\rho \boldsymbol{v}) + \nabla \cdot \left[\rho \boldsymbol{v}\boldsymbol{v}^T + \left(P + \frac{|\boldsymbol{B}|^2}{2}\right) I - \boldsymbol{B}\boldsymbol{B}^T\right] = \rho \boldsymbol{g}, \tag{2}$$

$$\frac{\partial \boldsymbol{B}}{\partial t} + \nabla \cdot \left(\boldsymbol{v}\boldsymbol{B}^T - \boldsymbol{B}\boldsymbol{v}^T\right) = 0, \tag{3}$$

$$\frac{\partial}{\partial t}(\rho E) + \nabla \cdot \left[\left(\rho E + P + \frac{|\boldsymbol{B}|^2}{2}\right) \boldsymbol{v} - \boldsymbol{B}(\boldsymbol{v} \cdot \boldsymbol{B})\right] = \rho \boldsymbol{g} \cdot \boldsymbol{v}, \tag{4}$$

$$E = \frac{|\boldsymbol{v}|^2}{2} + \frac{P}{(\gamma - 1)\rho} + \frac{|\boldsymbol{B}|^2}{2\rho}, \tag{5}$$

$$\boldsymbol{g} = -\frac{GM_\odot}{r^3}\boldsymbol{r}, \tag{6}$$

where ρ, $\boldsymbol{v}$, $\boldsymbol{B}$, E, P, $\boldsymbol{g}$, and I are the density, velocity, magnetic field, energy in the unit mass, pressure, gravity, and unit vector, respectively. The upper script T denotes a transposed matrix. The adiabatic index is assumed as $\gamma = 1.46$ [6]. The pressure P and temperature T are related by $P = (\rho/\mu)k_B T$, where k_B and $\mu (= 0.5)$ are the Boltzmann constant and mean molecular weight.

2.2. Computational Domain and Boundary Conditions

The computational domain is a cube with $x, y, z \in [-R_{\rm out}, R_{\rm out}]^3$, where $R_{\rm out}$ denotes the size of the computational domain. We set at $R_{\rm out} = 3000 R_\odot = 14$ au. The Sun is located at the origin (the center of the computational domain).

At the inner boundary condition, the solar wind is injected into the computational domain. The inner boundary condition is located at $r = R_{\rm in}$ in spherical coordinates (r, θ, φ). We set $R_{\rm in} = 25\,R_\odot$, following [6]. Note that the solar wind velocity should be super-Alfvénic at $r = R_{\rm in}$ in order to avoid inward propagation of the waves. The solar wind model imposed here is described in the next subsection.

The outer boundary condition is set at the surfaces $x, y, z = \pm R_{\rm out}$, the edges of the computational domain. The zero-gradient boundary condition is imposed there so that the wind goes out of the computational domain.

2.3. Solar Wind Model

The solar wind model is based on the space weather forecast system, SUSANOO[1] [7, 6]. The parameters of the solar wind are derived from observations of the solar magnetic field. Given time-series data of magnetogram synoptic maps, which are magnetic field distributions in two dimensions (θ, φ) at $r = R_\odot$, the model reproduces the solar wind as the inner boundary condition at $r = R_{\rm in}$. To date, the Global Oscillation Network Group (GONG) project[2] and the Wilcox Solar Observatory (WSO)[3] are available as input observational data. In this paper, we use the photospheric magnetograms provided by the GONG project.

Based on the magnetogram synoptic map, the magnetic fields in the corona of $R_\odot \le r \le R_s$ are reconstructed by using the potential field source surface (PFSS) model, where R_s denotes the source surface and is set at $2.5 R_\odot$. We thus obtain the radial component of the magnetic field at the source surface $B_r(R_s)$. The PFSS model assumes of the current-free magnetic field $(\nabla \times \boldsymbol{B} = 0)$, and the magnetic fields have only the radial component in the source surface $(B_\theta = B_\varphi = 0)$, connecting to the open magnetic fields.

The solar wind velocity at the inner boundary $(r = R_{\rm in})$ is obtained by the so-called WS formula as follows [8, 9],

$$v_r = 267.5 + \frac{410}{f_s^{0.4}}\,{\rm km\,s^{-1}}, \tag{7}$$

where f_s denotes the super-radial expansion factor defined as,

$$f_s = \left(\frac{R_\odot}{R_s}\right)^2 \frac{B_r(R_\odot)}{B_r(R_s)}. \tag{8}$$

The super-radial expansion factor expresses a change in magnetic field strength along each field line, and the magnetic field strength $B_r(R_\odot)$ is measured at the foot point of the field line of $B_r(R_s)$. We assume that $v_\theta = v_\varphi = 0$ at the inner boundary. The WSA model [10, 11] is an alternative model for v_r, and improvement of the solar wind speed was discussed using this model [12, 13]. The WSA model will be incorporated into our model in future work.

The magnetic field at the inner boundary is extended by assuming a simple relationship,

$$B_r(R_{\rm in}) = \left(\frac{R_{\rm in}}{R_s}\right)^{-2} B_r(R_s). \tag{9}$$

The azimuthal component of the magnetic field is set at,

$$B_\varphi(R_{\rm in}) = -B_r(R_{\rm in})\frac{R_{\rm in}\Omega_c \sin\theta}{v_r}, \tag{10}$$

[1] http://cidas.isee.nagoya-u.ac.jp/susanoo/
[2] https://gong.nso.edu
[3] http://wso.stanford.edu

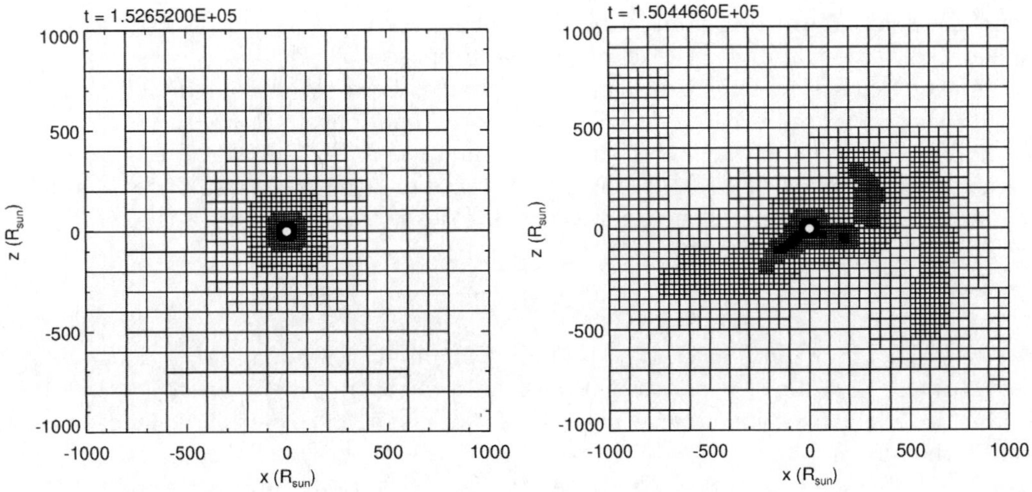

Figure 1. Grid configuration in the $y = 0$ slice plane. The squares with vertical and horizontal lines denote the AMR blocks. Each block contains 8^3 cells. The left panel shows the concentric nested grid according to equation (13). The right panel shows the refined grid according to detection of the HCS.

according to the Parker spiral, and $B_\theta = 0$, where $\Omega_c\,(= 2\pi/25.38\,\text{days})$ is the angular velocity of the Sun.

The density and temperature at $r = R_{\text{in}}$ are obtained by following [14],

$$n = \left(\frac{R_{\text{in}}}{50R_\odot}\right)^{-2} \left[62.98 + 866.4\left(\frac{v_r}{100\,\text{km}\,\text{s}^{-1}} - 1.549\right)^{-3.402}\right] \text{cm}^{-3}, \tag{11}$$

$$T = \left(\frac{R_{\text{in}}}{50R_\odot}\right)^{-2(\gamma-1)} \left(-0.455 + 0.1943\frac{v_r}{100\,\text{km}\,\text{s}^{-1}}\right) \times 10^6\,\text{K}. \tag{12}$$

Because the original formula[14] estimates the density and temperature at $r = 50R_\odot$, the equations scale them up to $r = R_{\text{in}}$.

The GONG project provides a daily magnetogram synoptic map with $1^\circ \times 1^\circ$ resolution in the $\theta - \varphi$ plane. By using the magnetogram synoptic maps and the prescription descried above, we derive the daily data of the two-dimensional distributions of $\boldsymbol{B}$, $\boldsymbol{v}$, n, and T at $r = R_{\text{in}}$ in the $\theta - \varphi$ plane with the same resolution of $1^\circ \times 1^\circ$. We performed a temporal linear interpolation of daily data in order to obtain the data at a specific time level for the simulation. The temporal interpolated data is remapped from the spherical grid to the Cartesian grid by performing a spatial linear interpolation. Because we calculate the heliosphere in the stationary frame, the inner boundary condition is shifted in the φ-direction in accordance with the rotation of the Sun.

The inner boundary condition described here does not take into account the divergence-free treatment of the magnetic field like [15]. Although a change in B_r at the inner boundary condition produces $\nabla \cdot \boldsymbol{B}$ in the computational domain, the hyperbolic divergence cleaning method [19] suppresses the instability which arises from the solenoidality of the magnetic field.

2.4. Grid Refinement
We used the MHD-AMR code, SFUMATO [16], which has been originally developed for star-formation problems. SFUMATO utilizes the block-structured grid with an octree arrangement.

The refinement is performed in units of blocks. Each block contains 8^3 cells.

We adopted AMR mainly for the following two purposes.

The first purpose is to set a grid suitable for radial flow. We adopted the concentrically nested grid shown in Figure 1 (left). This grid has the relationship $\Delta x \propto r$, which is the same as that of the azimuthal resolution of spherical coordinates. When the radial grid spacing is proportional to the azimuthal resolution, as is often done, the relationship $\Delta x \propto r$ holds for every dimension of the spherical coordinates. Each block is therefore refined up to the grid level of

$$\ell = \mathrm{int}\left[\log_2\left(\frac{R_{\mathrm{out}}}{d_{\mathrm{max}}}\right)\right] \tag{13}$$

where $\ell \in [0, \ell_{\mathrm{max}}]$ is the grid level, ℓ_{max} is the maximum grid level, and d_{max} is the maximum distance of the cells within that block from the origin. In this paper, we set $\ell_{\mathrm{max}} = 6$. Hereafter, this grid is called the fixed mesh refinement (FMR) because the resolution depends only on the location of the blocks.

The second purpose is to resolve the HCS with fine resolutions. In order to detect the HCS, we measure the anti-parallel magnetic field in the φ component and a high density as well. A block is refined when the following two criteria are satisfied,

$$\max_{\mathrm{block}}(B_\varphi) \times \min_{\mathrm{block}}(B_\varphi) < 0, \tag{14}$$

$$\max_{\mathrm{block}}(n)\left(\frac{r}{1\ \mathrm{au}}\right)^2 \geq 10\,\mathrm{cm}^{-3}, \tag{15}$$

where $\max_{\mathrm{block}}()$ and $\max_{\mathrm{block}}()$ mean the maximum and minimum values in a block, respectively. Equation (14) detects the blocks in which two adjacent cells have B_φ with different signs, e.g., $B_{\varphi,i,j,k}B_{\varphi,i+1,j,k} < 0$, showing that the HCS exists at the grid point $(i + 1/2, j, k)$. As shown in section 3.2, equation (14) detects noisy structures in the magnetic field as well. In order to exclude these structures, equation (15) is additionally evaluated. Finally, because of the geometrical effect of $B_\varphi \to 0$ as $\theta \to 0$, some blocks near the poles are satisfied with equation (14) although these block do not include the HCS. We therefore exclude the blocks near the poles after equations (14) and (15) are examined. If a block is satisfied with these criteria, the block is refined to two more levels than the FMR shown in Figure 1 (left). The resultant grid is shown in Figure 1 (right). Hereafter, this grid is called the AMR grid.

2.5. Scheme

The MHD method has third-order accuracy in space with MUSCL and second-order accuracy in time with the predictor-corrector method. We adopted the HLLD− Riemann solver [17] for the numerical flux. This solver is based on the HLLD Riemann solver [18], but it includes measures for curing the carbuncle phenomenon.

The hyperbolic divergence cleaning method [19] was adopted for suppressing numerical instability caused by $\nabla \cdot \boldsymbol{B}$. The method proposes several formulations, and the mixed EGLM formulation [19] was adopted here.

3. Results

3.1. Overall structure

We followed the evolution of the heliosphere from October 1, 2016 to June 1, 2017. Figure 2 shows the solar wind speed distribution in the heliosphere on March 1, 2017. This figure illustrates that the time-dependent solar wind is reproduced by the AMR grid.

The left panels of Figure 2 show that the Parker spiral exhibits complex structures, indicating that it is not in a steady state. In the middle-left panel, the fast winds catch up with the slow winds, producing the shock waves. These features correspond to the formation of co-rotating

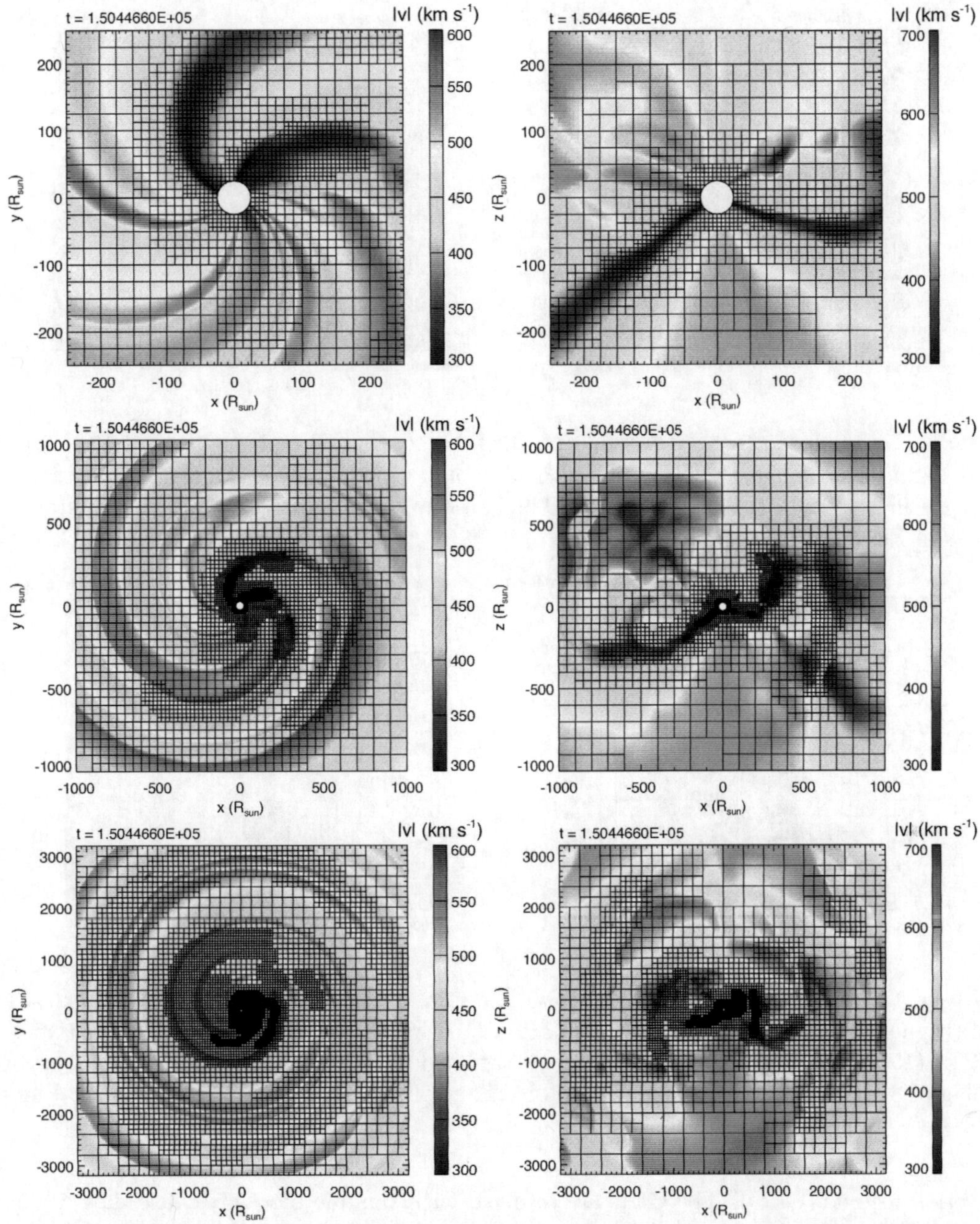

Figure 2. Velocity distributions of the heliosphere on March 1, 2017. Left and right panels are the cross sections in the $z = 0$ and the $y = 0$ planes, respectively, on larger scales from top to bottom. The color denotes the velocity. The squares with vertical and horizontal lines denote the AMR blocks. The white circle at the center denotes the inner boundary in each panel.

IOP Conf. Series: Journal of Physics: Conf. Series **1225** (2019) 012008 doi:10.1088/1742-6596/1225/1/012008

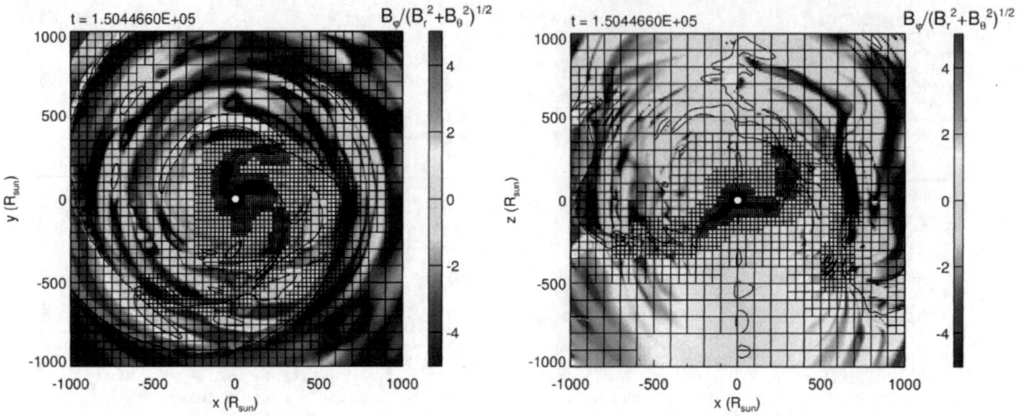

Figure 3. Ratios between the toroidal and poloidal components of magnetic fields, $B_\varphi/(B_r^2 + B_\theta^2)^{1/2}$ in the $z = 0$ plane (left) and the $y = 0$ plane (right). The squares with vertical and horizontal lines denote the AMR blocks. The contour curve denotes $B_\varphi = 0$. The stage is the same as in Figure 2.

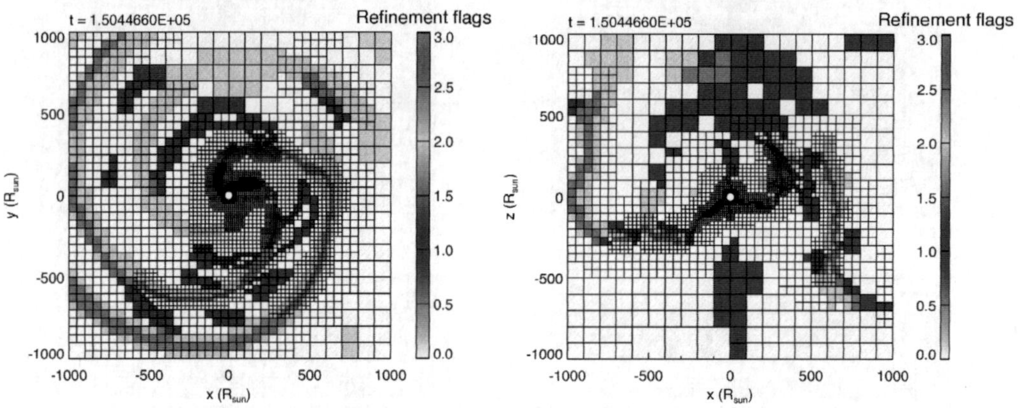

Figure 4. Distributions of the refinement flags in the $z = 0$ plane (left) and the $y = 0$ plane (right). The grids satisfied with equation (14) have 1 bit (blue), and those satisfied with equation (15) have 2 bits (green). The grids satisfied with both the criteria have 3 bits (red). The flag of 3 bits is a necessary condition for refinement. The stage is the same as in Figure 2.

interaction regions (CIRs). The CIRs are resolved with the fine blocks because they are located near the HCSs.

The right panels of Figure 2 clearly show that the HCSs are resolved with the fine blocks. The HCSs are associated with low velocity winds (blue regions). The ballerina skirt structures are clearly reproduced on all the scales.

3.2. Grid Refinement

Figure 3 shows distributions of the ratio between the toroidal and poloidal components of magnetic fields, $B_\varphi/(B_r^2 + B_\theta^2)^{1/2}$. The HCS exhibits boundaries between the positive (red) and negative (blue) toroidal component of the magnetic fields. Such boundaries exist not only near the mid-plane but also near the polar regions. The former corresponds to the HCSs, while the latter is caused by the geometrical effect. Only former regions are refined according to the

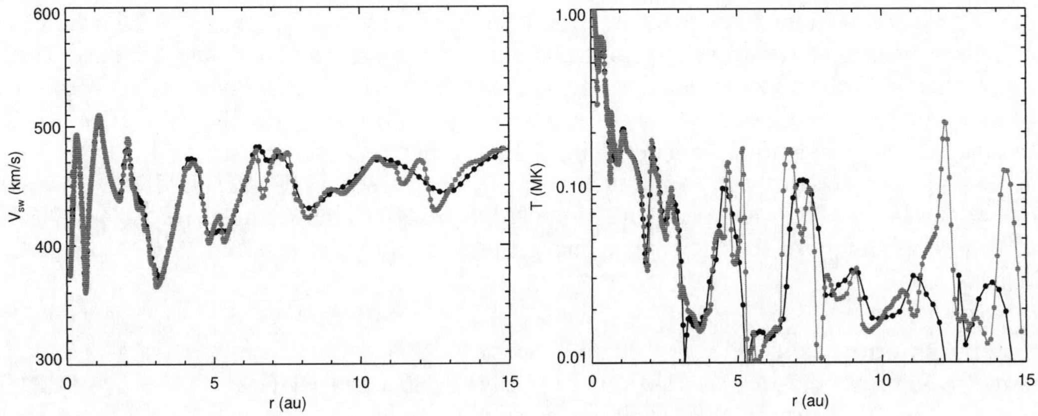

Figure 5. Solar wind speed (left) and temperature (right) distributions on the x-axis at the stage of June 1, 2017. The black and red lines denote the distributions solved by the FMR and AMR grids, respectively. The dots associated with lines denote the grid points.

criteria described in Section 2.4. Note that B_φ changes rapidly across the HCSs compared to the other boundaries.

Figure 4 illustrates how refinement criteria work. The grids satisfied with the refinement criterion of equation (14) have 1 bit flags (blue). These grids are distributed on the HCS and the polar regions. Because of the additional criterion of equation (15) (green), the refined grids (red) are selected appropriately except for the several grids in the northern polar region. In the right panel of Figure 4, the refined regions are not in contact with eacth other at $x \sim 500$ au because the blocks here are not satisfied with equation (15) (criterion with 2 bits). However, the 1-level coarser blocks safely cover this region because of the margin of refinement.

3.3. Effects of AMR

Figure 5 compares the solutions between the AMR and FMR grids. Both the solar wind speed and temperature in $r \lesssim 2$ au exhibit little differences because the CIRs are not developed within this radius. In contrast, in $r \gtrsim 2$ au, they exhibit significant differences between the AMR and FMR. The wind speed in the AMR exhibits sharp discontinuities, which corresponds to shock waves associated with the CIRs. The shock waves compressed the gas, leading to high temperature there.

4. Summary and Discussion

We developed a numerical model for the heliosphere in $r \leq 14$ au. The solar wind model is based on daily observations of solar magnetic fields, and the numerical model reproduces the time-dependent inner heliosphere. The AMR technique is adopted for resolving the HCS. A concentric nested grid is arranged, and the grid is refined further according to (1) the anti-parallel magnetic fields in the toroidal direction, (2) the high-density region or plasma sheet, and (3) exclusion of the polar region.

We examined an alternative refinement criteria as follows. We replaced the second criterion above by that on the low velocity, e.g., $|v| \leq 400\,\mathrm{km\,s^{-1}}$. Although the HCS has a low velocity, the amplitude of the v_r decreases as the wind flows outward, as shown in Figure 5. Figure 2 also shows a similar tendency. The criterion based on low velocity therefore does not work in the outer region.

In addition, we examined another criterion based on currents, e.g., $r^{-2}\nabla \times \boldsymbol{B}$, where r^{-2} is for scaling according to the distance from the Sun. Although the HCS has currents, they are weaker in the HCS than in the shock waves associated with CIRs. The shock waves sharply bend the magnetic field lines, leading a strong current. This suggests that the criterion based on the currents is more suitable for detecting CIRs rather than the HCS.

Our model does not take into account a coronal mass ejection (CME). CMEs have a strong impact on the heliospheric environment, especially near the maximum in the solar cycle. We reported here modeling of the heliosphere near the solar minimum, in which the role of CMEs is relatively small.

Acknowledgments
Numerical computations were carried out on Cray XC30 (ATERUI) and XC50 (ATERUI II) at the Center for Computational Astrophysics (CfCA), National Astronomical Observatory of Japan, and on FUJITSU FX100 (HOKUSAI-GreatWave) at the Advanced Center for Computing and Communication (ACCC), RIKEN. This research was supported by JSPS KAKENHI Grant Numbers 18H04449, 17K05671, 16H01182, 15H05813, 15H05816.

References

[1] Burger, R. A., & Sello, P. C. 2005, Advances in Space Research, 35, 643
[2] Miyake, S., Kataoka, R., & Sato, T. 2017, Space Weather, 15, 589
[3] Guo, X., & Florinski, V. 2014, Journal of Geophysical Research (Space Physics), 119, 2411
[4] Guo, X., & Florinski, V. 2016, ApJ, 826, 65
[5] Kataoka, R., Miyahara, H., & Steinhilber, F. 2012, Space Weather, 10, S11001
[6] Shiota, D., Kataoka, R., Miyoshi, Y., et al. 2014, Space Weather, 12, 187
[7] Kataoka, R., Ebisuzaki, T., Kusano, K., et al. 2009, Journal of Geophysical Research (Space Physics), 114, A10102
[8] Wang, Y.-M., & Sheeley, N. R., Jr. 1990, ApJ, 355, 726
[9] Arge, C. N., & Pizzo, V. J. 2000, JGR, 105, 10465
[10] Arge, C. N., Odstrcil, D., Pizzo, V. J., & Mayer, L. R. 2003, Solar Wind Ten, 679, 190
[11] Arge, C. N., Luhmann, J. G., Odstrcil, D., Schrijver, C. J., & Li, Y. 2004, Journal of Atmospheric and Solar-Terrestrial Physics, 66, 1295
[12] Riley, P., & Luhmann, J. G. 2012, Solar Physics, 277, 355.
[13] Riley, P., Linker, J. A., & Arge, C. N. 2015, Space Weather, 13, 154.
[14] Hayashi, K., Kojima, M., Tokumaru, M., & Fujiki, K. 2003, Journal of Geophysical Research (Space Physics), 108, 1102
[15] Hayashi, K. 2012, Journal of Geophysical Research (Space Physics), 117, A08105.
[16] Matsumoto, T. 2007, PASJ, 59, 905
[17] Miyoshi, T., & Kusano, K. 2007, AGU Fall Meeting Abstracts, SM41A-0311
[18] Miyoshi, T., & Kusano, K. 2005, Journal of Computational Physics, 208, 315
[19] Dedner, A., Kemm, F., Kröner, D., Munz, C.-D., Schnitzer, T., & Wesenberg, M. 2002, Journal of Computational Physics, 175, 645

IOP Conf. Series: Journal of Physics: Conf. Series **1225** (2019) 012009 doi:10.1088/1742-6596/1225/1/012009

Particle-in-Cell Simulations of Plasma Dynamics in Cometary Environment

Chaitanya Prasad Sishtla, Vyacheslav Olshevsky, Steven W. D. Chien, Stefano Markidis, Erwin Laure
KTH Royal Institute of Technology, Sweden
E-mail: {sishtla, slavik, wdchien, markidis, erwinl}@kth.se

Abstract. We perform and analyze global Particle-in-Cell (PIC) simulations of the interaction between solar wind and an outgassing comet with the goal of studying the plasma kinetic dynamics of a cometary environment. To achieve this, we design and implement a new numerical method in the iPIC3D code to model outgassing from the comet: new plasma particles are ejected from the comet "surface" at each computational cycle. Our simulations show that a bow shock is formed as a result of the interaction between solar wind and outgassed particles. The analysis of distribution functions for the PIC simulations shows that at the bow shock part of the incoming solar wind, ions are reflected while electrons are heated. This work attempts to reveal kinetic effects in the atmosphere of an outgassing comet using a fully kinetic Particle-in-Cell model.

1. Introduction

The simulation of interaction between solar wind and cometary plasma is a challenging task because of the peculiarities of the cometary plasma environment. In fact, comets consist of a nuclei composed of a mixture of ices, refractive materials and large organic molecules. These materials sublimate in the form of an outgas due to solar radiation. The extent of sublimation is dependent on the heliocentric distance to the comet. Photoionization and charge exchange with solar wind plasma results in the ionization of sublimated particles leading to the formation of a cometary plasma. The cometary ions are then assimilated by the solar wind through the $\mathbf{E}\times\mathbf{B}$ drift [1] and electromagnetic instabilities driven by the cometary ions [2].

The atmosphere of a comet is formed by its outgassing and interaction with solar wind. The innermost region of the comet is the comet nucleus which is an "icy dirt ball" from which sublimation of particles is happening. The outgassed neutral particles have a density variation $\sim 1/r^2$ [3]. Above the nucleus is the cometary atmosphere (coma) which consists of a "recombination" region where ionization and recombination of the sublimated particles happen [4]. This region exists inside the diamagnetic cavity [5] which is characterized by a decrease in magnetic field intensity to zero, and with plasma having a radial velocity inside the cavity. The coma is bound by the cometary ionopause. In the innermost regions of the coma where the neutral density is the largest, there are collisions between cometary ions and neutral particles which change the velocity of the cometary ion. For the plasma these collisions act as an ion-neutral friction force. The interaction between outgassed plasma and solar wind strongly depends on the outgassing rate and the direction of the interplanetary magnetic field. Mass

IOP Conf. Series: Journal of Physics: Conf. Series **1225** (2019) 012009 doi:10.1088/1742-6596/1225/1/012009

loading occurs due to the charge exchange between the newly acquired charged particles and the solar wind. If this exchange is substantial, it leads to the formation of a bow shock [6]. At the bow shock, picked up particles might exhibit unstable phase space distributions [7], leading to kinetic interaction of plasma with solar wind fluctuations and self-generated plasma waves. The goal of this work is to study the kinetic interaction between solar wind and a cometary plasma environment.

Previous modeling of cometary plasma processes have mainly focused on using either magnetohydrodynamic (MHD) simulations or hybrid simulations, which assume electrons behaving as a fluid and ions are treated as particles. Shou et al. [8] modeled the coma of the comet using a multi-fluid model with each cometary species treated by separate fluid equations. The simulation used the BATSRUS code [9, 10] and its results are verified against data collected on the 67P/Churyumov-Gerasimenko (CG) comet. MHD models have been developed for simulating comet interactions with the solar wind, and were validated against data collected by the Giotto spacecraft for the Halley comet [11] and the Rosetta mission for CG [12]. Koenders et al.[13] used a hybrid model based on the AIKEF code [14] to simulate the cometary atmosphere taking into account collisions, recombinations and ionizations. Huang et al. [15] uses a four-fluid model to simulate the comet solar wind interaction. However, a complete description of kinetic effects on electrons and ions is only possible via kinetic model. For example, the physical and production mechanisms of non-thermal electron distributions forund in the coma by the Ion and Electron Sensor (IES) onboard Rosetta are still unclear [16, 17]. The goal of our work is to use a Particle-in-Cell (PIC) model for studying the kinetic behavior of electrons and ions during the interaction of solar wind with cometary environment.

The first fully kinetic model of a cometary induced magnetosphere was presented by Deca et al. [18] using the iPIC3D code[19, 20], the authors have shown that the dynamics of ions and electrons in the atmosphere of a weakly outgassing comet could be represented, to first order, in terms of the four-fluid system with solar wind ions, solar wind electrons, cometary ions and cometary electrons each treated by separate fluid equations. Differently from this previous work, we develop an alternative approach for modeling cometary interaction with solar wind using iPIC3D. The key difference of our method is its outgassing mechanism. Instead of creating new plasma particles in the cometary environment at each computational cycle, we inject them from the "surface" of the comet according to the physical outgassing rate, i.e., we start from the first principles.

In this paper, we present the simulation approach and the results of our fully kinetic two-dimensional modeling of plasma dynamics in the outgassing comet plasma environment. The paper is organized as follows. Section 2 describes the governing equations of implicit method in the iPIC3D code. Then, Section 3 presents the simulation set-up and spatial and temporal scales that are involved in modeling the solar wind interaction with a cometary plasma environment. Section 4 presents the simulation results, such as evolution of electron and ions densities, macroscopic flows and velocity distribution functions. Finally, Section 5 summarizes the results, discusses limitations of this work and outlines future research.

2. Implicit Particle-in-Cell Method

We use the iPIC3D code [19] to model cometary plasma environment from first principles. The PIC method is based on the self-consistent solution of Maxwell's equations and the equations of motion of computational particles (superparticles). Maxwell's equations in iPIC3D are expressed in the second-order (Gaussian or curl-curl) formulation [21], with electric field evolution governed by

$$\nabla^2 \mathbf{E} - \frac{1}{c^2}\frac{\partial^2 \mathbf{E}}{\partial t^2} = \frac{4\pi}{c^2}\frac{\partial \mathbf{J}}{\partial t} + 4\pi \nabla \rho, \tag{1}$$

where $\mathbf{E}$ is the electric field, c is the speed of light, $\mathbf{J}$ is the current density, ρ is the charge density and CGS units are used. Once $\mathbf{E}$ is calculated, the magnetic field $\mathbf{B}$ is simply computed from the induction equation:

$$\frac{\partial \mathbf{B}}{\partial t} = -c\nabla \times \mathbf{E}. \tag{2}$$

The first two moments of the particle phase space distribution function provide the source terms for the above Maxwell's equations

$$\rho = \sum_s q_s \int f_s d\mathbf{v} \qquad \mathbf{J} = \sum_s q_s \int \mathbf{v} f_s d\mathbf{v} \tag{3}$$

where the index s denotes the plasma species (ions or electrons), q_s is the charge of a particle of the corresponding species, $\mathbf{v}$ is particle velocity, and $f_s(\mathbf{x}, \mathbf{v})$ is the phase space distribution function of species s. The distribution function of species s is then described as a collection of N_s computational particles labeled with index p.

The computational particles in the PIC method are used to sample phase space, i.e., each computational particle represents a cloud of real particles in phase space. The functional form of this cloud in iPIC3D is the Dirac's δ function in the velocity dimensions, and the zeroth order b-spline in the spatial dimensions

$$f_s(\mathbf{x}, \mathbf{v}, t) = b_0\left(\frac{x - x_p}{\Delta x}\right) b_0\left(\frac{y - y_p}{\Delta y}\right) b_0\left(\frac{z - z_p}{\Delta z}\right) \delta(\mathbf{v} - \mathbf{v}_p), \tag{4}$$

where $\mathbf{x_p}$ and $\mathbf{v_p}$ refer to the superparticle's position and velocity respectively. The b_0 spline is the hat-top function which is nonzero only in the interval $-0.5 \leq \zeta \leq 0.5$:

$$b_0(x) = \begin{cases} 0 & |\zeta| > 0.5 \\ 1 & |\zeta| \leq 0.5. \end{cases} \tag{5}$$

Particles and grid exchange information by interpolating fields $\mathbf{E}, \mathbf{B}$ (Lorentz force) from the grid to the particles, and by interpolating particle moments such as $\rho, \mathbf{J}$ from the particles to the grid. To evaluate the field $\mathbf{B}$ on the given particle p, the following sum is computed over the particle's neighbor grid points g

$$\mathbf{B}_p = \sum_g \mathbf{B}_g W(\mathbf{x} - \mathbf{x}_p). \tag{6}$$

To evaluate the moment ρ on the given grid point g the following sum is computed

$$\rho_g = \sum_s^{n_s} \sum_p^{N_s} q_s W(\mathbf{x} - \mathbf{x}_p), \tag{7}$$

and analogously for $\mathbf{J}$, following Eq. 3. In the above two equations, the key role is played by the interpolation function W which, in case of iPIC3D is the b-spline of order 1, $W(x - x_p) = b_1(x - x_p)/\Delta x$. The size of the particle in each spatial dimension corresponds to the grid step in this dimension, Δx, Δy or Δz. The b_1 spline is nonzero only in the interval $[-1; 1]$ where it is a triangular function with maximum $b_1(0) = 1$:

$$b_1(\zeta) = \begin{cases} 0 & |\zeta| > 1 \\ 1 - |\zeta| & |\zeta| \leq 1. \end{cases} \tag{8}$$

Particles move under the effect of an electromagnetic field according to Newton's equation

$$\frac{d\mathbf{x}_p}{dt} = \mathbf{v_p} \qquad \frac{d\mathbf{v}_p}{dt} = \frac{q_s}{m_s}\left(\mathbf{E}_p + \frac{\mathbf{v}_p \times \mathbf{B}_p}{c}\right), \tag{9}$$

where p is the particle's index, implying that $\mathbf{E}_p$ and $\mathbf{B}_p$ are interpolated from the grid to the particle's location.

The computational cycle of iPIC3D consists of the following four steps: (1) solve Maxwell's equations discretized on the grid; (2) interpolate new fields and compute the forces acting on each particle; (3) integrate the equations of particle motion 9; (4) interpolate new charge and current densities to the grid by integrating particle moments (Eq. 3) in each grid point's volume of control.

The distinctive feature of iPIC3D is the semi-implicit (implicit moment) scheme used to solve Maxwell's equations coupled to the equations of particle motion. In the implicit moment method, an approximate response of the particles to the fields is evaluated at the intermediate time moment at each computational cycle. Approximate values of $\hat{\rho}$ and $\hat{\mathbf{J}}$ are estimated from the Taylor expansion of the interpolation function W. This coupling of particle and field equations makes multiple interpolations between the grid and the particles at each computational cycle unnecessary, and dramatically reduces the compute cost compared to the fully implicit method [22]. This approximate solution, however, has been proven stable and accurate when the stability conditions are satisfied [23]. The stability conditions for implicit PIC methods are much more relaxed compared to the explicit PIC codes which must resolve the smallest physically meaningful scales, such as the Debye length in space and electron plasma frequency in time. Due to relaxed stability restrictions, iPIC3D is capable of modeling fluid-scale processes while retaining the full kinetic description of plasma.

3. Cometary Plasma Environment

The cometary plasma environment is formed as a result of the sublimation of heavy particles from the nucleus and their interaction with the solar wind. The sublimated particles are ionized by the extreme ultraviolet radiation and undergo inelastic collisions with plasma. The collisions account for the ion-neutral friction and electron cooling. This ionized part of the cometary atmosphere interacts with the incoming solar wind, forming plasma boundaries such as bow shock and ionopause [24]. The basic scales and properties of some comets are shown in Table 1. The scales for comets 1P Halley and 26P GS are reported at heliocentric distances of 0.89 AU and 1.01 AU respectively at the time of encounter during the Giotto mission [25, 26]: scales for 67P CG are reported before equinox at approximately 3 AU [27]. In the simulation presented outgassing is performed by injecting cometary ions and electrons instead of a neutral gas.

Comet	**Mean Diameter**	**Production Rate**($\times 10^{27}\mathrm{s}^{-1}$)	**Ion Gyroradius**
1P Halley	11km	690	10000km
67P CG	4km	0.2	1000km
26P GS	2.6km	7.5	4000km

Table 1: Mean diameter, production rates and cometary ion gyroradius for comets 1P/Halley, 67P/Churyumov-Gerasimenko and 26P/Grigg-Skjellerup [7] [28] [27] are shown. The production rate refers to the gas production rate due to sublimation processes which differ for each comet [7].

3.1. Simulation Setup

The iPIC3D simulation is carried out in a reduced 2D3V configuration, i.e., two spatial coordinates (x, z), and three coordinates for velocity, currents, electromagnetic fields, etc..

In iPIC3D, the simulation parameters are normalized with respect to ion scales as shown in Table 2, where ω_{pi} is the ion plasma frequency and $d_i = c/\ \omega_{pi}$ is the ion skin depth. The comet is placed at the center of the simulation domain measuring $50d_i \times 100d_i$, with solar wind injection from the left boundary at $x = 0$. The temperatures of solar wind ions and electrons are taken to be 7eV and 10eV respectively. The simulation assumes cometary and solar wind ions to be of equal mass to ensure faster assimilation of cometary ions into the solar wind plasma [29]. A reduced mass ratio of $m_i/m_e = 64$ is taken to ensure numerical stability, while retaining the scale separation between electron and ion dynamics [30]. The interplanetary magnetic field (IMF) is $\mathbf{B}_{IMF} = (0, 0, 0.01)\ m_i \omega_{pi}^{-1} e^{-1}$ where e is the elementary charge. Together with the IMF, a convective electric field $\mathbf{E} = -\mathbf{v_0} \times \mathbf{B}_{IMF}$ is initially imposed at $x = 0$, where $\mathbf{v_0}$ is the solar wind bulk velocity. Inflow boundary condition is imposed at $x = 0$ while other boundaries have outflow boundary conditions as discussed in [20, 31]. We note that the solar wind is super-sonic and super-Alfvénic.

Simulation Parameters	**Normalized Values**
ρ_{0e} , ρ_{0i}	1, 1
L_x, L_z	50 d_i, 100 d_i
N_x, N_z	400, 800
dx, dz	0.125 d_i, 0.125 d_i
$(e/m)_i$	1
$(e/m)_e$	-64
Radius of comet	1 d_i
Time step of simulation	0.15 / ω_{pi}
Electron thermal velocity	(0.045 c, 0.045 c, 0.045 c)
Ion thermal velocity	(0.0063 c, 0.0063 c, 0.0063 c)
Solar wind bulk velocity	(0.02 c, 0, 0)

Table 2: Main parameters of the simulation are shown. n_{0e} and n_{0i} are the electron and ion density. L_x, L_z are the box lengths in x and z directions respectively. Similarly N_x, N_z are the number of cells in x and z directions, and dx, dz are the spatial step lengths in x and z directions respectively. All values are normalized with respect to ion scales (ω_{pi}, d_i).

The main novelty of this paper is the development of a new numerical method to implement particle injection by an emitting boundary, rather than injecting particles throughout the atmosphere. Outgassing is performed by injecting new plasma particles from the "surface" of the comet. A spherical shell $5dx$ thick is defined inside the comet and acts as buffer zone similarly to buffer zones created for enabling open boundary conditions [20]. At each time step, new computational particles are created randomly and uniformly distributed inside this spherical shell with particle density $n_{0e} = n_{0i} = 1.0$. In this work, we inject 200 particles per time step. Each new particle receives a thermal speed with a random Gaussian distribution, as specified in Table 2, plus a radial outgassing velocity of 0.045 c. The radial outgassing velocity is chosen in a manner to ensure that the ejected cometary ions are fast enough to penetrate farther into the simulation domain within the time scales of the simulation. The strength of outgassing is controlled by the thermal velocity of the outgassed plasma. In the simulation thermal energy of cometary plasma is kept comparable to solar wind plasma to signify strong outgassing [29]. After creation, the particle position is advanced by one time step. If this particle remains inside the comet, it is deleted. Otherwise, it is retained and enters the simulation domain.

IOP Conf. Series: Journal of Physics: Conf. Series **1225** (2019) 012009 doi:10.1088/1742-6596/1225/1/012009

4. Results

In this section, we first show the macroscopic interaction of the solar wind with the cometary plasma environment, showing the electron and ion densities and flows. We then present the distribution functions revealing the microscopic dynamics of electrons and ions.

4.1. The Formation of the Cometary Plasma Environment

In order to create the cometary plasma environment, we initialize the simulation box with a uniform plasma, inject plasma at $x = 0$ and at the comet surface. Figures 1 and 2 show the contour plots of the densities of solar wind and cometary species superimposed with magnetic field lines at different times during the simulation. The empty circle in the middle of the domain shows the cometary body. When the solar wind flow reaches the comet ($t\omega_{pi} = 7.5$) the formation of a bow shock begins, seen in the density enhancement (roughly twice the initial density) in front of the comet in Figure 1. In Figure 2 the cometary ions and electrons are seen to be quickly picked up by the incoming solar wind, and are accelerated tailward via the $\mathbf{J}\times\mathbf{B}$ force [32]. The solar wind electron and ion density in the wake of the comet along the $z = 50d_i$ line starts decreasing while the cometary electron and ion densities starts increasing. As the simulation progresses and solar wind flow propagates further, the bow shock obtains its classical bell shape, and the low-density solar wind region and high-density cometary region in the wake extends towards the outflow boundary of the simulation domain. The simulated global structure of the solar wind-comet interaction is in accordance with results previously reported by hybrid and PIC simulations [13].

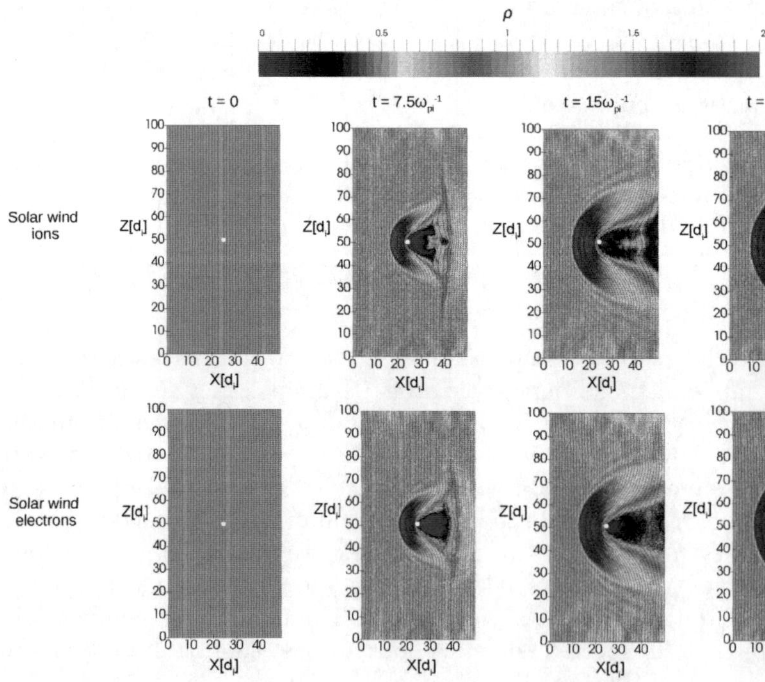

Figure 1: The evolution of solar wind ion and electron species as the simulation progresses is shown in different panels. The simulation starts at $t = 0$ and ends at $t = 22.5\omega_{pi}^{-1}$.

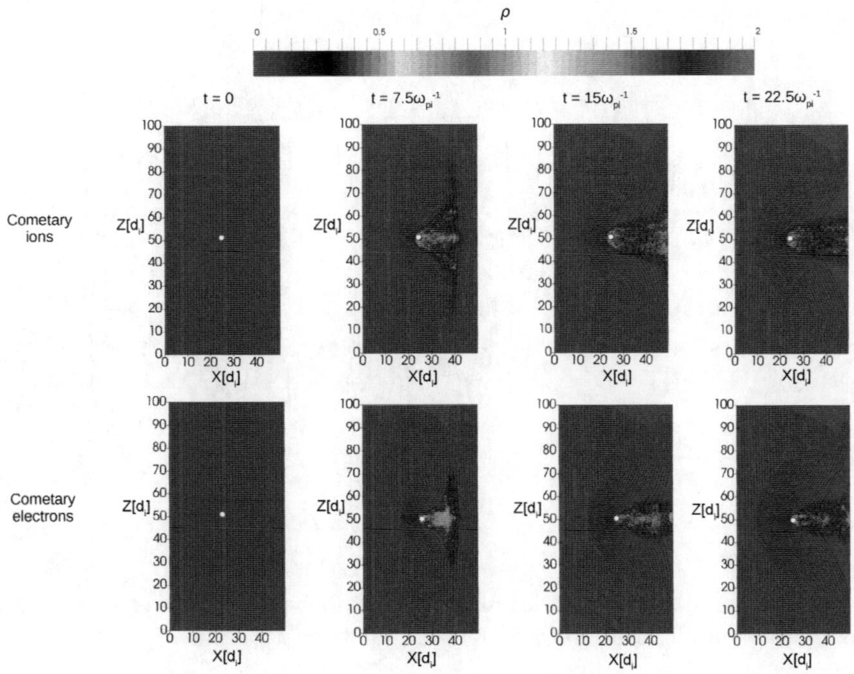

Figure 2: The evolution of cometary ion and electron species as the simulation progresses is shown in different panels. The simulation starts at $t = 0$ and ends at $t = 22.5\omega_{pi}^{-1}$.

4.2. Cometary Bow Shock

Our simulations show that the interaction of the solar wind and cometary plasma results in the formation of a bow shock. In order to investigate the location and thickness of the bow shock a line plot profiling the magnetosonic Mach number $M = u/(v_A^2 + c_s^2)$ can be used to observe the transition from super-magnetosonic to sub-magnetosonic flow. Here u is the total electron bulk speed, v_A is the Alfvén speed and $c_s = \sqrt{\gamma(T_e/m_i)}$ is the ion sound speed. Figure 3 shows a line plot at $z = L_z/2$ from $x = 0$ to $x = 24d_i$ of the z component of the magnetic field normalized to the IMF intensity, electron and ion densities, and the fast magnetosonic Mach number calculated using the total electron bulk speed at time $t = 22.5\omega_{pi}^{-1}$. The solar wind electron and ion densities overlap indicating very low extent of charge separation, but they both increase by a factor of 2 after the shock. The bow shock is located at $x \approx 8d_i$, and has a thickness of $\sim 1d_i$.

Figure 4 displays the contour plot of solar wind electron and ion flows in normalized units superimposed with a quiver plot showing the flow vectors at random points. Both the solar wind electrons and ions which enter the cometary environment are deflected in the z direction as they move around the comet.

An analysis of the distribution functions show the details of the microscopic dynamics of solar wind electrons and ions in proximity of the bow shock. We study particle distribution functions in the regions: upstream, bow shock front and downstream. Figure 5 shows the contour plots for solar wind electron and ion velocity distributions in three regions that are highlighted in the upper panels. The first region represents the pristine solar wind; the second region represents an interface between the solar wind and the coma; the third region is immersed into the coma. The solar wind flows in from the left boundary with a bulk speed of $0.02c$.

The top panel of Figure 5 shows the electron distribution function. Upstream the bow shock

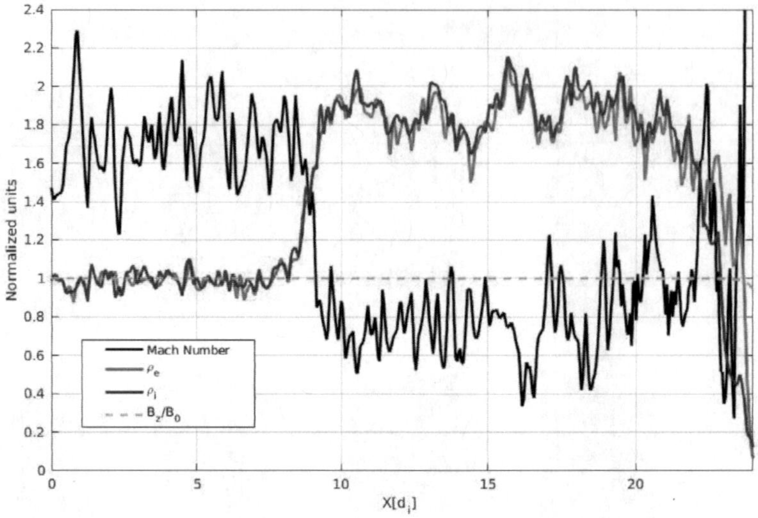

Figure 3: The fast magnetosonic Mach number, z component of the magnetic field normalized to the IMF intensity, and electron and ion densities at time $t = 22.5\ \omega_{pi}^{-1}$.

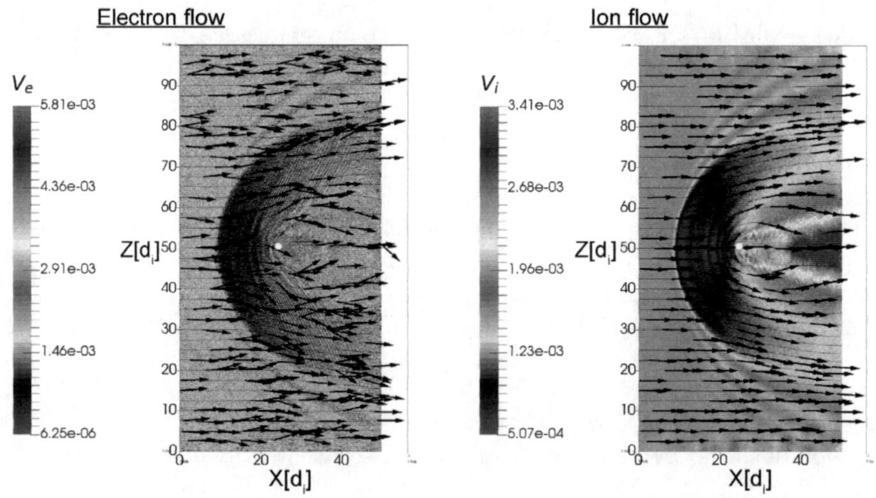

Figure 4: Solar wind electron and ion flows are shown. The figure presents the contour maps of the flow intensities, over which a quiver plot of the flow directions are shown.

in the pristine solar wind, the velocity distribution is purely Maxwellian and centered around $\mathbf{v} = (0.02c, 0, 0)$. As we move into the transition region between the solar wind and the coma, we see that the incoming electrons are heated especially in the z direction.

The bottom panel of Figure 5 shows the ion distribution function in the three regions. The distribution function in proximity of the bow shock (bottom center panel in Figure 5) is particularly interesting: crossing the bow shock, we can observe that the distribution function elongates towards the $-v_x$ axis reaching negative v_x velocity. The ion distribution function

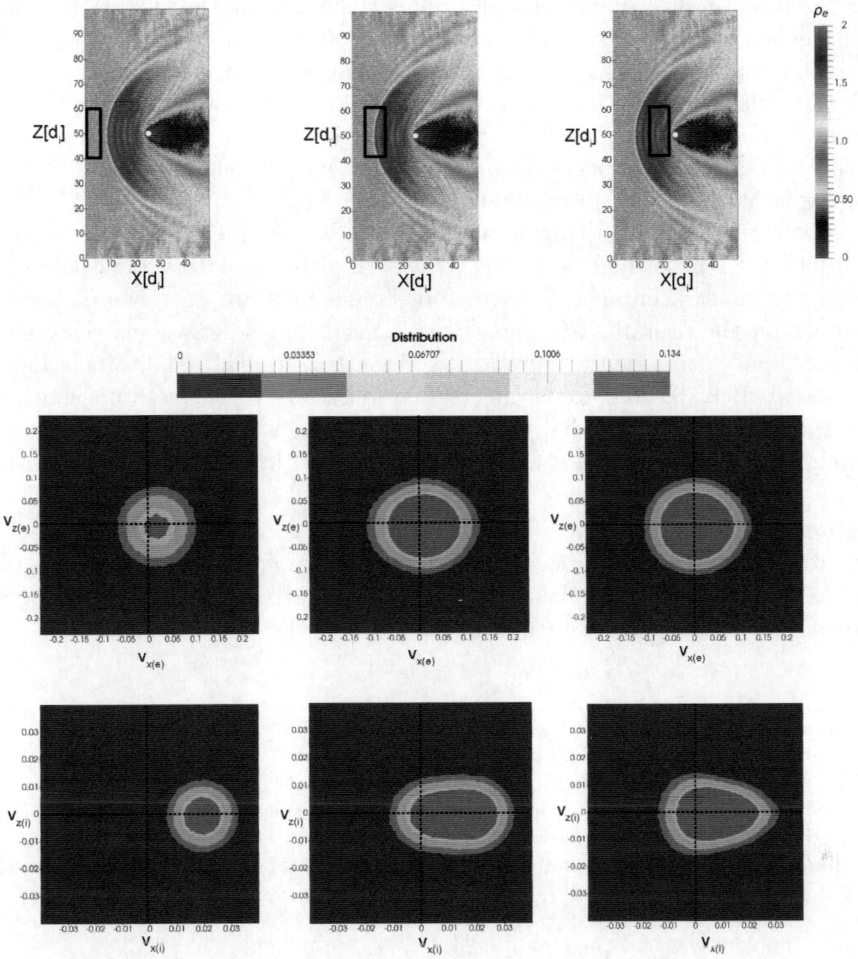

Figure 5: The solar wind electron and ion velocity distribution function in three distinct regions are shown. The first region is present completely in the solar wind, the second region is partly the solar wind and partly the coma while the third region is present completely inside the coma.

shows reflected ions at the bow shock. This is possibly due to the enhanced magnetic field at the bow shock [13].

5. Discussion and Conclusion

We have designed and developed a new fully-kinetic electromagnetic model of an outgassing comet interacting with the solar wind. We have implemented an empirical method for injection of cometary particles in the implicit PIC code iPIC3D: outgassing particles enter the simulation domain from the "surface" of the comet with the predefined radial velocity and a random thermal speed. These particles interact with the impinging solar wind and form a well-defined coma, tail, and a bow shock where the interplanetary magnetic field piles up, while the incoming solar wind flow is deflected and reflected. The overall structure of the simulated outgassing comet presented in this paper corresponds to observations and previous hybrid and PIC simulations [13]. The time needed to establish a stationary flow of the solar wind through the simulation domain is comparable to the solar wind transition time.

The incoming flow of the solar wind plasma is deflected in the direction parallel to the magnetic field (perpendicular to the solar wind speed). Closer to the comet, injected electrons mix with the solar wind and heat up, forming a broader distribution function. Solar wind ions mix with the outgassing particles and are reflected at the bow shock due to the enhanced magnetic field, forming an elongated velocity distribution in the v_x direction. This result is consistent with observations and previous simulations. Overall, our new model is promising for more advanced kinetic studies of the cometary atmospheres.

There are two main limitations to this work. First, we do not model collisions in iPIC3D. A collision model in iPIC3D would allow us to describe the cometary environment with high outgassing rates at higher accuracy. Second, our simulations are in a two dimensional space. This has an effect on the overall exchange of momentum and energy between solar wind and cometary plasma. Our future work will address these two limitations by including a collision model for plasma in iPIC3D and extending the simulations to three dimensions, potentially using iPIC3D and MHD-EPIC [33, 34] model. In addition, we plan to complete a parametric study for various solar wind velocities, IMF directions, and the comet outgassing rates.

Acknowledgments

This work was funded by Swedish Research (VR) Council project 2017-04508. Experiments were performed on resources provided by the Swedish National Infrastructure for Computing (SNIC) at PDC Center for High Performance Computing.

References

[1] Holzer T and Axford W 1971 *Journal of Geophysical Research* **76** 6965–6970
[2] Winske D, Wu C, Li Y, Mou Z and Guo S 1985 *Journal of Geophysical Research: Space Physics* **90** 2713–2726
[3] Haser L 1957 *Bulletin de la Societe Royale des Sciences de Liege* **43** 740–750
[4] Goldstein B, Altwegg K, Balsiger H, Fuselier S, Ip W H, Meier A, Neugebauer M, Rosenbauer H and Schwenn R 1989 *Journal of Geophysical Research: Space Physics* **94** 17251–17257
[5] Neubauer F, Marschall H, Pohl M, Glassmeier K H, Musmann G, Mariani F, Acuna M, Burlaga L, Ness N, Wallis M *et al.* 1993 *Astronomy and Astrophysics* **268** L5–L8
[6] Glassmeier K H 2017 *Phil. Trans. R. Soc. A* **375** 20160256
[7] Coates A J and Jones G H 2009 *Planetary and Space Science* **57** 1175–1191
[8] Shou Y, Combi M, Toth G, Tenishev V, Fougere N, Jia X, Rubin M, Huang Z, Hansen K, Gombosi T *et al.* 2016 *The Astrophysical Journal* **833** 160
[9] Powell K G, Roe P L, Linde T J, Gombosi T I and De Zeeuw D L 1999 *Journal of Computational Physics* **154** 284–309
[10] Tóth G, Van der Holst B, Sokolov I V, De Zeeuw D L, Gombosi T I, Fang F, Manchester W B, Meng X, Najib D, Powell K G *et al.* 2012 *Journal of Computational Physics* **231** 870–903
[11] Rubin M, Combi M, Daldorff L, Gombosi T, Hansen K, Shou Y, Tenishev V, Tóth G, van der Holst B and Altwegg K 2014 *The Astrophysical Journal* **781** 86
[12] Rubin M, Gombosi T, Hansen K, Ip W H, Kartalev M, Koenders C and Tóth G 2015 *Earth, moon, and planets* **116** 141–157
[13] Koenders C, Glassmeier K H, Richter I, Ranocha H and Motschmann U 2015 *Planetary and Space Science* **105** 101–116
[14] Müller J, Simon S, Motschmann U, Schüle J, Glassmeier K H and Pringle G J 2011 *Computer Physics Communications* **182** 946–966
[15] Huang Z, Tóth G, Gombosi T I, Jia X, Rubin M, Fougere N, Tenishev V, Combi M R, Bieler A, Hansen K C *et al.* 2016 *Journal of Geophysical Research: Space Physics* **121** 4247–4268
[16] Madanian H, Cravens T, Rahmati A, Goldstein R, Burch J, Eriksson A I, Edberg N J, Henri P, Mandt K, Clark G *et al.* 2016 *Journal of Geophysical Research: Space Physics* **121** 5815–5836
[17] Broiles T, Livadiotis G, Burch J, Chae K, Clark G, Cravens T, Davidson R, Eriksson A, Frahm R, Fuselier S *et al.* 2016 *Journal of Geophysical Research: Space Physics* **121** 7407–7422
[18] Deca J, Divin A, Henri P, Eriksson A, Markidis S, Olshevsky V and Horányi M 2017 *Physical review letters* **118** 205101
[19] Markidis S, Lapenta G and Rizwan-uddin 2010 *Mathematics and Computers in Simulation* **80** 1509–1519

[20] Peng I B, Markidis S, Vaivads A, Vencels J, Amaya J, Divin A, Laure E and Lapenta G 2015 The formation of a magnetosphere with implicit particle-in-cell simulations *ICCS2015, 2015* pp 1178–1187
[21] Jiang B N, Wu J and Povinelli L A 1996 *Journal of computational physics* **125** 104–123
[22] Markidis S and Lapenta G 2011 *Journal of Computational Physics* **230** 7037–7052
[23] Brackbill J U and Cohen B I 2014 *Multiple time scales* vol 3 (Academic Press)
[24] Nilsson H, Wieser G S, Behar E, Wedlund C S, Gunell H, Yamauchi M, Lundin R, Barabash S, Wieser M, Carr C *et al.* 2015 *Science* **347** aaa0571
[25] Krankowsky D, Lämmerzahl P, Herrwerth I, Woweries J, Eberhardt P, Dolder U, Herrmann U, Schulte W, Berthelier J, Illiano J *et al.* 1986 *Nature* **321** 326–329
[26] Johnstone A, Coates A, Huddleston D, Jockers K, Wilken B, Borg H, Gurgiolo C, Winningham J and Amata E 1993 *Astronomy and Astrophysics* **273**
[27] Hansen K C, Altwegg K, Berthelier J J, Bieler A, Biver N, Bockelée-Morvan D, Calmonte U, Capaccioni F, Combi M, De Keyser J *et al.* 2016 *Monthly Notices of the Royal Astronomical Society* **462** S491–S506
[28] Glassmeier K H, Boehnhardt H, Koschny D, Kührt E and Richter I 2007 *Space Science Reviews* **128** 1–21
[29] Omidi N and Winske D 1987 *Journal of Geophysical Research: Space Physics* **92** 13409–13426
[30] Lapenta G, Markidis S, Divin A, Goldman M and Newman D 2010 *Physics of Plasmas* **17** 082106
[31] Peng I B, Markidis S, Laure E, Johlander A, Vaivads A, Khotyaintsev Y, Henri P and Lapenta G 2015 *Physics of Plasmas* **22** 092109
[32] Whipple F L 2012 *Physics and chemistry of comets* (Springer Science & Business Media)
[33] Tóth G, Jia X, Markidis S, Peng I B, Chen Y, Daldorff L K, Tenishev V M, Borovikov D, Haiducek J D, Gombosi T I *et al.* 2016 *Journal of Geophysical Research: Space Physics* **121** 1273–1293
[34] Daldorff L K, Tóth G, Gombosi T I, Lapenta G, Amaya J, Markidis S and Brackbill J U 2014 *Journal of Computational Physics* **268** 236–254

IOP Conf. Series: Journal of Physics: Conf. Series **1225** (2019) 012010 doi:10.1088/1742-6596/1225/1/012010

Stochastic Propagation of Solar Energetic Particles in Coronal and Interplanetary Magnetic Fields

Ming Zhang, Lulu Zhao and Hamid K. Rassoul

Department of Physics and Space Sciences, Florida Institute of Technology, Melbourne, FL 32901, USA

E-mail: mzhang@fit.edu

Abstract. This paper describes a method of simulating solar energetic particle propagation through the magnetic fields of the solar corona and interplanetary medium. The simulation code is based on the focus transport equation of energetic particles in 3-d magnetic fields, which contains all the particle transport mechanisms, including streaming, convection, gradient/curvature drift, adiabatic focusing, pitch angle scattering by Alfvénic magnetic field fluctuations and perpendicular diffusion due to the random walk of field lines. In the simulation, particles are injected at their source in the corona, and their guiding center trajectories are calculated using stochastic differential equations. Because of the vastly different time scales of particle transport mechanisms included in the equation, we use the 4-th order Runge-Kutta method to integrate the particle streaming and adiabatic focusing terms, while the stochastic terms of pitch angle scattering and perpendicular diffusion are integrated with the Euler scheme. The model is applied to the 2017 September 10 solar energetic particle event. With perpendicular diffusion, we are able to explain SEP observations from Earth and STEREO-A. A pattern of SEP precipitation on the solar surface is also predicted.

1. Introduction

When a solar flare or a coronal mass ejection (CME) occurs in the corona, solar energetic particles (SEPs) up to many GeV can be produced and released into interplanetary space. The high-energy radiations travel to Earth or anywhere in the solar system, becoming a hazard to electronics on spacecraft or astronauts working in space. Therefore, we want to know how they are produced in the corona and how they propagate through the magnetic fields of the corona and interplanetary medium to Earth. We hope that, in future, we can reliably predict radiation dose using remote observations of solar events ahead of the arrival of radiation hazard.

Below the energy of a few GeV, the motion of SEPs in the coronal or interplanetary magnetic field is expected to strongly bound along magnetic field lines due to their small gyroradii compared to the domain size. Thus, we would expect that radiation exposure associated with SEPs only occur when one is directly connected by a magnetic field line to SEP sources near the sun. The size of a solar flare is small and even a large CME only covers a limit area of the solar surface, so SEPs should appear only at places magnetically connected to the solar event. However, multi-spacecraft measurements show that SEPs are often observed simultaneously over wide ranges of longitude, latitude, and radial distance. Wiedenbeck et al. (2013) reported that ^{3}He ions were detected by three spacecraft, ACE, STEREO-A (ST-A) and STEREO-B, spanning over 130° in longitude during the 2010 February 7 SEP event. Since ^{3}He is produced by

IOP Conf. Series: Journal of Physics: Conf. Series **1225** (2019) 012010 doi:10.1088/1742-6596/1225/1/012010

a stochastic particle acceleration process associated with magnetic reconnection in solar flares, the observation demonstrates that SEPs can spread widely across the longitude during their transport to 1 AU.

A number of theories have been proposed to explain the observation of widespread of SEPs. An extended or moving source of SEPs accelerated by a large CME shock could explain the widespread of SEPs observed in interplanetary space (Cliver et al. 1995). EUV waves are often seen to propagate very far on the solar surface from solar flare site (Klassen et al. 2000; Kienreich et al. 2009; Thompson & Myers 2009). If EUV waves are the footpoint of coronal CME shockwaves on the solar surface, one could argue that the CME shock can cover a much larger solar surface area than its interplanetary counterpart. Rouillard et al. (2012) hypothesized that the EUV waves can be used to track the expansion of a coronal shock, which is responsible for particle acceleration. Sometimes, a solar flare can trigger another or multiple eruptions in remote active regions that are linked by large-scale coronal magnetic fields or sympathetic solar flares (Wang et al. 2001; Schrijver & Title 2011; Shen et al. 2012; Schrijver et al. 2013). SEPs from multiple solar flares may be simultaneously observed by widely separated spacecraft. If the interplanetary magnetic field deviates significantly from the Parker spirals, locations of widely separated longitudes may be directly connected through those severely distorted interplanetary field lines (e.g., Richardson et al. 1991; Richardson & Cane 1996; Torsti et al. 2004; Chollet et al. 2010), but such an explanation only works for locations with a large separation in radial distance as well. The magnetic field in the solar coronal itself can connect over a wide range of longitudes and latitudes, so SEPs can spread laterally just by propagating along field lines in the corona (Schrijver & De Rosa 2003). Furthermore, magnetic field structures in the corona are extremely complex. In some places, magnetic reconnection can occur. This makes field lines from different regions get connected. SEPs following the reconnected field lines can easily access to a large volume of space. Lastly, energetic particles do not have to follow the magnetic field line all the time. Driven by magnetic field inhomogeneity and fluctuations, cross-field diffusion of particles can occur in both the corona and interplanetary medium. Marsh et al. (2013) and Dalla et al. (2013) suggest that drifts due to curvature and gradient of Parker spiral could be a considerable source of asymmetric cross-field transport for high-energy SEPs. Even in the absence of cross-field diffusion of the particles from the initial field lines, large-scale field line meandering can make an effective contribution to cross-field transport in the average magnetic field (Giacalone & Jokipii 2012; Kelly et al. 2012; Laitinen et al. 2013). The inclusion of particle cross-field transport is necessary to explain the SEP reservoir phenomenon. At the beginning of an SEP event, particle intensities at different locations could be very different depending on the magnetic connectivity to the SEP source near the sun, but after some time (typically one to several days), they reach a uniform level at all locations in the inner heliosphere up to a few AU radial distance (McKibben, 1972; Roelof et al. 1992; McKibben et al. 2001, Dalla et al. 2003; Zhang et al. 2003). This so-called SEP reservoir effect can only be explained by particle transport perpendicular to large-scale (averaged) magnetic fields in the corona or interplanetary medium.

Driven by the need to incorporate all the above possibilities of SEP transport modes, we have developed a computation code to calculate SEP propagation through the corona and interplanetary medium using a realistic magnetic field model. As the coronal magnetic field varies drastically over a time scale, from minutes to solar cycles, it is best to use a data-driven coronal magnetic field. In this paper, we present some calculation results of SEP transport in a potential field source surface coronal magnetic field model, which is derived from measurements of the photospheric magnetic field.

IOP Conf. Series: Journal of Physics: Conf. Series **1225** (2019) 012010 doi:10.1088/1742-6596/1225/1/012010

2. Model and computer code

We use the focus transport equation in the 3d spatial domain to describe SEP propagation from its source near the sun to anywhere in interplanetary space. The particle distribution f is gyrotropic and can be written as a time-dependent diffusion-drift equation in the 5-d phase space of position $\mathbf{r}$, momentum p and pitch-angle cosine μ. The equation is formulated as follows:

$$\frac{\partial f}{\partial t} = \nabla \cdot \boldsymbol{\kappa}_\perp \cdot \nabla f - (v\mu\hat{\mathbf{b}} + \mathbf{V} + \mathbf{V}_d) \cdot \nabla f + \frac{\partial}{\partial \mu} D_{\mu\mu} \frac{\partial f}{\partial \mu} - \frac{d\mu}{dt}\frac{\partial f}{\partial \mu} - \frac{dp}{dt}\frac{\partial f}{\partial p}. \tag{1}$$

The right-hand side of the equation includes the cross-field diffusion tensor $\boldsymbol{\kappa}_\perp$ and particle pitch-angle diffusion coefficient $D_{\mu\mu}$. These two diffusion terms contain the effect of magnetic field fluctuations. The terms of the first-order derivatives come from particle adiabatic motion in the ambient regular (averaged) large-scale magnetic field $\mathbf{B}$. These terms contain convection with the solar wind plasma velocity $\mathbf{V}$, streaming $v\mu\hat{\mathbf{b}}$ with particle speed v times pitch-angle cosine μ along the magnetic field direction $\hat{\mathbf{b}}$, gradient/curvature drift velocity $\mathbf{V}_d$, focusing $\frac{d\mu}{dt}$ and adiabatic cooling $\frac{dp}{dt}$, which can be written in terms of $\mathbf{B}$ and $\mathbf{V}$ as follows (e.g., Zhang 2006):

$$\begin{aligned}
\mathbf{V}_d &= \frac{cpv}{qB}\left\{\frac{1-\mu^2}{2}\frac{\mathbf{B}\times\nabla B}{B^2} + \mu^2\frac{\mathbf{B}\times[(\mathbf{B}\cdot\nabla)\mathbf{B}]}{B^3} + \frac{1-\mu^2}{2}\frac{\mathbf{B}(\mathbf{B}\cdot\nabla\times\mathbf{B})}{B^3}\right\}, \qquad (2)\\
\frac{d\mu}{dt} &= -\frac{(1-\mu^2)v}{2}\hat{\mathbf{b}}\cdot\nabla\ln B + \frac{\mu(1-\mu^2)}{2}(\nabla\cdot\mathbf{V} - 3\hat{\mathbf{b}}\hat{\mathbf{b}}:\nabla\mathbf{V}) - \frac{(1-\mu^2)p}{v}(\mathbf{V}\cdot\nabla\mathbf{V})\cdot\hat{\mathbf{b}},\\
\frac{dp}{dt} &= -\left[\frac{1-\mu^2}{2}(\nabla\cdot\mathbf{V} - \hat{\mathbf{b}}\hat{\mathbf{b}}:\nabla\mathbf{V}) + \mu^2\hat{\mathbf{b}}\hat{\mathbf{b}}:\nabla\mathbf{V}\right]p - \frac{\mu p}{v}(\mathbf{V}\cdot\nabla\mathbf{V})\cdot\hat{\mathbf{b}}.
\end{aligned}$$

2.1. Magnetic field model

We choose to use the potential field source surface (PFSS) model for the coronal magnetic field. The model assumes that there is no electric current. It could be a good approximation on the large scale, which is established after the plasma in the corona achieves mechanical equilibrium under the magnetic force. The PFSS magnetic field $\mathbf{B}$ can be calculated from a scalar potential Φ through $\mathbf{B} = -\nabla\Phi$. In a current free environment, the scalar potential obeys the Laplace equation with a radial field boundary condition at the solar wind source surface R_{ss}. The solution to the PFSS field model in the polar coordinates can be written as a summation of spherical harmonics (Altschuler & Newkirk, 1969):

$$\begin{aligned}
\Phi &= R_0 \sum_{l=0}^{\infty}\sum_{m=0}^{l} P_l^m(\cos\theta)\,(g_{lm}\cos m\phi + h_{lm}\sin m\phi) \qquad (3)\\
&\times \left[l+1+l\left(\frac{R_0}{R_{ss}}\right)^{2l+1}\right]^{-1}\left[\left(\frac{R_0}{r}\right)^{l+1} - \left(\frac{R_0}{R_{ss}}\right)^{l+1}\left(\frac{r}{R_{ss}}\right)^{l}\right].
\end{aligned}$$

In the above equation, P_l^m is the associated Legendre polynomial. The inner boundary is set at the solar surface $R_0 = 1R_s$ (photosphere), and the outer boundary is typically at $R_{ss} = 2.5R_s$ (the solar wind source surface). As we have measurements of the photospheric magnetic field at R_0, the spherical harmonic coefficients g_{lm} and h_{lm} can be determined from fits to photospheric magnetogram measurements. These coefficients along with photospheric magnetogram measurements are available at a few solar telescope observatories. We choose to use the products that are published on the National Solar Observatory (NSO) Global Oscillation

Network Group (GONG) website. The large-scale magnetic field structure in the corona does not change dramatically in a few days. It rigidly corotates with the sun. For this reason, in this paper, we use a stationary magnetic field model to study SEP propagation. The computation is performed in the corotating frame for convenience. The speed of plasma in the corona below R_{ss} is low compared to other particle transport mechanisms, so we assume $\mathbf{V} = 0$.

2.2. *Solar wind model*

Outside of R_{ss}, the effect of solar wind flow is not negligible. We use the model by LeBlanc (1998), in which the radial solar wind speed is prescribed as:

$$V_r = V_\infty \left[1 + k_4 \left(\frac{r}{R_0}\right)^{-2} + k_6 \left(\frac{r}{R_0}\right)^{-4}\right]^{-1}, \tag{4}$$

and the tangential speed in the corotating frame is

$$V_\phi = -\Omega_s (r - R_{ss}) \sin\theta, \tag{5}$$

where Ω_s is the sun's angular rotation speed. Note that V_ϕ is just a velocity component relative to a rotating coordinate, and it does not represent the true speed of the particles. The meridional plasma speed is set to $V_\theta = 0$ everywhere. The magnetic field in the solar wind can be calculated from the Parker model and the solar wind speed given its boundary condition at the solar wind source surface $B_r(R_{ss})$.

$$B_r = B_r(R_{ss}) \left(\frac{R_{ss}}{r}\right)^2, \quad B_\theta = 0, \quad \text{and} \quad B_\phi = B_r \frac{V_\phi}{V_r}. \tag{6}$$

2.3. *Diffusion coefficients*

We use the following form for the pitch-angle diffusion coefficient:

$$D_{\mu\mu} = D_0(r) p^{q-2} (1 - \mu^2)(|\mu|^{q-1} + h_0). \tag{7}$$

It mimics the result of particle scattering by an Alfvénic turbulence with a power-law spectrum of spectral index q. An additional parameter h_0 is used to phenomenologically describe the enhancement of scattering through $\mu = 0$ by either non-resonant scattering or non-linear effects. Unless h_0 is very small $\ll$0.05 or very large $\gg 0.5$, it cannot affect the calculation result significantly. The level of particle scattering is mainly controlled by the value of $D_0(r)$. It is left as a free parameter in our model because it is sensitive to solar activities and can change drastically from one event to another or from one region to another. We typically use the parallel mean free path to specify the level of particle pitch angle scattering, as they are related by the following equation (e.g., Hasselmann and Wibberenz, 1970):

$$\lambda_{||} = \frac{3v}{8} \int_{-1}^{1} d\mu \frac{(1 - \mu^2)^2}{D_{\mu\mu}}. \tag{8}$$

The spatial dependence of $D_0(r)$ reflects the spatial distribution of the magnetic field turbulence level. Following Bieber (1994), we set the radial mean free path to be constant so that $\lambda_{||}$ is proportional to $\cos^{-2}\psi$, where ψ is the spiral angle of the Parker magnetic field to the radial direction. In the corona, we take $\psi = 0$ for this part of the calculation although the magnetic field in the PFSS model is not radial everywhere.

The spatial diffusion perpendicular to the ambient magnetic field $\kappa_\perp$ mainly comes from the motion of particles that follow meandering or random walking magnetic field lines. The random walk of field line is not included in the large-scale PFSS field model, so we include its effect on particle transport in the diffusion coefficient. For a population of particles moving along the field lines with an average parallel speed of $v/2$, their perpendicular diffusion coefficient is:

$$\kappa_\perp = \frac{v}{2} D_{FLRW}, \tag{9}$$

where D_{FLRW} is the diffusion coefficient of field line random walk per unit length of the field line. The granular motion of plasma in the photosphere is the ultimate driver of field line random walk in the corona and interplanetary space. The footpoint of the field line in the photosphere is assumed to undergo a diffusive motion. As the field line is drawn out by the solar wind with speed V, D_{FLRW} can be estimated using the following equation:

$$D_{FLRW} = \frac{\langle \delta x^2 \rangle}{\langle \delta z \rangle} = k_{gd} \frac{B_0}{B} \frac{\langle \delta x_0^2 \rangle}{V \langle \delta t_0 \rangle} . \tag{10}$$

The typical size for supergranules is about $\delta x_0 = 3 \times 10^4$ km within a lifetime of $\delta t_0 = 24$ hours, and $k_{gd} < 1$ is a free parameter, indicating the coupling of the field line motion between the ionized corona and partially ionized photosphere. The ratio of magnetic field strength to the surface value B/B_0 describes the divergence effect of magnetic field flux tubes, which amplifies the extent of footpoint motion by that ratio because of the frozen-in magnetic field lines.

2.4. Stochastic differential equation solver

The focus transport equation (1) is a 5-dimensional diffusion-drift equation. Due to the complexity of magnetic field geometry and diffusion coefficients, the only way to obtain its solution is through numerical simulation. Regular ways to solve the partial differential equation, ie., finite difference or finite element methods, requires too many computer resources. Therefore, we resort to using stochastic differential equations to solve the focus transport equation. The Fokker-Planck diffusion equation (1) can be recast into 5 stochastic differential equations, which describe the time evolution of the guiding center position, pitch-angle cosine, and momentum of a stochastic process, which can be considered as a representation of real particles in this formulation of stochastic processes (e.g. Gardiner 1985):

$$\begin{aligned} d\mathbf{x}(t) &= \sqrt{2\,\boldsymbol{\kappa}_\perp} \cdot d\mathbf{w}(t) + (\nabla \cdot \boldsymbol{\kappa}_\perp + v\mu\hat{\mathbf{b}} + \mathbf{V} + \mathbf{V}_d)dt, \\ d\mu(t) &= \sqrt{2\max(D_{\mu\mu}, 0)}dw(t) + \left[\frac{\partial D_{\mu\mu}}{\partial \mu} + \frac{d\mu}{dt}\right] dt, \\ dp(t) &= \frac{dp}{dt} dt. \end{aligned} \tag{11}$$

In the above equation, $dw(t)$ is the Wiener process as a function of time t, and it can be generated by random numbers with a Gaussian distribution. The function max in (11) takes the maximum value of its arguments.

The stochastic differential equation is like a first-order ordinary differential equation. We solve the stochastic differential equations using an Euler scheme (Kloeden and Platen, 1992). As particle streaming along the magnetic field line and the focusing term in Equation (2), both of which are proportional to the particle speed v, are much faster than the other transport terms, we use a 4th-order Range-Kutta scheme (Press et al. 1991) to integrate those parts. In this way, particle trapping is correctly captured in our calculation. The high-order Range-Kutta integration accuracy is only necessary when the particle pitch-angle diffusion and perpendicular

IOP Conf. Series: Journal of Physics: Conf. Series **1225** (2019) 012010 doi:10.1088/1742-6596/1225/1/012010

spatial different are very weak, but normally with the typical values of the diffusion coefficients for most SEP events, accuracy higher than the 1-st order Euler scheme does not make much difference because, in the diffusion terms, $dw(t) \propto \sqrt{dt}$.

Equations (11) are time-forward stochastic differential equations that correspond to the focus transport equation (1). It is most suitable for the simulations in which particles are injected with a specific pattern. We typically inject particles at a solar flare site slightly above the solar surface or at a CME shock in the corona. As it is unlikely for particles to be effectively accelerated on closed field lines due to heavy loss in the solar atmosphere, and as it is also difficult to get the particles from closed field lines to interplanetary space, we only inject particles on open lines in the corona. The stochastic differential equations are integrated to follow the random trajectory of particles. Particle distribution, after some time of propagation, can be built by properly binning a large number of simulated particles.

3. An example of model results

The sun emits an intense X8.2 solar flare that peaks at 12:06 UT on 2017 September 10. It is a capstone event from Active Region 12673, which has been making solar flares since August. Viewed from Earth, the solar flare is located at S08W88. The solar flare triggers a fast halo CME. The center of the CME propagates out in the direction roughly perpendicular to the line of sight from Earth, so its speed and size can be more accurately determined from coronographs. The CME speed is greater than 3000 km/s, and its angular size is approximately 180° in latitude. Rapid increases in energetic particle fluxes can be seen at Earth immediately following the solar flare. At the time of the solar flare, the solar wind speed at Earth is 512 km/s. No shock is observed until the end of September 12, but the shock is not fast, being only less than 650 km/s. This suggests that the fast CME observed in the corona does not reach Earth either due to its size or strength.

ST-A spacecraft is located 128° west of Earth in longitude. The solar flare occurs on the opposite side of ST-A. In-situ measurements on ST-A show a slow wind of $\sim$ 400 km/s and do not detect any shock for several days. SEP intensities at ST-A increase gradually and they do not reach their peak intensity until 3 days later. This suggests that the observed particles could have reached ST-A through perpendicular transport in interplanetary space. ST-A gets into an SEP reservoir after observing a peak intensity 3 days after the solar flare.

We inject 30 MeV protons at the CME shock when it is at 1.5 R_s. The top panel of Figure 1 shows in the loci of open field lines at 1.5 R_s. We assume that the CME shock area is circular, extending 90° in each direction from the location of the solar flare. Particles are uniformly injected with an isotropic pitch-angle distribution over the open field lines within the CME shock. The radial mean free path is set to 5 R_s in the corona below 20 R_s and 20 R_s in interplanetary space outside of 20 R_s. Without perpendicular diffusion ($k_{gd} = 0$), a small fraction (1.7%) of particles enter into the loss cone and precipitate on the solar surface (Figure 1 middle panel). The bottom panel of Figure 1 shows the latitude-longitude distribution of escaped ($r > 2.5 R_s$) SEPs on the solar wind source surface. Once the particles reach 2.5 R_s, the chance to return to the solar surface is minimal in this simulation. The longitude and latitude of the particles are mapped along the Parker spirals on to the solar wind source surface at 2.5 R_s. The area coverage is very much the same as the CME shock although particles are only injected on open field lines with a smaller area at 1.5 R_s. The result is just a coincidence after mapping the field lines from 1.5 R_s to the solar wind source surface using the PFSS magnetic field of this particular event. On the other hand, this CME is so large that any distortion of its location during the field-line mapping could become less visible. The intensity distribution of the escaped particles also shows an effect of particle gradient/curvature drift, which is particularly clear near the neutral sheet. The footpoint of the magnetic field line to Earth (symbol E_m) is located directly within the SEP source regions. This explains why we see an immediate and fast

IOP Conf. Series: Journal of Physics: Conf. Series **1225** (2019) 012010 doi:10.1088/1742-6596/1225/1/012010

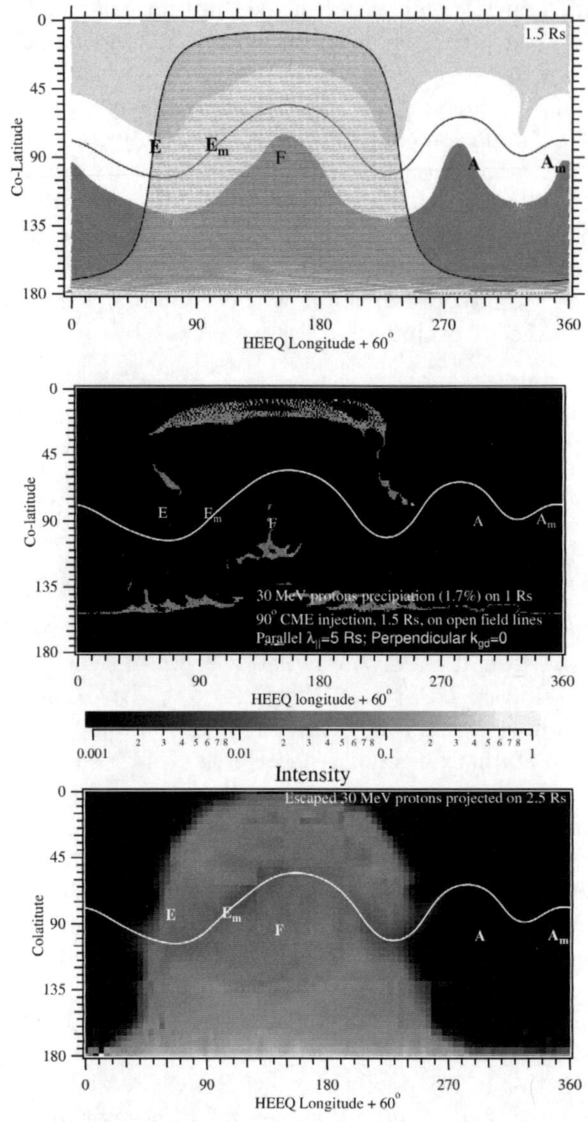

Figure 1. (Top) Loci of outward (green) and inward (red) open field lines as a function of solar latitude and longitude at 1.5 R_s. The black curve with the shaded area indicates an assumed CME shock size. (Middle) Loci of precipitating SEPs on the solar surface, and (Bottom) distribution of escaped SEPs projected on the solar wind source surface. The letter F shows the location of the solar flare. Letters E and A show the projection of Earth and ST-A. Letters E_m and A_m show the footpoint of magnetic field lines to Earth and ST-A. The blue and white curves are the neutral sheet from the PFSS coronal magnetic field model. Perpendicular diffusion is turned off in the calculation.

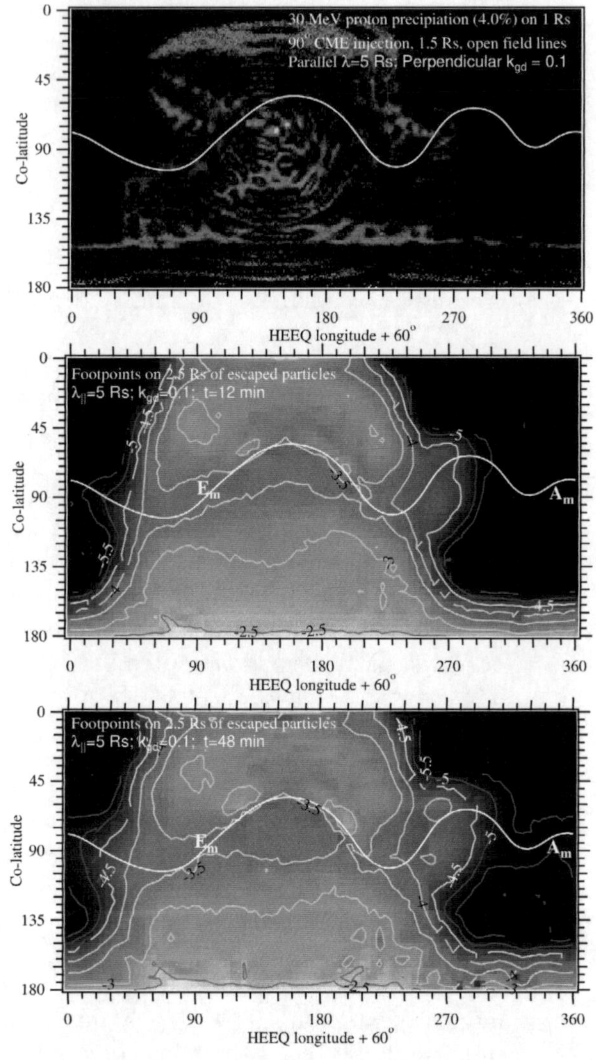

Figure 2. Loci of precipitating SEPs on the solar surface (Top), and (Middle at t =12 min and Bottom at t = 48 min) the latitude-longitude intensity distribution of escaped SEPs on the solar wind source surface at 2.5 R_s when SEP transport has a perpendicular diffusion equal to 10% the rate derived from supergranular diffusion. The contours are in the $\log_{10}$ scale with an arbitrary intensity unit. The white curve shows the neutral sheet of the PFSS magnetic field model. Letters E_m and A_m indicate the footpoint of magnetic field lines to Earth and ST-A.

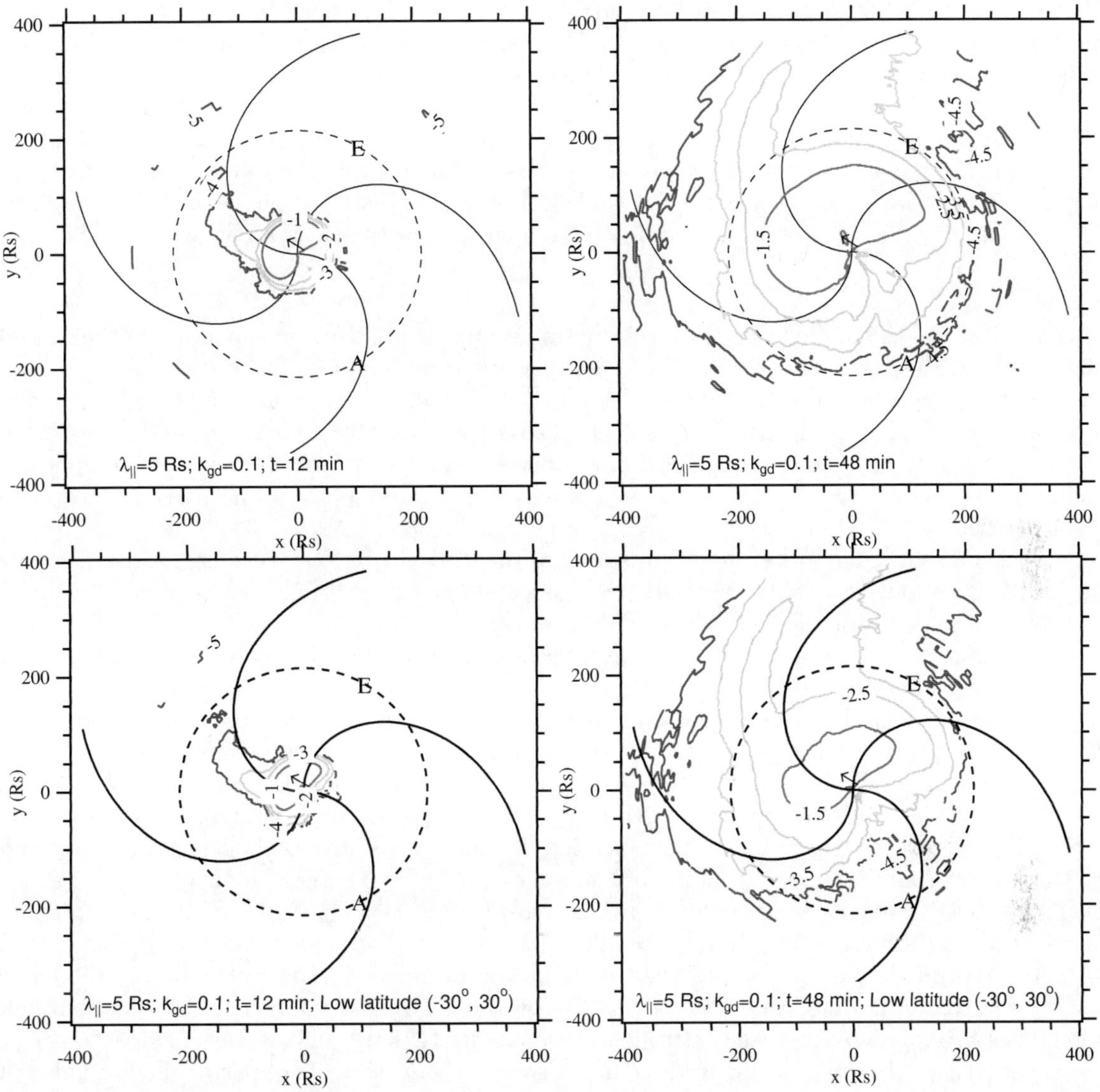

Figure 3. Distribution of escaped SEPs projected latitudinally on the solar equatorial plane at two time intervals t =12 min and t = 48 min. The contours are in the $\log_{10}$ scale with an arbitrary intensity unit. The top two panels include particles from all latitudes, while the bottom two panels only include particles at low latitudes between $-30°$ and $30°$

increase of SEPs after the solar flare. There are no SEP source regions on magnetic field lines that connect to ST-A, which is consistent with ST-A observations.

Figure 2 shows calculation results of the latitude-longitude distribution of precipitating (top) and escaped (middle and bottom) SEPs with a finite $k_{gd} = 0.1$ level of perpendicular diffusion. The percentage of SEPs precipitating on the solar surface has increased to 4.0%, and the precipitation covers more solar surface area. This is because particles can diffuse across the magnetic field into close field lines. The longitudinal spread of escaped SEPs can be seen by comparing the distribution at 12 min and 48 min in the middle and bottom panels of Figure 2. The footpoint of the magnetic field line to ST-A does not lie within the source initially. The

SEPs spread in longitude, which occurs particularly more easily near the neutral sheet due to the combined effects of enhanced particle drift and perpendicular diffusion in weak field regions. This could be used to explain the gradual increase of SEP fluxes at ST-A. Figure 3 shows the intensity distributions of SEPs projected on the solar equatorial plane at 12 min and 48 min. It demonstrates that Earth is inside the regions of immediate radiation exposure, whereas the radiation will take a much longer time to reach ST-A. The bottom two panels only include the particles at low latitudes between $-30°$ and $30°$. Particle distribution spreads less in radial distance, mostly due to the tighter winding of Parker magnetic field lines at low latitudes.

4. Summary

We have described a computer code for conducting simulations of SEP propagation in the coronal and interplanetary magnetic fields. We use stochastic differential equations to solve the focus transport equation in 3-d. The transport equation essentially includes all the particle transport mechanisms in regular and turbulent magnetic fields. Because of the flexibility and simplicity of solving stochastic differential equations, our computer model can incorporate any complicated magnetic field geometry derived from magnetogram measurements of the solar photospheric magnetic field.

We have used the computer model to simulate the 2017 September 10 solar flare SEP event. Its results show our capability to predict SEP escape from the corona and precipitation on to the solar surface. A small fraction of SEP produced by the CME shock in the high corona can precipitate on the solar surface. These precipitating particles may produce high-energy, non-thermal gamma-rays or neutrons, some of which could be detected at Earth. The emission can be far away from the site of the solar flare that emits mostly in soft x-rays. Currently, we are conducting further simulations to investigate whether SEPs can be confined in the corona for many hours so that we can use the model to explain long-duration gamma-ray emissions detected by the Fermi telescope (Omodei et al. 2018). This situation is highly plausible, given the high levels of magnetic field turbulence left behind by the CME shock. The majority of SEPs produced and injected at the CME shock can escape to interplanetary space. Particle mirroring by the magnetic field is the major reason for the asymmetry of particle intensity between precipitation and escape . Although particles are only injected on open field lines with a limited latitude-longitude coverage, the distribution of escaped SEPs is essentially the entire range of latitude and longitude that is covered by the CME. Perpendicular diffusion can help spread the SEPs in longitude and latitude a little bit, particularly into weaker field regions near the neutral sheet. However, compared to the size of the CME, the role of perpendicular diffusion transport in the spread of particles within an hour is not that important. Subsequent particle transport in interplanetary space is needed to further spread the particles to locations where magnetic field lines are far away from the CME. The gradual increase of SEPs observed by ST-A is consistent with perpendicular particle transport in the interplanetary space.

5. Acknowledgments

This work was supported in part by NASA Grants NNX15AN72G, NNX15AB76G, and 80NSSC18K0644. We highly appreciate discussions with the SEP connectivity focus science team members of the NASA Living With the Star Program.

6. References

[1] Altschuler, M. D., & Newkirk, G., 1969 *Solar Phys.* **9** 131
[2] Bieber, J.W.,Matthaeus, W. H., Smith, C.W.,Wanner, W., Kallenrode, M.,& Wibberenz, G. 1994, *Astrophys. J* **420** 294
[3] Chollet, E. E., Mewaldt, R. A., Cummings, A. C., et al. 2010 *J. Geophys. Res.* **115** 12106
[4] Cliver et al. 1995, *Proc. ICRC* **4**, 57

[5] Dalla, S., Balogh, A., Krucker, S., et al., 2003, *Geophys. Res. Lett.* **30** 8035
[6] Dalla, S., Marsh, M. S., Kelly, J., & Laitinen, T. 2013, *J. Geophys. Res.* **118** 5979
[7] Gardiner, C. W. 1985 *Handbook of Stochastic Methods for Physics, Chemistry and the Natural Sciences*, Berlin: Springer-Verlag
[8] Giacalone, J., & Jokipii, J. R. 2012 *Astrophys. J. Lett.* **751** L33
[9] Hasselmann, K., & Wibberenz, G. 1970 *Astrophys. J.* **162** 1049
[10] Kelly, J., Dalla, S., & Laitinen, T. 2012 *Astrophys. J. Lett.* **750** 47
[11] Kienreich, I. W., Temmer, M., & Veronig, A. M. 2009, *Astrophys. J Lett* **703** L118
[12] Klassen, A., Aurass, H., Mann, G., & Thompson, B. J. 2000 *Astron & Astrophys. Supp* **141** 357
[13] Kloeden, P. E., Platen, E. 1992, *Numerical Solution of Stochastic Differential Equations* Berlin: Springer-Verlag
[14] Laitinen, T., Dalla, S., & Marsh, M. S. 2013 *Astrophys. J. Lett.* **773**, L29
[15] LeBlanc, Y., Dulk, G. A., & Bougeret, J., 1998 *Solar Phys.* **183** 165
[16] Marsh, M. S., Dalla, S., Kelly, J., & Laitinen, T. 2013 *Astrophys. J.* **774** 4
[17] McKibben, R.B., *J. Geophys. Res.* **77** 3975.
[18] McKibben, R.B., Lopate, C., & Zhang, M., 2001 *Space Sci. Rev.* **97** 257
[19] Omodei, N., Pesce-Rollins, M., Longo, F, Allafort, A., Krucker, S., 2018, *Astrophys. J. Lett.* **submitted** arXiv:1803.07654
[20] Press, William H.; Teukolsky, Saul A.; Vetterling, William T.; Flannery, & Brian P. 1992, *Numerical recipes in FORTRAN. The art of scientific computing* Cambridge: University Press
[21] Richardson, I. G., & Cane, H. V. 1996 *J. Geophys. Res.* **101** 27521
[22] Richardson, I. G., Cane, H. V., & von Rosenvinge, T. T. 1991 *J. Geophys. Res.* **96** 7853
[23] Roelof, E. C., Gold, R. E., Simnett, G. M., Tappin, S. J., Armstrong, T. P., & Lanzerotti, L. J., 1992 *Geophys. Res. Lett.* **19** 1243
[24] Rouillard, A. P., Sheeley, N. R., Tylka, A., et al. 2012, *Astrophys. J.* **752** 44
[25] Schrijver, C. J., & De Rosa, M. L. 2003 *Solar Phys.* **212** 165
[26] Schrijver, C. J.,& Title, A. M. 2011 *J. Geophys. Res.* **116** 4108
[27] Schrijver, C. J., Title, A. M., Yeates, A. R., & DeRosa, M. L. 2013 *Astrophys. J.* **773** 93
[28] Shen, Y., Liu, Y., & Su, J. 2012 *Astrophys. J.* **750** 12
[29] Thompson, B. J., & Myers, D. C. 2009, *Astrophys. J. Supp* **183** 225
[30] Torsti, J., Riihonen, E., & Kocharov, L. 2004 *Astrophys. J. Lett.* **600** L83
[31] Wang, H., Chae, J., Yurchyshyn, V., et al. 2001 *Astrophys. J.* **559** 1171
[32] Wiedenbeck, M. E., Mason, G. M., Cohen, C. M. S., Nitta, N. V., Gómez-Herrero, R., & Haggerty, D. K. 2013, *Astrophys. J.* **1762** 54
[33] Zhang, M., McKibben, R. B., Lopate, C., Jokipii, J. R., Giacalone, J., Kallenrode, M.-B., & Rassoul, H. K., 2003 *J. Geophys. Res.* **108** 1154
[34] Zhang, M., 2006 *J. Geophys. Res.* **111** 4208

IOP Conf. Series: Journal of Physics: Conf. Series **1225** (2019) 012011 doi:10.1088/1742-6596/1225/1/012011

The relativistic implicit Particle-in-Cell method

Fabio Bacchini, Jorge Amaya, and Giovanni Lapenta

Centre for mathematical Plasma Astrophysics, Department of Mathematics, KU Leuven, Celestijnenlaan 200B, B-3001 Leuven, Belgium

E-mail: fabio.bacchini@kuleuven.be

Abstract. Relativistic Particle-in-Cell (PiC) methods are among the most reliable methods for the investigation of plasma phenomena at the particle scale. Standard explicit and semi-implicit PiC methods are affected by numerical instabilities that restrain the range of admissible simulation parameters, and prevent their application to large domains over long time scales. Here, we present a three-dimensional, fully-implicit algorithm for relativistic PiC simulations that conserves energy exactly (to machine precision) and eliminates numerical instabilities, allowing long-term calculations. We describe the full numerical solution procedure and test the algorithm with several one- and two-dimensional case-studies of kinetic instabilities, comparing the results with theoretical predictions.

1. Introduction

High-energy astrophysical environments where plasmas reach relativistic speeds are a natural laboratory where the interplay of strong electromagnetic fields and the dynamics of particles can be studied. Such scenarios are considered realistic for plasma flows in the magnetosphere of compact objects (black holes and neutron stars, see e.g. [1]), pulsar wind nebulae (PWNe, see [2]), shock fronts of relativistic jets [3], and many others. The modeling of relativistic plasmas can be efficiently carried out with numerical codes, simulating the relevant dynamics on a global scale with magnetohydrodynamic (MHD) approaches, or at the particle scale with a fully-kinetic description.

Particle-in-Cell (PiC) methods are the most successful tool for the simulation of the kinetic dynamics of plasmas. PiC algorithms in a broad sense represent a class of numerical methods for the solution of partial differential equations in contexts where advection of physical quantities plays an important role in the relevant phenomena. As such, PiC methods can be applied to the modeling of fluids [4], to solid mechanics problems [5] and, most importantly, to the kinetic description of plasma processes. The latter context is by far the most widely renowned field of application of PiC codes, hence the common association of the term "PiC" with kinetic physics. PiC codes exist in a broad variety of formulations both for simulations in the nonrelativistic limit [6][7][8], and for relativistic applications [9][10][11].

Despite their success, commonly employed relativistic PiC codes are based on the explicit solution of particle and field equations, which is well-known for exhibiting spurious heating of the lighter particles due to numerical instabilities [12]. In the most sever cases, such instabilities can completely invalidate the numerical results, which become unphysical and often cause the failure of the simulation run. Early semi-implicit schemes, based on the implicit solution of the field equations (with the particles evolved explicitly), relax the constraints imposed by numerical instabilities, but are still prone to destabilization when specific criteria for the chosen spatial and temporal resolutions are not met. In practice, one is often bound to choose a grid spacing that resolves the electron Debye length $\lambda_{De} = \sqrt{\varepsilon_0 kT_e/(e^2 n_e)}$, calculated from the electron temperature kT_e, electron charge e, and number density n_e. For electron-positron or electron-ion plasmas with low electron temperatures, however, the dynamics of specific phenomena such as magnetic reconnection can take place over the much larger length and time scales of the (electron or ion) skin depth [8]. The requirement of resolving the unimportant Debye length scales imposes a detrimental increase in the computational effort required by explicit codes.

Energy-conserving PiC algorithms are not subjected to this issue, due to the elimination of instabilities related to the spurious non-conservation of energy. The exact conservation of global energy is attained via specific choices of discretization of the grid and particle equations. Energy-conserving semi- and fully-implicit methods have been proposed and successfully applied for plasma simulations in the nonrelativistic limit [13][6][8]. In the relativistic framework, only fully-implicit algorithms allow for attaining exact energy conservation, due to the implicit nature of the momentum equation for the particles [14].

In this work, we present a multi-dimensional, fully implicit algorithm for relativistic plasma simulations that conserves energy exactly (to machine accuracy), exhibiting superior stability properties. The new scheme is based on that presented by [14], but differs in the solution procedure that is adapted to handle coarse temporal and spatial resolutions, as well as realistic ion-to-electron mass ratios, and achieves affordable computational costs in multi-dimensional runs. In the next Section, we present in full the numerical scheme and two alternative solution strategies for the discretized Vlasov-Maxwell system of equations. Then, we test our algorithm against analytic calculations by simulating several types of kinetic instabilities and comparing to the predictions from linear theory. Finally, we provide an example application to a magnetic reconnection case-study, involving large separation of scales between the ion and electron dynamics.

2. Relativistic implicit Particle-in-Cell

2.1. The Vlasov-Maxwell system and the finite particle approach

The physics of plasmas at the microscopic scale is described by the system composed of Maxwell's equations for the electromagnetic fields **E**, **B**,

$$\frac{\partial \mathbf{E}}{\partial t} = \nabla \times \mathbf{B} - \mu_0 \mathbf{J}, \tag{1}$$

$$\frac{\partial \mathbf{B}}{\partial t} = -\nabla \times \mathbf{E}, \tag{2}$$

$$\nabla \cdot \mathbf{E} = \frac{\rho}{\varepsilon_0}, \tag{3}$$

$$\nabla \cdot \mathbf{B} = \mathbf{0}, \tag{4}$$

and the Vlasov equation for the distribution function $f_s(\mathbf{x}, \mathbf{u}, t)$ of each particle species s,

$$\frac{\partial f_s}{\partial t} + \frac{d\mathbf{x}}{dt} \cdot \frac{\partial f_s}{\partial \mathbf{x}} + \frac{d\mathbf{u}}{dt} \cdot \frac{\partial f_s}{\partial \mathbf{u}} = 0. \tag{5}$$

Here, all quantities are expressed in SI units, with the speed of light as a normalization constant.

A most successful numerical approach for solving the Vlasov-Maxwell system of equations above is the so-called kinetic Particle-in-Cell (PiC) method [12]. This class of numerical methods employs a finite number N_s of "superparticles" for each species s, with the fundamental assumption that the function f_s can be sampled by the superposition

$$f_s(\mathbf{x}, \mathbf{u}, t) = \sum_p^{N_s} S(\mathbf{x} - \mathbf{x}_p)\delta(\mathbf{u} - \mathbf{u}_p), \tag{6}$$

where S is a shape function characterizing each superparticle p. Because of the linearity of the superposition, the evolution of f_s from equation (6) is still governed by the Vlasov equation (5). PiC methods evolve the position and momentum, $\mathbf{x}_p$ and $\mathbf{u}_p$ of each superparticle in time. The particle position and momentum obey the equations of motion,

$$\frac{d\mathbf{x}_p}{dt} = \frac{\mathbf{u}_p}{\gamma_p}, \tag{7}$$

$$\frac{d\mathbf{u}_p}{dt} = \frac{q_p}{m_p}\left(\mathbf{E}(\mathbf{x}_p) + \frac{\mathbf{u}_p}{\gamma_p} \times \mathbf{B}(\mathbf{x}_p)\right), \tag{8}$$

where $\gamma_p = \sqrt{1 + u_p^2}$ is the Lorentz factor, and the right-hand side of equation (8) is the Lorentz force for a particle of mass m_p and charge q_p. All the computational (super)particles are evolved by solving equations (7) and (8).

The evolution of the electromagnetic fields is calculated by solving finite-difference approximations of Maxwell's equations (1)-(4) on a computational grid of N_g elements. Each computational cell has volume V_g. Because the shape function S defines the finite size of each particle, the average electric and magnetic field acting on a single particle are defined as

$$\mathbf{E}(\mathbf{x}_p) = \int_{V_g} \mathbf{E}(\mathbf{x})S(\mathbf{x} - \mathbf{x}_p)d\mathbf{x} = \sum_g \mathbf{E}_g W(\mathbf{x}_g - \mathbf{x}_p), \tag{9}$$

$$\mathbf{B}(\mathbf{x}_p) = \int_{V_g} \mathbf{B}(\mathbf{x})S(\mathbf{x} - \mathbf{x}_p)d\mathbf{x} = \sum_g \mathbf{B}_g W(\mathbf{x}_g - \mathbf{x}_p), \tag{10}$$

where the electric and magnetic fields on each grid element $\mathbf{E}_g$ and $\mathbf{B}_g$ contribute to the particle according to the value assumed by the interpolation function $W(\mathbf{x}_g - \mathbf{x}_p)$. The interpolation function is related to the shape function by

$$W(\mathbf{x}_g - \mathbf{x}_p) = \int_{V_g} S(\mathbf{x} - \mathbf{x}_p)d\mathbf{x}. \tag{11}$$

In one dimension, a typical choice for the shape function S is the zeroth-order b-spline, and as a consequence the interpolation function W simply results in the first-order b-spline. In multi-dimensional simulations, the shape function and interpolation function are simply taken as the convolution of b-splines along each spatial dimension. Finally, the interpolation function defines the contribution of the particles to the source terms of Maxwell's equations at each grid point,

$$\rho_g = \frac{1}{V_g}\sum_s \sum_p^{N_s} q_p W(\mathbf{x}_g - \mathbf{x}_p) \tag{12}$$

$$\mathbf{J}_g = \frac{1}{V_g}\sum_s \sum_p^{N_s} q_p \frac{\mathbf{u}_p}{\gamma_p} W(\mathbf{x}_g - \mathbf{x}_p). \tag{13}$$

IOP Conf. Series: Journal of Physics: Conf. Series **1225** (2019) 012011 doi:10.1088/1742-6596/1225/1/012011

2.2. Implicit time discretization

Implicit PiC schemes employ an implicit time discretization of Maxwell's equations and of the particle equations of motion. Our choice of discretization, in the most general case, consists of a θ-scheme for the fields,

$$\frac{\mathbf{E}_g^{n+1} - \mathbf{E}_g^n}{\Delta t} = \nabla \times \mathbf{B}_g^{n+\theta} - \mu_0 \tilde{\mathbf{J}}_g, \tag{14}$$

$$\frac{\mathbf{B}_g^{n+1} - \mathbf{B}_g^n}{\Delta t} = -\nabla \times \mathbf{E}_g^{n+\theta}, \tag{15}$$

which eliminates the restrictions on the time step Δt imposed by the CFL condition when $\theta > 1/2$ [7]. The particle equations of motion are discretized with a midpoint scheme [14],

$$\frac{\mathbf{x}_p^{n+1} - \mathbf{x}_p^n}{\Delta t} = \frac{\overline{\mathbf{u}}_p}{\overline{\gamma}_p}, \tag{16}$$

$$\frac{\mathbf{u}_p^{n+1} - \mathbf{u}_p^n}{\Delta t} = \frac{q_p}{m_p}\left(\mathbf{E}^{n+\theta}(\overline{\mathbf{x}}_p) + \frac{\overline{\mathbf{u}}}{\overline{\gamma}} \times \mathbf{B}^{n+\theta}(\overline{\mathbf{x}}_p)\right). \tag{17}$$

Here, quantities at time level $n+\theta$ are taken as a linear combination of values between time levels n and $n+1$, e.g. $\mathbf{E}^{n+\theta} = \theta\mathbf{E}^{n+1} + (1-\theta)\mathbf{E}^n$, with the decentering parameter $\theta = [1/2, 1]$. For the case $\theta = 1/2$, the scheme above is of second-order accurate in time. Barred quantities thus represent values at $n + 1/2$, e.g. $\overline{\mathbf{u}}_p = (\mathbf{u}_p^{n+1} + \mathbf{u}_p^n)/2$. A special case is represented by the average current $\tilde{\mathbf{J}}$, which we define as

$$\tilde{\mathbf{J}}_g = \sum_s \sum_p^{N_s} q_p \frac{\overline{\mathbf{u}}_p}{\overline{\gamma}_p} W(\overline{\mathbf{x}}_p - \mathbf{x}_g). \tag{18}$$

Note that in the system above, we do not consider the two divergence equations (3) and (4) for simplicity. It can be shown that the divergence-free condition for the magnetic field is automatically satisfied by our choice of formulation. Gauss's law for the electric field, instead, can be enforced by either applying a divergence cleaning step [7] or by subcycling during the current deposition step (as demonstrated in [6] in the Newtonian limit). Note that the former technique, however standard, violates the energy conservation properties of the scheme [14]; the latter, instead, ensures energy conservation at the cost of an increased computational cost. Alternatively, one can rely on charge-conserving deposition schemes [15], which however mutually exclude the energy-conserving formulation of the current (18). Very recently a second alternative approach was proposed for nonrelativistic, semi-implicit energy-conserving PiC methods that enforces Gauss's law by modifying the particle position at each time step [16]. Whether this is applicable in our algorithm is an interesting possibility that will be considered in the future. Despite the absence of a method to enforce charge conservation, in the results of the tests presented here we did not detect severe unphysical features or spurious oscillations in the field quantities.

For the remainder of this work, we focus on the case $\theta = 1/2$. With our choice, the system of equations to be solved is

$$\frac{\mathbf{E}_g^{n+1} - \mathbf{E}_g^n}{\Delta t} = \nabla \times \overline{\mathbf{B}}_g - \mu_0 \tilde{\mathbf{J}}_g, \tag{19}$$

$$\frac{\mathbf{B}_g^{n+1} - \mathbf{B}_g^n}{\Delta t} = -\nabla \times \overline{\mathbf{E}}_g, \tag{20}$$

$$\frac{\mathbf{x}_p^{n+1} - \mathbf{x}_p^n}{\Delta t} = \frac{\overline{\mathbf{u}}_p}{\overline{\gamma}_p}, \tag{21}$$

$$\frac{\mathbf{u}_p^{n+1} - \mathbf{u}_p^n}{\Delta t} = \frac{q_p}{m_p}\left(\overline{\mathbf{E}}(\overline{\mathbf{x}}_p) + \frac{\overline{\mathbf{u}}}{\overline{\gamma}} \times \overline{\mathbf{B}}(\overline{\mathbf{x}}_p)\right). \tag{22}$$

This choice makes the overall scheme globally energy-conserving, as shown in the original work by Lapenta & Markidis [14].

2.3. Solution procedure

The numerical scheme proposed above is inherently implicit in both the field and particle variables. Compared to the relatively simple time-marching algorithm of explicit PiC schemes, implicit methods involve a higher degree of difficulty in the solution procedure. A comparison between the two solution algorithms is sketched in Figure 1. Explicit PiC codes (left) typically employ a four-step sequential scheme to interpolate the fields at the particle position, advance the particle quantities, gather the source term for the fields from the particles, and advance the field quantities on the grid. Implicit PiC schemes (right), instead, must solve particle and field equations simultaneously, clearly involving a higher computational cost. Additionally, for the relativistic case considered here, the particle equations are also nonlinear, due to the presence of the Lorentz factor in equation (22). Because the field equations are coupled implicitly to the particle equations, the whole system of particle and field equations inherits the nonlinear character, and thus is typically solved iteratively.

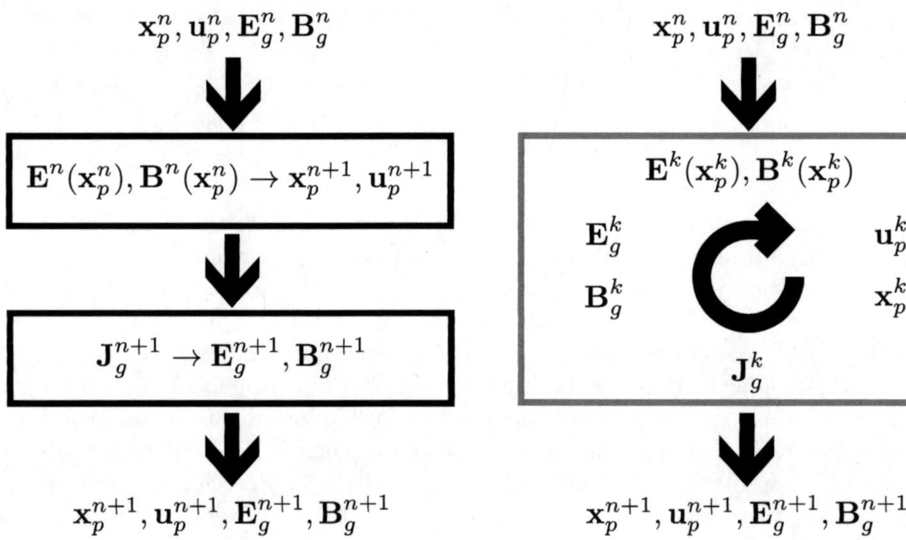

Figure 1: Schematic comparison between the typical computational cycle of an explicit PiC method (left) and an implicit method (right). Black boxes indicate explicit (non-iterative) operations, while red boxes indicate the use of an iterative solution process.

The most direct method for solving the implicit equations consists of handling the full system (19)-(22) with any preferred nonlinear solver. This solution was shown to be applicable for one-dimensional systems in the original work [14]. This strategy, however, is not suitable for realistic three-dimensional simulations. In this case, the full system consists of $6N_g + 6N_p$ coupled nonlinear equations. For typical PiC runs where the number of particles N_p is of order millions or billions, the computational cost of such a direct approach is simply prohibitive.

Here, we consider instead two alternative strategies that reduce the complexity of the system to $3N_g$ coupled linear or nonlinear equations for the fields, and $3N_p$ independent nonlinear equations for the particles. The implicit solution of such a system involves a computational cost that surely exceeds that of an explicit scheme, but that can still be handled by state-of-the-art linear/nonlinear solvers, contrary to the original $6N_g + 6N_p$ nonlinear system. The workflow of the two algorithms is compared in Figure 2. In the next Sections, we describe the two strategies in detail and discuss their features.

2.3.1. System reduction As a first step, we can reduce the total size of the system of equations from $6N_g + 6N_p$ by a factor 2. The field equations can be reduced with a standard procedure employed in implicit PiC codes [13] as follows. First, substitute the electric and magnetic fields at time level $n+1$ with the definitions in terms of the corresponding barred quantities into equations (19) and (20),

$$\frac{\overline{\mathbf{E}}_g - \mathbf{E}_g^n}{\Delta t/2} = \nabla \times \overline{\mathbf{B}}_g - \mu_0 \tilde{\mathbf{J}}_g, \tag{23}$$

$$\frac{\overline{\mathbf{B}}_g - \mathbf{B}_g^n}{\Delta t/2} = -\nabla \times \overline{\mathbf{E}}_g. \tag{24}$$

Then, take the curl of equation (24) and substitute it in equation (23) to obtain a second-order equation in $\overline{\mathbf{E}}$ only,

$$\frac{\overline{\mathbf{E}}_g - \mathbf{E}_g^n}{\Delta t/2} = \nabla \times \mathbf{B}_g^n - \frac{\Delta t}{2} \nabla \times \nabla \times \overline{\mathbf{E}}_g - \mu_0 \tilde{\mathbf{J}}_g. \tag{25}$$

With such a manipulation, the field part of the system is reduced to $3N_g$ equations. The magnetic field can then be computed with equation (24).

The particle equations can be reduced in two ways. A first, straightforward approach is to employ a leap-frog scheme for the position and momentum, such that the position equation (21) is written

$$\frac{\mathbf{x}_p^{n+1/2} - \mathbf{x}_p^{n-1/2}}{\Delta t} = \frac{\mathbf{u}_p^n}{\gamma_p^n}. \tag{26}$$

In this way, the average position $\overline{\mathbf{x}}_p \equiv \mathbf{x}_p^{n+1/2}$ does not depend on the unknown $\overline{\mathbf{u}}_p$. A second, more consistent approach is to substitute for $\overline{\mathbf{x}}_p$ into equation (21), such that

$$\frac{\overline{\mathbf{x}}_p - \mathbf{x}_p^n}{\Delta t/2} = \frac{\overline{\mathbf{u}}_p}{\overline{\gamma}_p}. \tag{27}$$

Then, by substituting this expression into equation (22), we are left with a system of equations in the unknown $\overline{\mathbf{u}}_p$ only ($\mathbf{u}_p^{n+1}$ can also be substituted by inverting $\overline{\mathbf{u}}_p = (\mathbf{u}_p^{n+1} + \mathbf{u}_p^n)/2$). Either approach reduces the system size to $3N_p$ nonlinear equations.

2.3.2. Field hiding The field hiding (FH) strategy (Figure 2, left) proposed recently for nonrelativistic implicit PiC [17] decouples the field and particle parts of the initial set of equations (19)-(22). The particle equations of motion (21)-(22) are solved independently for each particle with an iterative process, until convergence below a prescribed tolerance. With an initial guess for $\overline{\mathbf{u}}$, the current, $\tilde{\mathbf{J}}$, is computed on the grid. Then, equations (25) and (24) are solved for the electromagnetic fields, $\overline{\mathbf{E}}$ and $\overline{\mathbf{B}}$. With the new values of the fields, the momentum $\overline{\mathbf{u}}$ is computed from equation (22), and the iteration is restarted.

The fundamental assumption of the FH strategy is that the motion of each particle can be resolved while neglecting the motion of all other particles. This dependence is in fact "hidden" in the field terms. In practice, when solving equation (22) for $\overline{\mathbf{u}}$, one neglects the dependence $\overline{\mathbf{E}}_g(\overline{\mathbf{u}})$ and $\overline{\mathbf{B}}_g(\overline{\mathbf{u}})$ (arising from the expression of $\tilde{\mathbf{J}}_g$), and the field terms are essentially considered fixed for the current iteration. This effectively reduces the global problem to the solution of $6N_p$ ($3N_p$

IOP Conf. Series: Journal of Physics: Conf. Series **1225** (2019) 012011 doi:10.1088/1742-6596/1225/1/012011

after the reduction outlined above) *independent* equations. Should this dependence be taken into account, the system of equations to be solved would consist instead of $6N_p$ ($3N_p$) *coupled* equations, which is clearly prohibitive, given the large number of particles employed in typical three-dimensional PiC simulations.

The strategy above implies, as a secondary assumption, that when solving equation (25) for $\overline{\mathbf{E}}$ the source term, $\tilde{\mathbf{J}}$, is kept fixed. This simplification makes equation (25) linear, such that it can be solved with any preferred solver for linear systems.

In summary, in the FH approach it is sufficient to solve a system of $3N_g$ coupled linear equations for the fields, and $3N_p$ independent nonlinear equations for the particles, at each iteration step. The reduction in computational cost, compared to the initial system size, is of orders of magnitude. The resulting system of equations can be easily handled by modern linear solvers. However, neglecting the nonlinear dependence that characterizes the global system of equations can have a detrimental effect on the results. The FH algorithm is indeed not suitable for all applications; for specific cases, e.g. electron-proton plasmas with realistic mass ratios, it requires a critical reduction of the time step, thus neglecting the advantage of the implicit solution procedure. On the other hand, if the FH scheme can be applied, it allows for neglecting the nonlinear coupling between all the particles, which introduces a much higher complexity in the overall algorithm (millions or billions of coupled nonlinear equations).

2.3.3. Kinetic enslavement An alternative approach to the FH strategy consists of retaining the nonlinearity of the initial system, but confining it to the field equations instead of the particle equations. With the kinetic enslavement (KE) approach (Figure 2, right), an iterative solution is carried out on the field quantities only. At each time step, an initial guess is given for $\overline{\mathbf{E}}_g$ and $\overline{\mathbf{B}}_g$. Then, the particles are evolved independently of each other, by solving equation (22), and the current $\tilde{\mathbf{J}}_g$ is gathered on the grid. With the new values of the current, the fields are recalculated by solving equation (25). The iteration is carried out until convergence of the fields below a prescribed tolerance. Such a technique was first developed for nonrelativistic implicit PiC methods [18], and successfully applied in the Newtonian regime [6][13][19]. To the best of our knowledge, here we show for the first time the application of the KE strategy to relativistic implicit PiC simulations, which also implies a substantially different approach for the solution of the momentum equation (see next Section).

Here, the key concept is that the nonlinear character of equation (25) is retained by considering the nonlinear dependence $\tilde{\mathbf{J}}_g(\overline{\mathbf{E}}_g)$ (arising via the particle equations). As a consequence, one needs to solve a system of $3N_g$ coupled *nonlinear* equations, contrary to the FH strategy where the system is linear. Therefore, a nonlinear solver must be employed for the field advancing step. This approach shows much higher robustness for typical implicit PiC applications, at a cost that does not outmatch dramatically the FH strategy. In summary, the overall complexity of this strategy involves solving $3N_g$ (after reduction) coupled nonlinear equations for the fields, and $3N_p$ independent nonlinear equations for the particles.

2.3.4. Solution of the momentum equation The discrete equation of motion (22) differs from typical choices of discretization employed in relativistic particle simulations. Typical schemes are formulated such that the discretized momentum equation, though implicit, can be formally inverted in order to carry out the momentum update explicitly (see e.g. [20]). The main difference between these schemes lies in the definition of the "average velocity" used in the cross product term of the Lorentz force. In our formulation of equation (22), the average velocity is

$$\frac{\overline{\mathbf{u}}_p}{\overline{\gamma}_p} = \frac{\mathbf{u}_p^{n+1} + \mathbf{u}_p^n}{\gamma_p^{n+1} + \gamma_p^n} \tag{28}$$

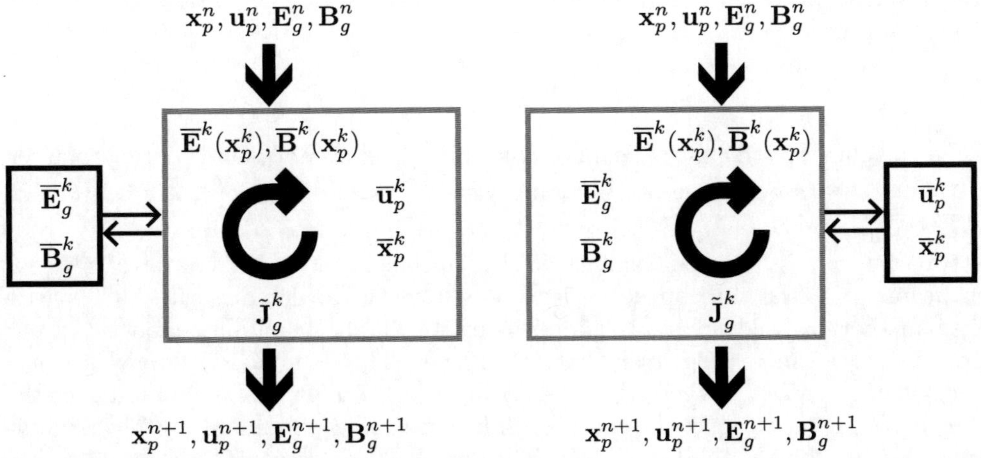

Figure 2: Schematic comparison between the FH and KE strategy for the implicit solution of the nonlinear discretized equations. The FH approach (left) requires a nonlinear iteration (red box) in the particle momentum, while the fields are solved for in a linear system (black box). The KE approach (right) requires instead a nonlinear iteration in the field quantities, while the momentum update is carried out particle by particle outside of the iteration.

where the barred quantities are defined above as the mean value between time steps n and $n+1$. With this choice, the momentum equation becomes more complicated to handle, and a formal inversion is not possible. Here, we consider an iterative solution for the numerical update of $\mathbf{u}_p$. One can employ a simple root-finding algorithm (e.g. Picard's or Newton's method) for the nonlinear solution of equation (22) in the variable $\mathbf{u}_p^{n+1}$. This approach has the advantage of being applicable also in case the system size is not reduced with the leap-frog approach. In typical simulations, the number of iterations required remains relatively limited (around 4-5 in the examples of the next Sections). The solution of equation (22) is carried out by iteratively searching for the roots of the residual equation

$$\mathbf{F}(\mathbf{u}_p) = \mathbf{u}_p - \mathbf{u}_p^n - \frac{q_p \Delta t}{m_p} \left(\overline{\mathbf{E}}(\overline{\mathbf{x}}_p) + \frac{\mathbf{u}_p + \mathbf{u}_p^n}{\gamma_p + \gamma_p^n} \times \overline{\mathbf{B}}(\overline{\mathbf{x}}_p) \right), \tag{29}$$

where $\gamma_p = \sqrt{1 + u_p^2}$ is a function of the iteration variable $\mathbf{u}_p$. If the leap-frog approach is not used for the position equation, then also $\overline{\mathbf{x}}_p = \mathbf{x}_p + (\Delta t/2)(\mathbf{u}_p + \mathbf{u}_p^n)/(\gamma_p + \gamma_p^n)$ is a function of $\mathbf{u}_p$. This procedure is employed in all numerical experiments in the next Sections.

3. Validation tests

3.1. Counter-streaming instabilities

As is customary for kinetic PiC code, a set of simple one-dimensional tests is run in order to check the accuracy of the new algorithm. Here we choose to quantitatively assess the performance of the relativistic implicit solver both in the electrostatic and in the full electromagnetic limits, by simulating counter-streaming instabilities in pair plasmas.

For the electrostatic limit, we consider the two-stream instability, a classical benchmark for PiC codes in both Newtonian and relativistic regimes. The simplest such a setup consists of two counter-streaming electron-positron beams of opposite initial speed $\pm v_0$, flowing parallel to the wave-vector $\mathbf{k} = k\hat{\mathbf{x}}$. The system is unstable to specific perturbation modes, and the beams tend to interact, converting beam bulk energy into thermal energy. The instability is characterized by a maximum growth rate

$$\Gamma_{\max} = \frac{\omega_p^0}{2\gamma_0^{3/2}}, \tag{30}$$

for symmetric counter-streaming beams with initial velocity v_0 and Lorentz factor $\gamma_0 = (1 - v_0^2)^{-1/2}$. The associated most unstable wave-vector is $k_{\max} = \sqrt{3}\omega_p^0/(2v_0\gamma_0^{3/2})$, where $\omega_p^0 = \sqrt{n_0 e^2/(\epsilon_0 m_e)}$.

The two-stream instability is simulated by initializing two counter-streaming electron-positron beams at initial drift speed v_0 in a one-dimensional domain with 1000 particles per cell. A small thermal velocity v_{th} is added to excite all the instability modes. In our set of runs, we consider an increasing initial $\gamma_0 = 2, 10, 50$. The domain is enlarged accordingly as $L = 128d_e^0, 1280d_e^0, 12800d_e^0$ (where $d_e^0 = c/\omega_p^0$) in order to allow the maximum growth rate to dominate. For each γ_0, we vary the spatial resolution from a minimum of 32 to a maximum of 256 total cells. In all cases the Debye length is not resolved, exceeding the stability constraint that characterizes explicit schemes. While the highest resolutions suffice, in some cases, to resolve the nonrelativistic skin depth d_e^0, the *relativistic* skin depth $d_e = c/\omega_p = c\sqrt{\gamma_0}/\omega_p^0$ is never resolved. For all values of Δx we keep the CFL ratio $\Delta t/\Delta x = 0.25$ fixed and we run until $t = 100(\omega_p^0)^{-1}$. The results for the $\gamma_0 = 10$ run are shown in Figure 3 (left panel), where we compare the growth of the electric energy associated to the most unstable wavenumber $k_{\max}$ with the analytic growth rate from equation (30). For almost all resolutions and γ_0, we observe excellent agreement between numerical results and theoretical predictions. For the lowest resolution of 32 grid points, the numerical results show worse agreement with the analytic prediction, which is expected since the spatial scale of $k_{\max}$ is not accurately resolved. We also show the conservation of total energy (right panel), which is kept to machine accuracy at all times, as expected.

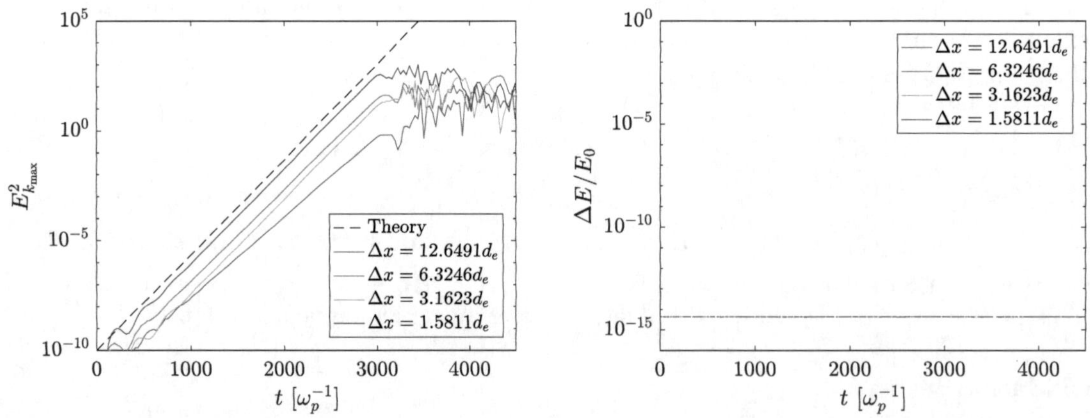

Figure 3: Relativistic two-stream instability for an electron-positron plasma with initial $\gamma_0 = 10$ for several grid resolutions and CFL ratio 0.25. Left: evolution of the electric field energy associated to the most-unstable wave-vector, compared to linear theory. Right: evolution of the relative error in the conservation of the total energy. In all cases the relativistic skin depth $d_e = c/\omega_p$ is underresolved.

In order to test the code in the electromagnetic limit, we consider the filamentation instability, another classical benchmark for PiC codes. The simplest one-dimensional setup consists again of two counter-streaming electron-positron beams, with initial velocity $\mathbf{v}_0 \perp \mathbf{k}$, where $\mathbf{k} = k\hat{\mathbf{x}}$. The filamentation instability generally dominates over the parallel modes, when the two beams have equal density, in the relativistic regime [21]. The maximum growth rate saturates at large k to the value

$$\Gamma_{\max} = \omega_p^0 v_0 \sqrt{\frac{2}{\gamma_0}}. \tag{31}$$

We test our algorithm by considering variable spatial and temporal resolutions in the simulation of two symmetric, counter-streaming electron-positron beams. The initial drift velocity is set to v_0 along the y-direction. A small thermal velocity is added to seed the instability. We initialize the system with increasing $\gamma_0 = 2, 10, 50$, enlarging the simulation box accordingly to $L = 12.8d_e^0, 25.6d_e^0, 51.2d_e^0$. We keep a CFL ratio of 0.125 and vary the resolution from 32 to 256 total cells, with 1000 particles per cell and periodic boundary conditions. Since the growth rate is insensitive of k, we can measure the total magnetic energy for comparison with linear theory. The results for $\gamma_0 = 10$ are shown in Figure 4 (left panel), where we find excellent agreement between our simulations and the expected growth rate of the instability (left panel). We note that satisfactory agreement with the analytic growth rate from equation (31) is achieved only for well-resolved runs where the grid spacing is below the relativistic skin depth d_e. This is not unexpected, since this corresponds to the fastest-growing wavelength for the filamentation instability. However, the nonrelativistic skin depth can be underresolved, without loss of accuracy or stability of the calculation. Figure 4 (right panel) shows the effect of a variation in the CFL ratio for the $\gamma_0 = 10$ case, demonstrating that the algorithm remains stable and accurate even in the extreme situation where $\Delta t \simeq \Delta x$, which exceeds by far the limit of explicit algorithms.

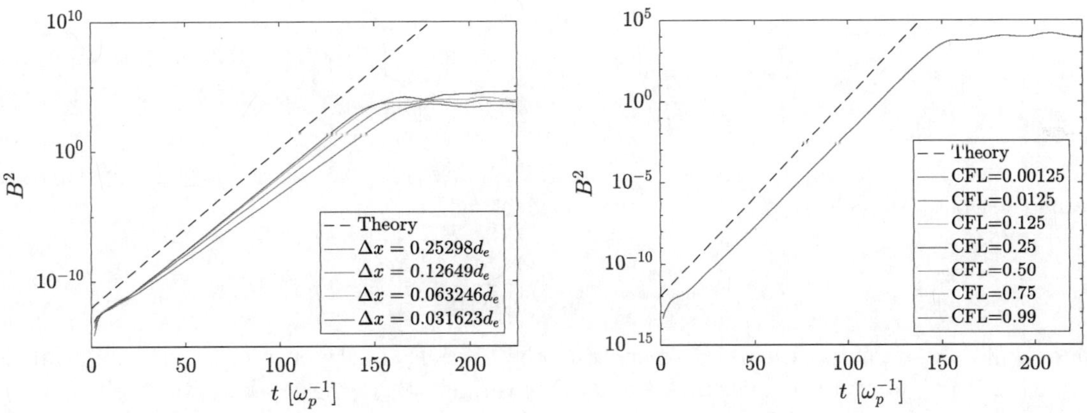

Figure 4: Relativistic filamentation instability for an electron-positron plasma with initial $\gamma_0 = 10$ for several grid resolutions and CFL ratio 0.125. Left: evolution of the total magnetic field energy, compared to linear theory. Right: effect of the variation of the CFL number, up to the extreme case where $\Delta t \simeq \Delta x$.

The results show an extreme robustness of the algorithm in both the electrostatic and electromagnetic regimes, with exact energy conservation at all times even for simulations in which both the relativistic and the nonrelativistic skin depth are heavily underresolved. The performance of the implicit solution procedure greatly surpasses that of explicit algorithms, allowing for much larger Δx and Δt without loss of stability.

For all runs we employ the kinetic enslavement technique explained in Section 2.3.3. The field hiding approach proves unreliable as it produces acceptable results only for highly resolved runs where Δx and Δt do not exceed specific limiting values. When this is not verified, the iteration fails to converge and the results become unphysical. The KE technique, instead, allows for less demanding simulation parameters such has large Δt and Δx, while still retaining the physical correctness of the results.

3.2. Relativistic tearing instability

In order to test the implementation of the implicit algorithm more extensively, we apply the method to a relativistic reconnection setup. A classic benchmark for this purpose is the onset of tearing modes in a two-dimensional, relativistic Harris equilibrium. Following [11], a double current sheet is initialized in a periodic two-dimensional domain by choosing the magnetic field profile

$$\mathbf{B}(x) = B_0 \left[-1 - \tanh\left(\frac{x - x_{\text{cs}}}{\Delta}\right) + \tanh\left(\frac{x + x_{\text{cs}}}{\Delta}\right)\right] \hat{\mathbf{y}}, \tag{32}$$

where B_0 is the magnetic field strength away from the current sheet and $\pm x_{\text{cs}}$ are the locations of the current sheets. The half-width Δ characterizes the sheet thickness. The magnetic field is sustained by a population of particles flowing at velocity $\pm U_{s,\text{cs}}$ (sign depending on the charge of each species s). The associated particle density profile in the current sheet reads

$$n_s(x) = n_{s,\text{cs}} \left[\cosh^{-2}\left(\frac{x + x_{\text{cs}}}{\Delta}\right) - \cosh^{-2}\left(\frac{x - x_{\text{cs}}}{\Delta}\right)\right], \tag{33}$$

where $n_{s,\text{cs}}$ is the density at the center of the current sheet, measured in the simulation frame.

The typical Harris equilibrium is obtained upon ensuring that the magnetic pressure at the center of the current sheet is balanced by thermal pressure. For a pair plasma of electrons and positrons, neutral ($n_{s,\text{cs}} = n_{\text{cs}}$ for all species), and populated by equally hot particles (with temperature $T_{s,\text{cs}}/(m_s c^2) = \Theta_{s,\text{cs}} = \Theta_{\text{cs}}$) this condition reads [22]

$$\frac{B_0^2}{2\mu_0} = 2 n_{\text{cs}} \Theta_{\text{cs}} c^2 m_e. \tag{34}$$

A second condition, deriving deriving from Ampère's law, ensures that the particle current sustains the magnetic field profile at the current sheet, as

$$\gamma_{s,\text{cs}} U_{s,\text{cs}} = -2 \frac{\Theta_{\text{cs}} c^2 m_e}{q_s B_0 \Delta}. \tag{35}$$

Upon deriving the temperature and drift speed of the particles in the current sheet, we initialize such particles at the current sheet locations by loading the corresponding Maxwell-Jüttner distribution according to the technique presented by Melzani et al. [11]. An additional, background population of cold particles is initialized with a nonrelativistic Maxwellian distribution.

The electron-positron equilibrium of a Harris current sheet as described above is subject to tearing instability modes. Magnetic field lines in the vicinity of the sheet are pushed close to each other, eventually undergoing reconnection, that converts magnetic energy into particle kinetic energy. The growth rate of the tearing instability, for pair plasmas, is given by the solution of the dispersion relation as given by [22].

In order to test our algorithm, we run a series of simulations in a two-dimensional periodic domain of size $L_x \times L_y$ discretized with 1024×512 cells in the x and y directions, respectively. The cell spacing is uniform in both directions and equal to $0.1 d_e^0$. We choose an initial plasma $\beta = 4\mu_0 n_0 T_0 / B_0^2$ of the background equal to 0.01, and we vary the magnetization $\sigma = B_0^2/(2 n_0 \mu_0 c^2) = 5, 10, 15, 20$ while keeping an overdensity ratio $\eta = n_{\text{cs}}/n_0 = 3$

constant. In all cases the initial current sheets have a half-width $\Delta = 1d_e^0$ and are located at $x_{\rm cs} = L_x/4, 3L_x/4$. Finally, the time step is taken to be equal to the grid spacing, i.e. $\Delta t/\Delta x = 1$, which is the maximum value allowed by our algorithm. Since these runs only involve pair plasmas with a sufficiently high resolution, we have employed the FH strategy, which is less computationally expensive than the more robust KE scheme.

Figure 6 shows the results of the $\sigma = 20$ run, with the magnitude of the z-component of the current J_z^2 and the magnetic energy B^2 (with selected field lines) shown at several times during and after the linear phase of the instability. After an initial equilibrium phase, thermal noise triggers the tearing instability, which initiates magnetic reconnection. Magnetic islands start growing and merging along the current sheets, in a process that converts magnetic energy into particle kinetic energy.

The conversion of magnetic energy into particle kinetic energy is shown in Figure 5 (left panel) for $\sigma = 20$, where the interplay between particles and fields appears clearly. In order to quantitatively confirm the accuracy of our scheme, we also show the comparison (right panel) between the evolution of the magnetic field energy in the $x-$direction and the theoretical growth rate from [22]. In all cases, we find excellent agreement between the two.

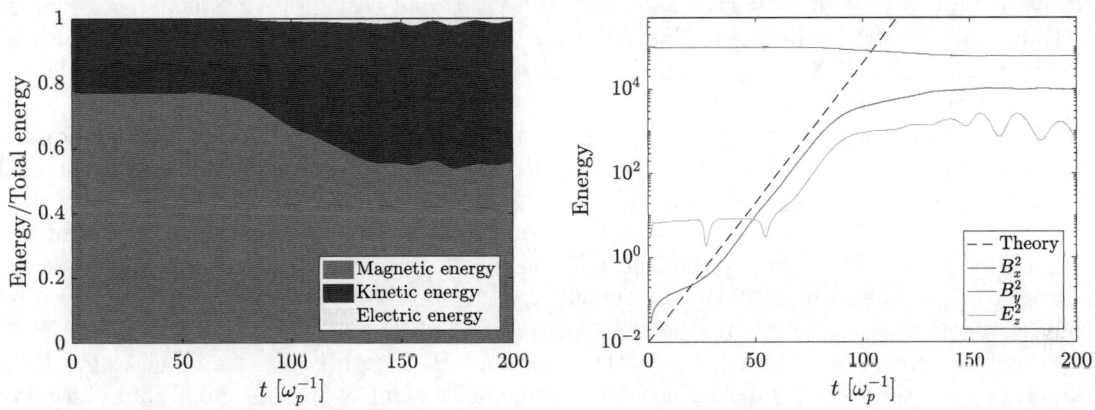

Figure 5: Energy analysis for the $\sigma = 20$ run. Left: time evolution of the magnetic, kinetic, and electric energy, showing the process of conversion of magnetic energy into particle energy. Right: evolution of the magnetic field energy in the x-direction, compared to the maximum growth rate from [22]. The magnetic energy in the y-direction and the electric energy in the z-direction are also shown for comparison.

4. Conclusions

We presented a three-dimensional implementation of the relativistic implicit Particle-in-Cell method for kinetic simulations of plasmas. The algorithm is based on the method presented in [14], but introduces two new strategies for the solution of the implicit equations, that allow for higher-dimensional simulations at accessible computational costs. The algorithm is tested by simulating one- and two-dimensional kinetic instabilities, and shows excellent agreement between analytic predictions for the growth rates and numerical results.

The method conserves energy to machine precision at all times, achieving high physical accuracy of the results. At the same time, this effectively eliminates the constraints imposed on the spatial and temporal resolution by numerical instabilities that affect explicit and semi-implicit PiC methods. The superior stability of the algorithm allows for long-term calculations without compromising the accuracy of the results. In all cases where explicit methods are bound to resolve the Debye length, in order to control numerical heating of the particles, our algorithm allows instead for much larger Δx and Δt, which can result in dramatic reductions of the computational cost.

While we refer to a future, more detailed work for a full analysis of computational costs and performances of the various strategies here presented, here we provide a first estimate of the cost of the implicit algorithm, compared to a typical explicit method. In particular, we have considered the two-dimensional reconnection setup from Section 3.2, with the typical production run parameters used for the tearing instability test. First, we have run the setup with the FH strategy (used also for the test) in the implicit implementation; then, we have re-run the same setup with a standard explicit algorithm (namely a Boris push an no iteration on the calculation of the current, which characterises the implicit cycle) implemented in the same code framework. By comparing the runtime of the two algorithms, we have found that a good estimate of the ratio between costs of the implicit and explicit runs is given by the number of nonlinear iterations performed by the latter. This holds essentially because the main cost of one time step is determined by the current gathering step, which is performed only once in the explicit time update. In the implicit cycle, at each time step, a certain number of nonlinear iterations are carried out, each requiring one current gathering. Since the number of nonlinear iterations is typically around 4-5, the resulting computational cost of the implicit run becomes $\sim$4-5 times that of the explicit run. We confirmed this with actual measures of the runtime: the explicit run took $\sim$7900 s to perform 2000 time steps, while the implicit run took $\sim$29200 s, slightly less than 4 times as long.

The high accuracy and reliability of relativistic implicit PiC methods allow for applications in several contexts. Large-scale simulations of cold plasmas, where it is desirable to employ spatial resolutions much coarser than the Debye length (see e.g. [23]), are the ideal field of application of this method. At the same time, long-term calculations where energy errors usually prevent the application of explicit methods, are instead allowed by the energy conservation property of the new scheme (e.g in the case of relativistic turbulence, see e.g. [24]). A final example is provided by coupled algorithms where PiC methods are complemented by magnetohydrodynamic (MHD) calculations (see e.g. [25]). In such cases, the typical MHD length-scales are much larger than the kinetic scales, therefore rendering necessary the employment of PiC methods that can relax the simulation parameters as much as possible, in order to avoid excessive computational costs.

Acknowledgements
FB is supported by projects GOA/2015-014 (2014-2018 KU Leuven) and the Interuniversity Attraction Poles Programme by the Belgian Science Policy Office (IAP P7/08 CHARM). FB is also supported by the Research Fund KU Leuven and Space Weaves RUN project. The computational resources and services used in this work were provided by the VSC (Flemish Supercomputer Center), funded by the Research Foundation Flanders (FWO) and the Flemish Government - department EWI

IOP Conf. Series: Journal of Physics: Conf. Series **1225** (2019) 012011 doi:10.1088/1742-6596/1225/1/012011

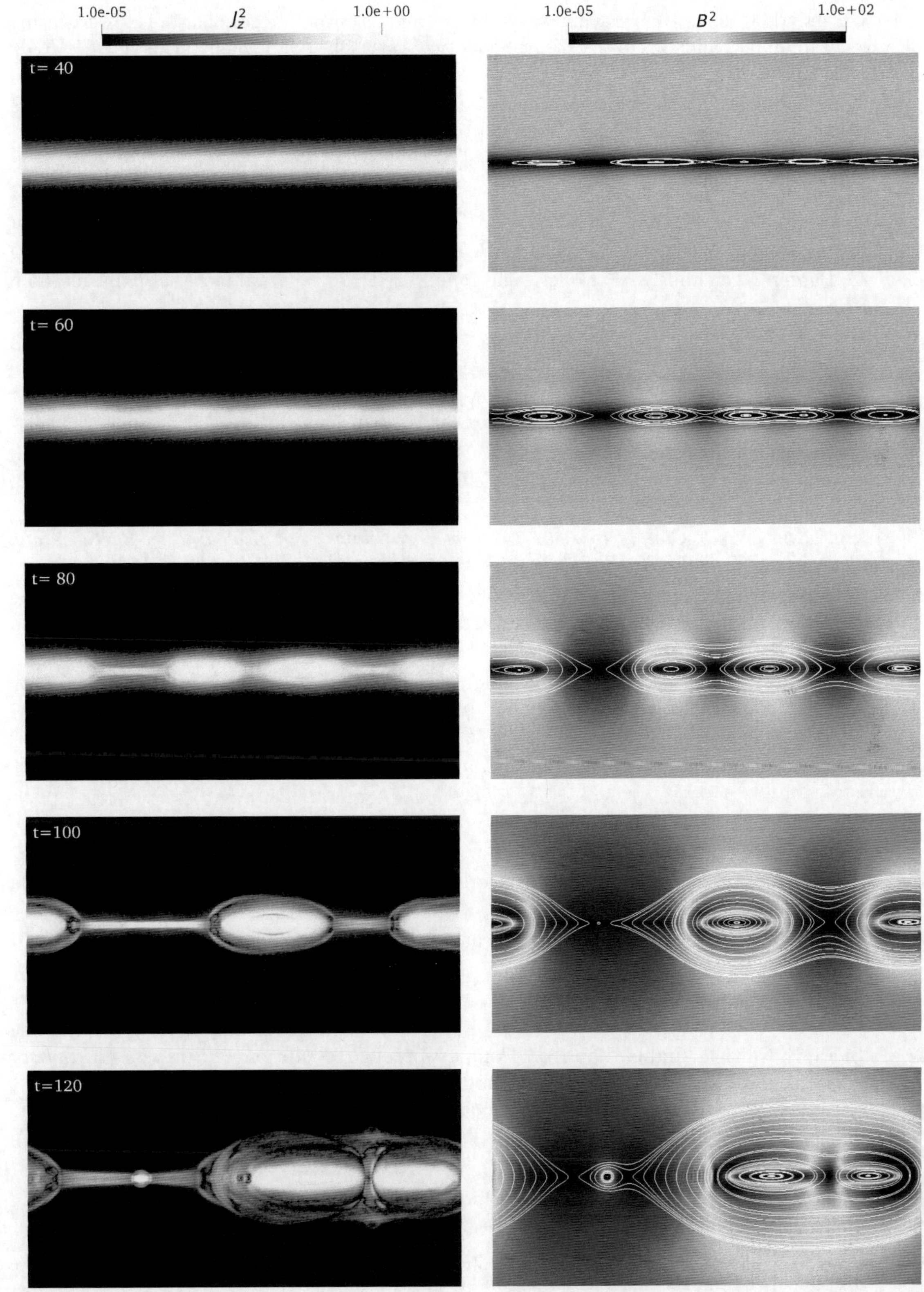

Figure 6: Spatial distribution of the current density (left column) and magnetic energy (right column) at several times during and after the linear growth of the tearing instability, for one of the two current sheets in the simulation. Tl g and merging of primary and secondary magnetic islands appears clearly, together wi 111 connection of magnetic field lines.

IOP Conf. Series: Journal of Physics: Conf. Series **1225** (2019) 012011 doi:10.1088/1742-6596/1225/1/012011

References

[1] A. Chael, M. Rowan, R. Narayan, M. Johnson, and L. Sironi. The role of electron heating physics in images and variability of the Galactic Centre black hole Sagittarius A*. *MNRAS*, 478, 4, 2018.

[2] L. Sironi and A. Spitkovsky. Acceleration of particles at the termination shock of a relativistic striped wind. *ApJ*, 741, 39, 2011.

[3] A. Spitkovsky. Particle acceleration in relativistic collisionless shocks: Fermi process at last? *ApJL*, 682, L5, 2008.

[4] F. Bacchini, V. Olshevsky, S. Poetds, and G. Lapenta. A new particle-in-cell method for modeling magnetized fluids. *CPC*, 210, 79, 2017.

[5] S. G. Bardenhagen and E. M. Kober. The generalized interpolation material point method. *CMES*, 5, 6, 2004.

[6] G. Chen, L. Chacón, and D. C. Barnes. An energy- and charge-conserving, implicit, electrostatic particle-in-cell algorithm. *JCP*, 230, 18, 2011.

[7] S. Markidis, G. Lapenta, and Rizwan-uddin. Multi-scale simulations of plasma with iPIC3D. *Mathematics and Computers in Simulation*, 80, 7, 2010.

[8] G. Lapenta, D. Gonzalez-Herrero, and E. Boella. Multiple-scale kinetic simulations with the energy conserving semi-implicit particle in cell method. *J. Plasma Phys.*, 83, 705830205, 2017.

[9] A. Spitkovsky. *AIP Conf. Proc. 801, Astrophysical Sources of High Energy Particles and Radiation, ed. T. Bulik, B. Rudak, and G. Madejski (Melville, NY: AIP)*, 345, 2005.

[10] B. Cerutti, G. R. Werner, D. A. Uzdensky, and M. C. Begelman. Simulations of particle acceleration beyond the classical synchrotron burnoff limit in magnetic reconnection: An explanation of the crab flares. *ApJ*, 770, 147, 2013.

[11] M. Melzani, C. Winisdoerffer, R. Walder, D. Folini, J. M Favre, S. Krastanov, and P. Messmer. `Apar-T`: code, validation, and physical interpretation of particle-in-cell results. *A&A*, 558, A133, 2013.

[12] C.K. Birdsall and A.B. Langdon. *Plasma physics via computer simulation.* IoP Publishing, Bristol, 1991.

[13] S. Markidis and G. Lapenta. The energy conserving particle-in-cell method. *J. Comp. Phys.*, 230, 18, 2011.

[14] G. Lapenta and S. Markidis. Particle acceleration and energy conservation in particle in cell simulations. *Phys. Plasmas*, 072101, 18, 2011.

[15] J. Villasenor and O. Buneman. Rigorous charge conservation for local electromagnetic field solvers. *CPC*, 69, 306, 1992.

[16] Y. Chen and G. Tóth. Gauss's Law Satisfying Energy-Conserving Semi-Implicit Particle-in-Cell Method. *arXiv:1808.05745*, 2018.

[17] L. Siddi, E. Cazzola, and G. Lapenta. Comparison of Preconditioning Strategies in Energy Conserving Implicit Particle in Cell Methods. *arXiv:1806.04454*, 2018.

[18] W. Taitano. Development of a Jacobian-free Newton-Krylov method with kinetic enslavement to implicitly solve Vlasov-Poisson system in plasma physics. *PhD Thesis*, 2010.

[19] G. Chen and L. Chacón. A multi-dimensional, energy- and charge-conserving, nonlinearly implicit, electromagnetic VlasovDarwin particle-in-cell algorithm. *CPC*, 197,73, 2015.

[20] B. Ripperda, F. Bacchini, J. Teunissen, C. Xia, O. Porth, L. Sironi, G. Lapenta, and R. Keppens. A comprehensive comparison of relativistic particle integrators. *ApJS*, 235:21, 2018.

[21] A. Bret, L. Gremillet, and M. E. Dieckmann. Multidimensional electron beam-plasma instabilities in the relativistic regime. *Physics of Plasmas*, 17(12), 2010.

[22] J. Pétri and J. G. Kirk. Growth rates of the weibel and tearing mode instabilities in a relativistic pair plasma. *Plasma Phys. Control. Fusion*, 49, 1885, 2007.

[23] M. André, A. Vaivads, Y. V. Khotyaintsev, T. Laitinen, H. Nilsson, G. Stenberg, A. Fazakerley, and J. G. Trotignon. Magnetic reconnection and cold plasma at the magnetopause. *GRL*, 37, 22, 2010.

[24] V. Zhdankin, D. A. Uzdensky, G. R. Werner, and M. C. Begelman. Numerical investigation of kinetic turbulence in relativistic pair plasmas i. turbulence statistics. *MNRAS*, 474, 2, 2018.

[25] K. Makwana, R. Keppens, and G. Lapenta. Two-way coupled mhd-pic simulations of magnetic reconnection in magnetic island coalescence. *J. Phys.: Conf. Ser.*, 1031, 012019, 2018.

IOP Conf. Series: Journal of Physics: Conf. Series **1225** (2019) 012012 doi:10.1088/1742-6596/1225/1/012012

GPU Acceleration of an Established Solar MHD Code using OpenACC

R. M. Caplan, J. A. Linker, Z. Mikić, C. Downs, T. Török, and V. S. Titov

Predictive Science Inc., 9990 Mesa Rim Road Suite 170, San Diego, CA 92121

E-mail: caplanr@predsci.com, linkerj@predsci.com, mikicz@predsci.com, cdowns@predsci.com, ttorok@predsci.com, titovv@predsci.com

Abstract. GPU accelerators have had a notable impact on high-performance computing across many disciplines. They provide high performance with low cost/power, and therefore have become a primary compute resource on many of the largest supercomputers. Here, we implement multi-GPU acceleration into our Solar MHD code (MAS) using OpenACC in a fully portable, single-source manner. Our preliminary implementation is focused on MAS running in a reduced physics "zero-beta" mode. While valuable on its own, our main goal is to pave the way for a full physics, thermodynamic MHD implementation. We describe the OpenACC implementation methodology and challenges. "Time-to-solution" performance results of a production-level flux rope eruption simulation on multi-CPU and multi-GPU systems are shown. We find that the GPU-accelerated MAS code has the ability to run "zero-beta" simulations on a single multi-GPU server at speeds previously requiring multiple CPU server-nodes of a supercomputer.

1. Introduction

Hardware accelerators (e.g. graphics processing units (GPUs)) have significantly impacted the state-of-the-art of high-performance computing (HPC) in the fields of AI and machine learning, as well as in traditional scientific applications [1]. GPU-accelerated codes can achieve performance normally only available on super computers using a single 'in-house' system with a modest number of accelerator cards (often at much lower cost). GPUs exhibit drastic improvements in power efficiency – a key reason why many of the largest HPC systems in the world are based around them. In some cases, they can provide necessary performance that cannot be achieved with traditional CPU systems.

Originally, the only way for software to take advantage of GPU accelerators was to write/rewrite code using APIs such as CUDA [2] and OpenCL [3]. However, for legacy scientific applications, rewriting code can be far too difficult in terms of development time, effort, and education of the user base. Additionally, because portability and longevity is of critical concern, the wide-scale adoption of accelerated programming into such applications has been relatively slow.

Programming for accelerators through the use of directives addresses the above concerns. Accelerator directives are expressed as special comments in the code. This allows a single source code to be compiled for both accelerators (using a directive-supporting compiler), and CPUs (using any CPU compiler). For NVIDIA GPUs, the most common directive-based API

IOP Conf. Series: Journal of Physics: Conf. Series **1225** (2019) 012012 doi:10.1088/1742-6596/1225/1/012012

in production is the open standard OpenACC[1] [4, 5]. Implementations of OpenACC can also compile to other device types [5] including FPGAs, multi-core CPUs, Power9, and SunWay processors [6]. The most complete implementation of OpenACC is currently in the PGI compiler from the Portland Group[2]. There are also open-source OpenACC implementations, including the GNU compiler[3], but these currently lack some features. In this paper, we will use PGI's OpenACC implementation to accelerate our code for use on NVIDIA GPUs. We note that OpenMP[4] has also recently added accelerator support into its API, but we opted for the maturity and relative simplicity of OpenACC for this work.

In Figure. 1, we illustrate the coding advantages of OpenACC over traditional GPU language extensions, by converting a simple SAXPY code segment for use on NVIDIA GPUs using both CUDA and OpenACC. We see that the OpenACC implementation requires much less additional

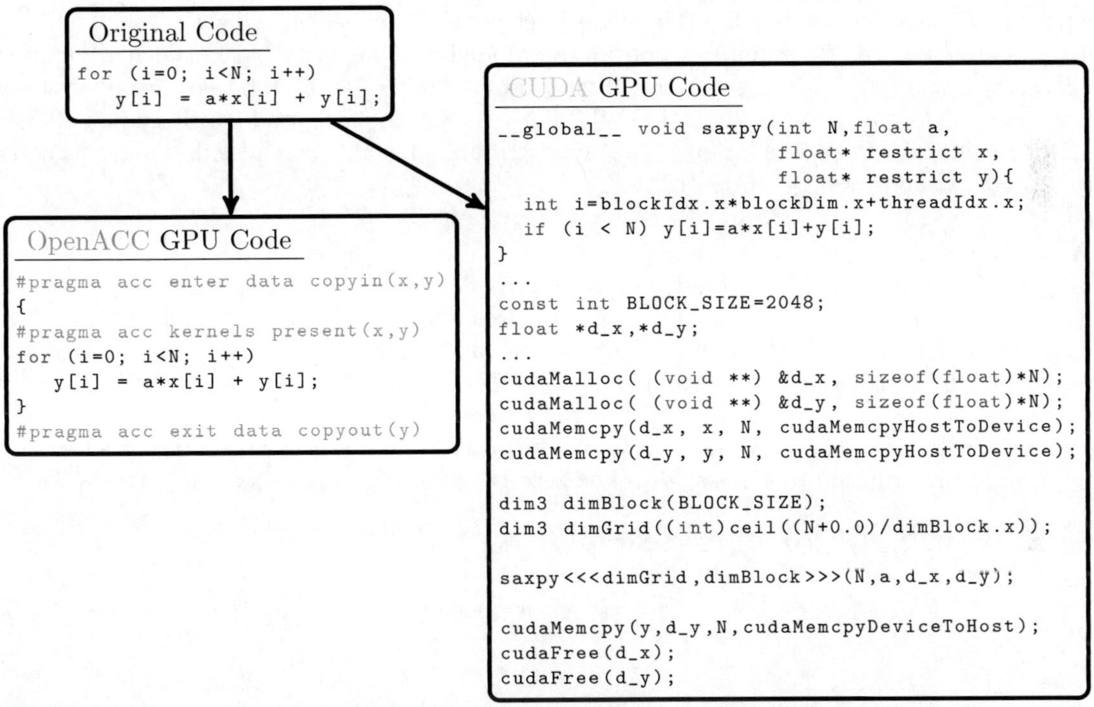

Figure 1. Example GPU implementations of a SAXPY algorithm using CUDA and OpenACC. The OpenACC implementation is simpler and maintains portibility.

code, preserves the original code, and is backwards compatible with the original code.

In this paper, we show our initial implementation of OpenACC into our Solar Magnetohydrodynamic (MHD) code MAS (described in Sec. 2), allowing it to run on multiple NVIDIA GPUs. Our goal in porting MAS to GPUs is to allow for multiple small-to-medium sized simulations to be run on an in-house multi-GPU server, greatly benefiting development and parameter studies. In addition, very large simulations using MAS take up a substantial portion of our HPC allocations. Having the code run on multiple GPU nodes can decrease the time-to-solution using the same number of nodes and/or reduce the total allocation usage.

[1] http://www.openacc.org
[2] http://www.pgroup.com
[3] https://gcc.gnu.org/wiki/OpenACC
[4] http://www.openmp.org

IOP Conf. Series: Journal of Physics: Conf. Series **1225** (2019) 012012 doi:10.1088/1742-6596/1225/1/012012

The paper is organized as follows: in Sec. 2 we introduce the MAS code and describe its use, models, and algorithms. We then describe the production-level test simulation we use for timing results in Sec. 3. The OpenACC implementation, including algorithm analysis, difficulties encountered, and an overall summary of the effort, is discussed in Sec. 4. In Sec. 5 we show extensive timing results across multiple algorithms and hardware models (both GPUs and CPUs), and we conclude in Sec. 6.

2. The MAS Code

Magnetohydrodynamic Algorithm outside a Sphere (MAS) is an in-production MHD code with over 15 years of ongoing development used extensively in Solar physics research [7, 8, 9, 10, 11, 12, 13, 14]. It was used to predict the coronal structure during the August 21st, 2017 total solar eclipse [15] and has demonstrated the ability to reproduce qualitative and quantitative properties of solar eruptions [16]. The code is currently included as a part of the CORHEL suite [17] hosted at the Community Coordinated Modeling Center (CCMC)[5] which allows users to generate background MHD solutions of the corona and heliosphere based on observational boundary conditions. MAS is written in FORTRAN ($\sim 50,000$ lines) and parallelized with MPI. It can run simulations containing over two hundred million grid cells and exhibits performance scaling to thousands of CPU cores [18].

2.1. Models

MAS solves a well-featured thermodynamic resistive MHD model with numerous physics and algorithm options [19, 18]. However, not all features and model equations are used for every simulation. The so-called 'zero-beta' mode freezes the density and sets the pressure to zero, essentially eliminating several evolution equations. Zero-beta simulations are applicable in low-beta regions of the corona (i.e. where the magnetic forces dominate the dynamics) and can have many uses including the generation of initial solutions for solar storm magnetic structures and analyzing initial eruption dynamics [16]. The zero-beta model is described as

$$\begin{aligned}
\frac{\partial \mathbf{A}}{\partial t} &= \mathbf{v} \times (\nabla \times \mathbf{A}) - \frac{c^2 \eta}{4\pi} \nabla \times \nabla \times \mathbf{A} \\
\frac{\partial \mathbf{v}}{\partial t} &= -\mathbf{v} \cdot \nabla \mathbf{v} + \frac{1}{\rho}\left[\frac{1}{c}\mathbf{J} \times \mathbf{B}\right] + \frac{1}{\rho} \nabla \cdot (\nu \rho \nabla \mathbf{v}) + \frac{1}{\rho} \nabla \cdot \left(S\,\rho\,\nabla \frac{\partial \mathbf{v}}{\partial t}\right) \\
p &= 0 \\
\rho &= \rho_0(\mathbf{r})
\end{aligned} \tag{1}$$

where $\mathbf{A}$ is the magnetic vector potential, $\mathbf{B} = \nabla \times \mathbf{A}$ is the magnetic field, $\mathbf{J} = \frac{c}{4\pi} \nabla \times \mathbf{B}$ is the current density, ρ is the plasma density, p is the plasma pressure, and $\mathbf{v}$ is the plasma velocity, η is the coefficient of resistivity, ν is the coefficient of kinematic viscosity, and S is a semi-implicit factor defined as $S = (\Delta t^2\, \tilde{k}^2)^{-1}\,(C_w^2/(1 - C_f)^2 - 1)$, where $\tilde{k}^2$ is the upper limit of the combined inverse grid spacing, C_f is the flow CFL, and C_w is the fast magnetosonic wave CFL [18]. The resistivity and viscosity operators are added in order to avoid the potential emergence of oscillatory behavior due to the algorithms in MAS not guaranteeing monotonicity. The semi-implicit operator is added in order to be able to advance the code at the flow CFL time-step stability limit which is much larger than the fast magneto-sonic wave limit (see Ref. [20] for details).

Since the computations involved in running the zero-beta model contain all the major algorithms and data structures used in the fully-featured mode of MAS, the zero-beta model makes an ideal first step for our OpenACC implementation of MAS and is the focus of this paper.

[5] `http://ccmc.gsfc.nasa.gov`

IOP Conf. Series: Journal of Physics: Conf. Series **1225** (2019) 012012 doi:10.1088/1742-6596/1225/1/012012

2.2. Algorithms
MAS solves the MHD equations on a non-uniform logically-rectangular staggered grid using finite difference algorithms. Several time integration schemes are used for various parts of the MHD model. Advective and reactive terms are integrated with an explicit predictor-corrector upwinding scheme [20], the semi-implicit term is solved in the predictor-corrector framework implicitly using two preconditioned conjugate gradient (PCG) solves, while diffusive-like terms are either solved implicitly with PCG solves or explicitly with a Runge-Kutta Legendre super time-stepping (STS) scheme [21]. The PCG solver in MAS has two communication-free preconditioning options. The first (PC1) uses the inverse diagonal of the matrix (point-Jacobi), while the second (PC2) is a block-Jacobi using a zero-fill incomplete LU factorization [22]. For 'easy' solves (low condition number), PC1 typically runs faster than PC2, while for 'hard' solves (high condition number) PC2 can be much faster than PC1. The resistivity solve is typically always 'easy' and therefore always uses PC1. For the semi-implicit and viscosity solves in coronal simulations, PC2 is usually faster than PC1, but PC1 scales more efficeintly to high numbers of processors than PC2. The STS algorithm for viscosity typically runs as fast (often faster) than PCG+PC2, and scales better than PCG even with PC1. However, it can suffer from accuracy-breakdown in certain problem configurations (see Ref. [18] for details). For the implementation and testing in this paper, we define the following four algorithmic choices:

1) Semi-Implicit: PCG+PC1, Viscosity: PCG+PC1
2) Semi-Implicit: PCG+PC2, Viscosity: PCG+PC2
3) Semi-Implicit: PCG+PC1, Viscosity: STS
4) Semi-Implicit: PCG+PC2, Viscosity: STS

3. Real-world Test Case
In order to provide useful timings, we select a full production run of MAS. We use a global zero-beta simulation of the onset of a solar eruption with a TDm flux rope [23] structure (see Ref. [24] for similar simulations). The domain extends all around the Sun (avoiding the need to impose artificial boundary conditions) from the solar surface out to 10 solar radii. It uses an extremely non-uniform grid of ~11 million cells, with very small cell sizes in the erupting region, coarsening greatly with distance. We therefore employ a large, uniform, viscosity coefficient sufficient to ensure stability in the very coarse regions. The simulation covers a physical evolution of ~200 seconds requiring ~700 non-uniform time steps. Figure 2 shows visualizations of the run. The semi-implicit PCG solves take an average of ~70 iterations per time step with the PC1 preconditioner and between ~30–50 iterations using PC2 (the PC2 preconditioner degrades in effectiveness as the number of MPI ranks increases). The viscosity solve is more difficult (due

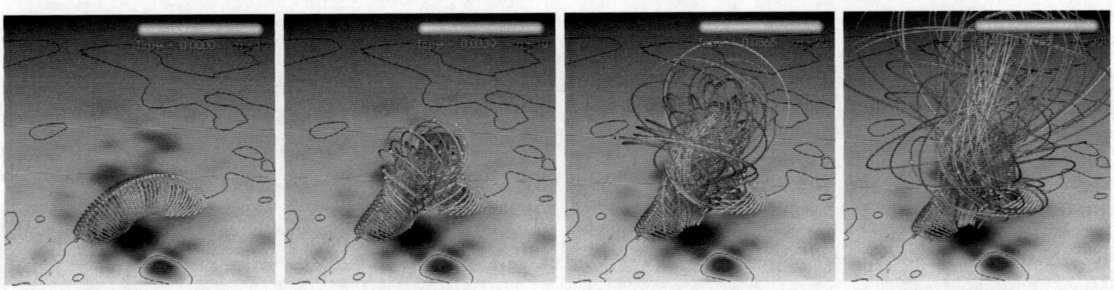

Figure 2. Magnetic field evolution from the simulation used to test code performance. The orange and cyan lines represent magnetic field lines, while the colormap represents the radial magnetic field values on the solar surface.

to the large viscosity coefficient), requiring an average of ∼550 iterations using PCG+PC1, and between ∼100–200 iterations using PCG+PC2. Advancing viscosity with the explicit STS algorithm requires an average of ∼200 sub-steps per time-step.

3.1. Validation

As mentioned in Sec. 2.2, using the STS algorithm for the viscosity operator is often faster than using PCG+PC2. In Ref. [18], we showed that the STS algorithm was over twice as fast as PC2+PCG in a thermodynamic MHD coronal relaxation simulation. However, in that case, the STS algorithm exhibited a break-down in accuracy, causing nonphysical oscillations in high resolution areas. Therefore, we must carefully validate the solution when deciding whether or not to use the STS method. In Figure. 3 we show values of the total magnetic and kinetic energies, and the total $|J \times B|$ and $|J \cdot B|$ for the test simulation of Sec. 3 using both the PCG+PC2 and STS algorithms, along with a side-by-side comparison of visualizations. We see that the two simulations yield very similar results and the solutions appear nearly indistinguishable. Therefore, for this example problem, we can safely include STS as a valid choice of algorithm for viscosity.

Our GPU runs described below were also validated with the original CPU runs. The GPU runs matched the CPU runs quantitatively to within the PCG solver tolerance (due to floating-point variations).

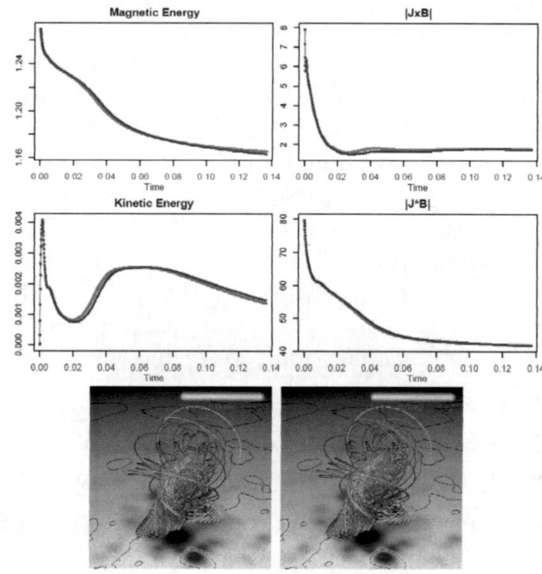

Figure 3. Comparison of the test simulation using PCG+PC2 and STS for viscosity. Top two rows: Volume integrated quantities (in code units) between PCG+PC2 (red) and STS (blue). Bottom row: Magnetic field visualization of the run with viscosity solved by PCG+PC2 (left) and STS (right).

4. Implementing OpenACC in MAS

Here we summarize our implementation of OpenACC into MAS. Further details, including specific code examples, strategies and development tips, can be found in Ref. [25].

4.1. Algorithm Analysis

GPUs are very efficient for a variety of algorithms, but not all. Therefore it is important to analyze the algorithms and determine if they are suitable for GPU acceleration. It is key that an algorithm is vectorizable (i.e. that it can perform a single operation on a contiguous section of an array in memory (SIMD)). Basic array operations (e.g. AXPY) are vectorizable, as are explicit finite difference algorithms (such as the stencil operations in the advection operators and the STS algorithm). The core operators in our PCG solvers consist of sparse matrix-vector products (essentially stencil operations), basic array operations, and dot products, all of which are vectorizable to varying degree. The PC1 preconditioner consists of a simple array operation, but the PC2 preconditioner relies on a local triangular matrix solve computed using a standard backward-forward sequential algorithm which is not vectorizable. To address this issue, NVIDIA has developed the CUDA-based cuSparse library containing advanced (and vectorizable) triangular sparse solvers for GPUs. Although OpenACC codes are able to link to CUDA libraries (like `cuSparse`), doing so breaks portability as the libraries often do not contain CPU versions of the algorithms. To test if this loss of portability is outweighed by the performance advantage of PC2, we implemented `cuSparse` into our OpenACC GPU-accelerated potential field solver POT3D [26]. The results showed that solving our symmetric matrices (which have few off-diagonal non-zero entries) on modern GPUs with PC2 yielded similar performance as using PC1. Therefore, we choose to prioritize portability and do not GPU-accelerate the PCG+PC2 algorithm in MAS (limiting the algorithm choices in Sec. 2.2 to (1) and (3) for use on GPUs).

4.2. Examples of OpenACC

A full description of the OpenACC directives used to port the MAS code is beyond the scope of this paper. We refer the reader to the many available books and tutorials on OpenACC programming [4, 5]. Here, we describe some of the more fundamental concepts through selected examples.

One key aspect of efficient GPU programming is carefully managing data movement. GPUs have their own local memory and data needs to be first transferred to the GPU for computation, and then retrieved if needed by the CPU. This transfer has very high latency and needs to be avoided as much as possible. For MAS, we use OpenACC's unstructured data regions to manually send the data to the GPU, only transferring it back to the CPU for I/O. This is shown schematically in Figure. 4.

```
      allocate/initialize "y" on CPU ...
!$acc enter data copyin (y)
      GPU version of "y" now available for
      use in OpenACC compute regions ...
!$acc update self (y)
      CPU version of "y" updated
      (for I/O, etc.) ...
!$acc exit data delete(y)
```

Figure 4. Outline for using unstructured data regions in OpenACC.

In order to avoid data transfers except in the case of I/O, we needed to apply OpenACC directives to every part of the code that uses the data, resulting in a large number of added directives. In Figure 5, we show examples of adding OpenACC directives to two basic code loops - an AXPY operator and a summation operator. The `default(present)` and `present()` directives indicate that the required data has already been transferred to the GPU and are available. In the summation loop, `total` is needed to be declared as a reduction variable along with the reduction operator (in this case "+").

```
!$acc kernels loop default(present)
   do i=1,n
     y(i) = a*x(i) + y(i)
   enddo
!$acc end kernels
```

```
!$acc kernels loop present(y) reduction(+:total)
   do j=1,m
     total = total + y(j)
   enddo
!$acc end kernels
```

Figure 5. Basic examples of applying OpenACC to an AXPY operator loop (left) and a summation (right).

For large problems, it is critical to be able to compute on multiple GPUs - both within a server node and between nodes on a cluster. The MAS code splits the domain of the simulation and gives each MPI rank (typically assigned to one CPU core) a section of the domain to work on. When one rank requires data from other ranks, MPI communications are used to send and receive the required data. To use multiple GPUs, we assign MPI ranks so that there is one CPU rank per GPU. We then use OpenACC's `set device_num` directive to tell the run-time to use the desired GPU. For this to work across a cluster, the code needs to know the local MPI ranks on a given node. To do this we use MPI-3's shared communicator feature as shown in Figure 6. Figure 6 also shows an example of how we avoid CPU-GPU transfers when using MPI communication by using the `host_data use_device` clause in OpenACC to call MPI routines with GPU data directly (this requires a 'CUDA-aware' MPI library).

```
  call MPI_Comm_split_type (MPI_COMM_WORLD,MPI_COMM_TYPE_SHARED,0,
&                           MPI_INFO_NULL,comm_shared,ierr)
  call MPI_Comm_size (comm_shared, nprocsh, ierr)
  call MPI_Comm_rank (comm_shared, iprocsh, ierr)
  igpu = MODULO(iprocsh, nprocsh)
!$acc set device_num(igpu)
```

```
!$acc host_data use_device(y) if_present
  call MPI_Allreduce (MPI_IN_PLACE,y,n,MPI_DOUBLE,MPI_SUM,MPI_COMM_WORLD,ierr)
!$acc end host_data
```

Figure 6. Top: Example approach to selecting the GPU device to use for a particular MPI rank. Bottom: Example of using CUDA-aware MPI to avoid CPU-GPU transfers for MPI communications.

These are just a few examples of the OpenACC directives needed for the MAS code. We refer the reader to Ref. [25] for many more detailed examples.

4.3. Challenges

Implementing OpenACC comes with its fair share of challenges. As a relatively new programming API, there are often delays in clear documentation for new features, delays in compilers implementing the latest features, as well as inevitable compiler bugs. We encountered all of these during our development of OpenACC into MAS, adding difficulty to the otherwise fairly straight-forward implementation. For example, in order to work around a few bugs in the PGI compiler, we had to modify more CPU code than otherwise required. We also tried using very new language features that offered even simpler coding, but in the end, needed to use standard features for the code to function properly with the compiler.

In addition to the challenges posed by compiler implementations, we also encountered difficulties running the resulting code on the HPC resources available to us. For example, certain new features of OpenACC that were required for MAS were not available in the older versions of the compiler installed on the cluster, and updating the compiler took some time. Also, some MPI libraries were not up-to-date, not allowing some required GPU features. Although we were able to get everything working eventually through the help of the centers' staff, these problems added non-trivially to the development time. We note that HPC systems specifically designed for GPU computing (such as the Summit system at ORNL [27]), are not likely to exhibit these issues.

4.4. Summary of code modifications

The current implementation of OpenACC into MAS for zero-beta runs required modifying less than 5% of the code. The total number of added OpenACC comments amounted to less than 1.5% the length of the original code. Certain compiler bugs and currently lacking features of OpenACC account for much of the remaining code modifications, along with optional CPU code optimizations implemented during the development (since only the zero-beta features of MAS were accelerated, an implementation of the full feature-set of MAS will necessarily require additional OpenACC directives). The accelerated code only uses directives, keeping the code fully portable and able to be compiled with any previously used CPU compiler.

IOP Conf. Series: Journal of Physics: Conf. Series **1225** (2019) 012012 doi:10.1088/1742-6596/1225/1/012012

5. Timing Results

Comparing CPU and GPU performance of a given code in a 'fair' manner can be difficult since the optimal implementation of the algorithms are often unique for each hardware (either explicitly in the code, or implicitly through the compiler). In addition, different algorithms may perform better on one hardware than another, (e.g. the PC2 preconditioner of our PCG solvers is more suited for a CPU, while PC1 is better suited for a GPU). However, our goal is not to benchmark hardware, but rather to test the effective time-to-solution for a given problem using GPU acceleration versus the original CPU code. Therefore, we first test each available algorithm on each hardware configuration using the best suited compiler and select the fastest running algorithm for each. Implementation of new algorithms could potentially improve performance on the CPU and GPU, but here we want to focus on what can be achieved adding GPU acceleration to our current code base in a portable way.

To measure the effective time-to-solution, we record the absolute wall clock time of our runs including all CPU-code setup, I/O, CPU-GPU transfers, etc. While we do not include the time spent waiting in the scheduling queue of HPC clusters, it should not be overlooked when evaluating the advantage of an 'in-house' solution. The timings are recorded using calls to the MPI library's timing API routines, and all simulations are computed in double-precision (FP64).

5.1. HPC environment

In order to give a broad picture of performance, we run the simulations on multiple generations of hardware. For the CPU code, we run on five generations of Intel processors (from Sandy Bridge (rel. 2012) up to Skylake (rel. 2017)) on NASA's NAS Pleiades and Electra systems. For the GPU code, we run on three generations of NVIDIA GPUs (Kepler (rel. 2014) up to Volta (rel. 2017)) on SDSC's Comet and NVIDIA's PSG machines. Each node on the CPU systems are dual-socket (2 CPUs per node), while each GPU node has four GPUs per node (the Kepler K80 GPU is a dual-GPU card and therefore each node has two K80 GPU cards to make up the four GPUs). We also test each code using an in-house desktop PC with an Intel Broadwell CPU and a NVIDIA TitanXP GPU. Detailed specifications for all CPU and GPU hardware and software configurations (including compiler versions and flags, driver versions, etc.) are given in the Appendix.

5.2. Algorithm comparisons

Here, we test each available algorithm option from Sec. 2.2 for the code running on CPUs and GPUs to determine the most efficient for each. We use the best available model of CPU and GPU for the timings (in this case, the Skylake CPU and Volta GPU) and record the performance and scaling using the test run of Sec. 3. The results are shown in Fig. 7. The CPU results show that using only the PCG solver with PC2 runs almost twice as fast as when using PC1 for low number of CPUs. However, for the highest numbers of CPUs, the improved scaling of the PC1 preconditioner makes the two solvers run in similar time. Using STS for viscosity exhibits perfect scaling, and the run time of using PC1 or PC2 for the semi-implicit solves yields similar results. However, since the PC2+STS does exhibit slightly faster times overall than PC1+STS, we will use PC2+STS for our CPU comparison tests.

Both algorithm choices (PC1+PC1 and PC1+STS) for the GPU show similar scaling, with PC1+STS being about twice as fast as PC1+PC1. Therefore, we choose the PC1+STS algorithm for our GPU comparison runs. There is a noticeable reduction in scaling past 4 GPUs (although the runs still reduces in time with more GPUs). Since each compute node contains 4 GPUs, this reduction appears to be an inter-node scaling problem. While it is possible that the test problem is simply too small to scale to many GPUs, we note that the PSG cluster and Comet do not have direct RDMA GDR enabled for our choice of MPI library. This means that even when using CUDA-aware MPI, all communications require the transfer of data to and from the GPU.

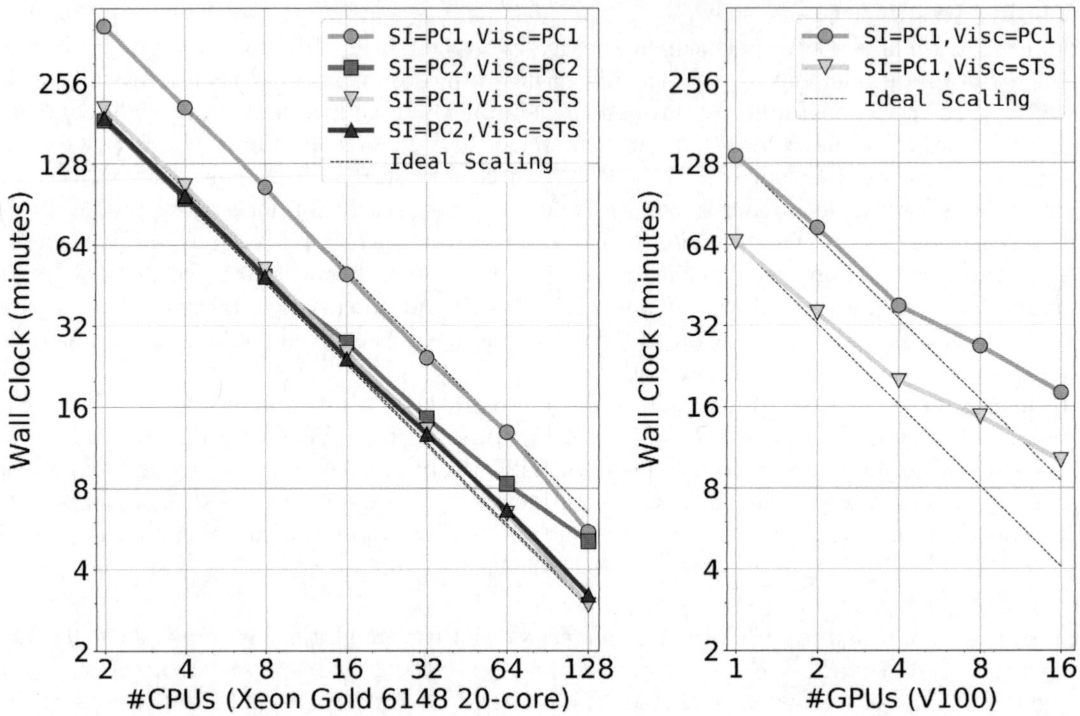

Figure 7. Comparison of run times and scaling for the algorithm choices in Sec. 2.2 for the test run of Sec. 3 using CPUs (left) and GPUs (right). The CPU timings were performed on server nodes with 2 Skylake 20-core CPUs per node, while the GPU timings used servers with 4 V100 GPUs per node (see Appendix for details).

Since the amount of data transferred in the MPI calls is much larger in these runs compared to comparably-timed CPU runs (due to much fewer MPI ranks leading to larger grid halo sizes), the lack of RDMA is likely a contributing factor to the poor scaling of the inter-node runs.

We have chosen two unique algorithms for our CPU and GPU comparison timings. While this may seem like an unfair comparison, as mentioned, we are only interested in the effective 'time-to-solution' we can achieve on both platforms, not in an apples-to-apples comparison of the hardware. However, since the run times of the PC2+STS and PC1+STS algorithms on the CPUs are quite similar, the CPU-GPU comparisons do happen to be close to a direct comparison in this case.

5.3. Hardware comparisons

Using the best chosen algorithms for the CPU and GPU, we compare wall clock times for several generations of CPU and GPU architectures. Since a single GPU server could be configured with up to 16 GPUs and a single CPU server could be configured with up to 4 CPU multi-core processors, it is difficult to directly compare timing results. In line with our practical approach, we choose to compare timing results as a function of the number of compute nodes used on the HPC systems (for CPUs, dual-socket configurations (2 CPU processors per node) and for GPUs, 4 GPUs per node). Figure 8 shows the results for 5 models of CPU and 3 models of GPU. The CPU runs all scale nearly perfectly as expected from the tests in Sec. 5.2, while all three GPU models show similar scaling problems between nodes. However, even with the reduced scaling, for a given number of compute nodes, the GPU runs are much faster than those of the

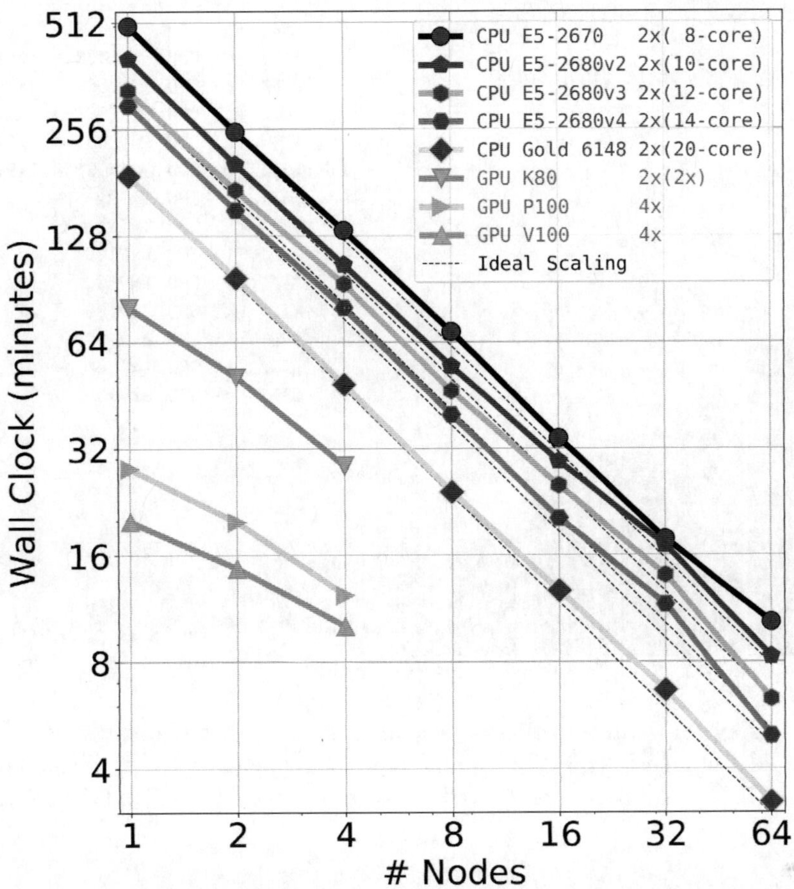

Figure 8. Comparison of wall clock times for the test run using various generations of CPU and GPU hardware. The CPU runs use the PC2+STS algorithm, while the GPU runs use the PC1+STS algorithm (see Sec. 5.2). Details of the hardware and software environment used is given in the Appendix.

CPUs. Allocation limits prevented us from testing past 4 GPU-enabled nodes. Using just 1 node of 4xV100s yields a run time that would require more than 8 40-core SkyLake CPU nodes (or nearly 32 SandyBridge nodes) to achieve. Thus, the GPU nodes exhibit excellent compact performance.

5.4. Single System results

One of the goals for adding GPU acceleration to MAS is to allow small-to-medium simulations to run 'in-house' that otherwise would require the use of an HPC CPU cluster. To illustrate this, in Figure 9 we show timing results for CPUs and GPUs on a single server (limited to those available to us). We also add CPU and GPU timings for a consumer-level desktop PC equipped with a consumer-level NVIDIA GPU for comparison. All hardware and software details are listed in the Appendix. We see that while a dual-socket Skylake server runs the test problem over 5 times faster than our local desktop PC, a single server of 4xV100 GPUs runs the same test almost 50 times faster (9.5x faster than the Skylake node). When considering that one is able to configure a single server with up to 16 GPUs, the advantages of the GPU implementation

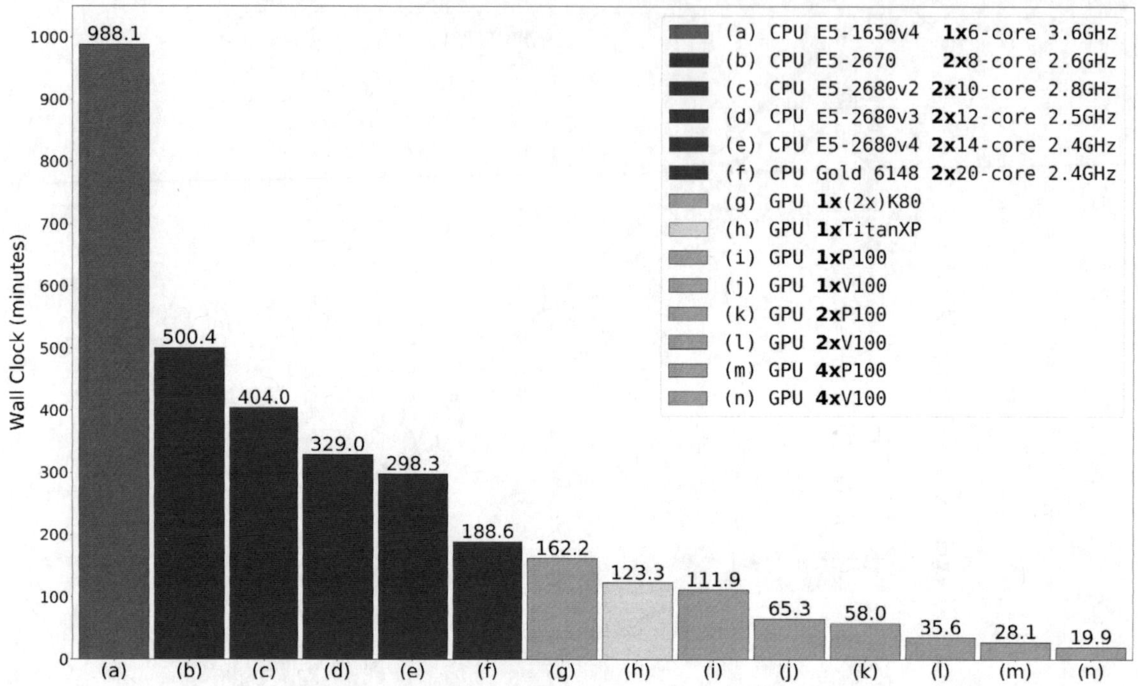

Figure 9. Single system wall clock comparisons for the test run on various CPU and GPU hardware. The advantages of the GPU implementation is clearly seen. Although we only had access to single GPU systems up to 4 GPUs, we note that up to 16xGPU single servers are available.

are compelling. We also see that the consumer-level GPU (TitanXP) performs almost as well as the server counterpart of the same architecture (P100). This may seem surprising at first, since, while the TitanXP has a clock rate $\sim$15% faster than the P100, it contains 15x less double-precision compute cores. We can explain the result by noting that the memory bandwidth of the two GPUs are within 25% of each other. Since the algorithms in MAS are all 'memory-bound' (and never reach close to the theoretical maximum FLOP/s on any architecture), it is not too surprising that the performance of the two cards are within 10% of each other. This performance similarity opens the door for very inexpensive GPU computing with the MAS code (or other codes with memory-bound algorithms), greatly increasing the relevance of the single-server results.

6. Conclusions

We have successfully added GPU acceleration to the MAS code running in zero-beta mode in a fully portable manner using OpenACC. The use of OpenACC allows us to maintain a single source code that can be compiled to either multi-CPU or multi-GPU platforms, with minimal modifications to the original code. This avoids the steep learning curves and development interference many associate with GPU programming.

Using the best algorithm choice for each hardware, we showed that one can achieve performance using a few GPUs on one server that requires numerous CPU servers to match. This opens the door for running small-to-medium resolution simulations 'in-house', greatly assisting development and research. The OpenACC implementation will also allow the MAS code to utilize the latest state-of-the-art HPC multi-GPU systems (e.g. the Summit machine at ORNL

[27]).

Efforts are underway to continue the OpenACC implementation, with the goal of eventually supporting MAS's full set of features for use with a wide range of solar physics and space weather applications.

Acknowledgments

Work at Predictive Science was supported by AFOSR, NASA, and NSF. Computational resources provided by the NSF-supported XSEDE program for use of San Diego Supercomputer Center's COMET system, NASA NAS for use of the Pleiades and Electra systems, and NVIDIA Corporation for use of their PSG test system.

References

[1] Matthews D 2018 *Nature* **562** 151–152
[2] Cook S 2012 *CUDA Programming: A Developer's Guide to Parallel Computing with GPUs* (Morgan Kaufmann Publishers Inc.)
[3] Kaeli D R, Mistry P, Schaa D and Zhang D P 2015 *Heterogeneous Computing with OpenCL 2.0* (Morgan Kaufmann Publishers Inc.)
[4] Farber R 2016 *Parallel Programming with OpenACC* (Morgan Kaufmann, Elsevier) ISBN 9780124103979
[5] Chandrasekaran S and Juckeland G 2017 *OpenACC for Programmers Concepts and Strategies* (Addison-Wesley Professional) ISBN 9780134694283
[6] Fu H, Liao J, Yang J, Wang L, Song Z, Huang X, Yang C, Xue W, Liu F, Qiao F *et al.* 2016 *Science China Information Sciences* **59** 072001
[7] Mikić Z, Linker J A, Schnack D D, Lionello R and Tarditi A 1999 *Phys. Plasmas* **6** 2217
[8] Linker J A, Mikić Z, Biesecker D A, Forsyth R J, Gibson S E, Lazarus A J, Lecinski A, Riley P, Szabo A and Thompson B J 1999 *Journal of Geophysical Research: Space Physics* **104** 9809–9830
[9] Downs C, Linker J A, Mikić Z, Riley P, Schrijver C J and Saint-Hilaire P 2013 *Science* **340** 1196–1199
[10] Lionello R, Riley P, Linker J A and Mikić Z 2005 *The Astrophysical Journal* **625** 463–473
[11] Lionello R, Linker J A, Mikić Z and Riley P 2006 *The Astrophysical Journal Letters* **642** L69
[12] Linker J A, Lionello R, Mikić Z, Titov V S and Antiochos S K 2011 *The Astrophysical Journal* **731** 110
[13] Linker J A, Mikić Z, Lionello R, Riley P, Amari T and Odstrcil D 2003 *Phys. Plasmas* **10** 1971
[14] Lionello R, Downs C, Linker J A, Török T, Riley P and Mikić Z 2013 *The Astrophysical Journal* **777** 76
[15] Mikić Z, Downs C, Linker J A, Caplan R M, Mackay D H, Upton L A, Riley P, Lionello R, Török T, Titov V S *et al.* 2018 *Nature Astronomy* 1
[16] Török T, Downs C, Linker J A, Lionello R, Titov V S, Mikić Z, Riley P, Caplan R M and Wijaya J 2018 *The Astrophysical Journal* **856** 75
[17] Riley P, Linker J A, Lionello R and Mikic Z 2012 *Journal of Atmospheric and Solar-Terrestrial Physics* **83** 1–10
[18] Caplan R M, Mikić Z, Linker J A and Lionello R 2017 *Journal of Physics: Conference Series* **837** 012016
[19] Lionello R, Linker J A and Mikić Z 2009 *The Astrophysical Journal* **690** 902
[20] Lionello R, Mikić Z and Linker J A 1999 *Journal of Computational Physics* **152** 346–358
[21] Meyer C D, Balsara D S and Aslam T D 2014 *Journal of Computational Physics* **257** 594–626
[22] Saad Y 2003 *Iterative methods for sparse linear systems* (Siam)
[23] Titov V, Török T, Mikic Z and Linker J A 2014 *The Astrophysical Journal* **790** 163
[24] Mikić Z, Török T, Titov V, Linker J A, Lionello R, Downs C and Riley P 2013 *AIP Conference Proceedings* vol 1539 (AIP) pp 42–45
[25] Caplan R, Linker J, Downs C, Török T, Mikić Z, Lionello R, Titov V, Riley P and Wijaya J 2018 Using OpenACC to port solar storm modeling code to GPUs `https://devblogs.nvidia.com/solar-storm-modeling-gpu-openacc/`
[26] Caplan R M, Mikić Z and Linker J A 2017 *ArXiv e-prints* (*Preprint* 1709.01126)
[27] Hines J 2018 *Computing in Science Engineering* **20** 78–82 ISSN 1521-9615

IOP Conf. Series: Journal of Physics: Conf. Series **1225** (2019) 012012 doi:10.1088/1742-6596/1225/1/012012

Appendix: Hardware and software details

The hardware and software details used for the CPU tests are given in Table 1 and those for the GPU tests in Table 2. The compiler flags used for each test are shown in Table 3.

	NASA NAS Pleiades & Electra					Local Desktop
Compiler	Intel 2018.3.222					GNU 5.4.0
MPI Library	SGI MPT 2.15r20					OpenMPI 2.1.1
CPU Family	Sandy Bridge	Ivy Bridge	Haswell	Broadwell	Skylake	Broadwell
Instruction Set	AVX		AVX2		AVX512	AVX2
CPU Model	E5-2670	E5-2680v2	E5-2680v3	E5-2680v4	Gold 6148	E5-1650v4
CPU Clock Rate	2.6 GHz	2.8 GHz	2.5 GHz	2.4 GHz	2.4 GHz	3.6 GHz
#Sockets x #Cores	2x8	2x10	2x12	2x14	2x20	1x6
Memory Bandwidth	51.2 GB/s	59.7 GB/s	68 GB/s	76.8 GB/s	128 GB/s	76.8 GB/s

Table 1. System hardware and software configuration details for CPU results.

	SDSC COMET		NVIDIA PSG	Local Desktop
Compiler	PGI 18.4		PGI 18.3	PGI 18.4
MPI Library	OpenMPI 2.1.2		OpenMPI 1.10.7	OpenMPI 2.1.2
CUDA Version	CUDA 9.1			
Driver Version	367.48		396.26	410.48
GPU Family	Kepler	Pascal	Volta	Pascal
GPU Model	K80	P100	V100	TitanXP
#GPUs/node	2x2	4	4	1
GPU Clock Rate	0.56 GHz	1.33 GHz	1.38 GHz	1.58 GHz
#FP64 Cores/GPU	2x832	1792	2560	120
Memory Bandwidth/GPU	2x240 GB/s	732 GB/s	900 GB/s	547.6 GB/s

Table 2. System hardware and software configuration details for GPU results.

Compiler	Flags
Intel 2018.3.222 (CPU)	`-O3 -heap-arrays -fp-model precise -axCORE_AVX512 -xSSE4.2 -DNDEBUG`
GNU 5.4.0 (CPU)	`-O3 -mtune=native -DNDEBUG`
PGI 18.3 & 18.4 (GPU)	`-O3 -ta=tesla:cuda9.1,cc60,cc70 -DNDEBUG`

Table 3. Compiler flags used to compile MAS for the test runs.

IOP Conf. Series: Journal of Physics: Conf. Series **1225** (2019) 012013 doi:10.1088/1742-6596/1225/1/012013

thornado-transport: IMEX schemes for two-moment neutrino transport respecting Fermi-Dirac statistics

Ran Chu

Department of Physics and Astronomy, University of Tennessee, Knoxville, TN 37996-1200

E-mail: rchu@vols.utk.edu

Eirik Endeve

Computational and Applied Mathematics Group, Oak Ridge National Laboratory, Oak Ridge, TN 37831 USA
Department of Physics and Astronomy, University of Tennessee, Knoxville, TN 37996-1200
Joint Institute for Computational Sciences, Oak Ridge National Laboratory, Oak Ridge, TN 37831 USA

E-mail: endevee@ornl.gov

Cory D. Hauck

Computational and Applied Mathematics Group, Oak Ridge National Laboratory, Oak Ridge, TN 37831 USA
Department of Mathematics, University of Tennessee, Knoxville, TN 37996-1320

E-mail: hauckc@ornl.gov

Anthony Mezzacappa

Department of Physics and Astronomy, University of Tennessee, Knoxville, TN 37996-1200
Joint Institute for Computational Sciences, Oak Ridge National Laboratory, Oak Ridge, TN 37831 USA

E-mail: mezz@utk.edu

Bronson Messer

Scientific Computing and Theoretical Physics Groups, Oak Ridge National Laboratory, Oak Ridge, TN 37831 USA
Department of Physics and Astronomy, University of Tennessee, Knoxville, TN 37996-1200

E-mail: bronson@ornl.gov

Abstract. We develop implicit-explicit (IMEX) schemes for neutrino transport in a background material in the context of a two-moment model that evolves the angular moments of a neutrino phase-space distribution function. Considering the upper and lower bounds that are introduced by Pauli's exclusion principle on the moments, an algebraic moment closure based on Fermi-Dirac statistics and a convex-invariant time integrator both are demanded. A finite-volume/first-order discontinuous Galerkin(DG) method is used to illustrate how an algebraic moment closure based on Fermi-Dirac statistics is needed to satisfy the bounds. Several algebraic closures are compared with these bounds in mind, and the Cernohorsky and Bludman closure, which satisfies the bounds, is chosen for our IMEX schemes. For the convex-invariant time integrator, two IMEX schemes named PD-ARS have been proposed. PD-ARS denotes a convex-invariant IMEX Runge-Kutta scheme that is high-order accurate in the streaming limit, and works well in the diffusion limit. Our two PD-ARS schemes use second- and third-order, explicit, strong-stability-preserving Runge-Kutta methods as their explicit part, respectively, and therefore are second- and third-order accurate in the streaming limit, respectively. The accuracy and convex-invariance of our PD-ARS schemes are demonstrated in the numerical tests with a third-order DG method for spatial discretization and a simple Lax-Friedrichs flux. The method has been implemented in our high-order neutrino-radiation hydrodynamics (`thornado`) toolkit. We show preliminary results employing tabulated neutrino opacities.

1. Introduction

Core-collapse supernovae (CCSNe) are the explosions of massive stars that end their lives. They are directly or indirectly responsible for the lion's share of elements heavier than oxygen and play important roles in many astrophysical phenomena, such as neutron star and black hole formation. Furthermore, these explosions occur at energies and densities relevant to address fundamental questions in nuclear, particle, and gravitational physics. A solid theoretical framework for the CCSN explosion mechanism may help answer important questions in fundamental physics [1].

One essential part of the explosion mechanism is neutrino transport. Neutrino energy deposition is believed to be the major driver of CCSN explosions, except in peculiar cases where rapid rotation is present and magnetohydrodynamic effects may dominate (for reviews, see [2, 3, 4, 5]). Ideally, neutrino transport would be modeled by the Boltzmann transport equation, which is an integro-partial-differential equation evolving a phase-space distribution function f (e.g., see [6, 7, 8, 9, 10, 11, 12, 13, 14, 15, 16, 17, 18] for studies of CCSN with Boltzmann transport in various approximate settings). Simulating the neutrino transport implies finding a solution of the Boltzmann equation for a specific domain and period, with acceptable accuracy.

However, solving the Boltzmann transport equation with sufficient phase-space resolution and full weak interaction physics is at present too expensive. To balance physical fidelity and computational expediency, an approximate method called the two-moment method has been adopted (e.g., see [19, 20, 21, 22]). Using the two-moment method, the evolved variables are the zeroth and first angular moments of the distribution function f – the spectral particle density $\mathcal{J}$ and flux $\boldsymbol{\mathcal{H}}$, respectively. However, the equation of $\boldsymbol{\mathcal{H}}$ includes the second angular moment $\boldsymbol{\mathcal{K}}$. Knowledge of $\boldsymbol{\mathcal{K}}$ is needed to close the two-moment system. Therefore, a closure that gives $\boldsymbol{\mathcal{K}}$ consistent with $\mathcal{J}$ and $\boldsymbol{\mathcal{H}}$ is needed. The better the closure predicts $\boldsymbol{\mathcal{K}}$, the more accurate the two-moment method will be. The two-moment method has been widely applied in the CCSN modeling community with different algebraic closures, such as the Minerbo [23] closure (e.g. O'Connor and Couch [24], Pan and et al. [25], Glas et al. [26], and Just et al. [21]) and the Levermore [27] closure (e.g. Vartanyan et al. [22], Cabezon et al. [28], and Kuroda et al. [19]).

Applying the two-moment method does simplify the problem, but doesn't guarantee an affordable solution. To be precise, how to discretize the continuous system of equatio ns given by the two-moment method and solve the discretized system efficiently remains a question. In fact, the time scales of neutrino interactions with the background (they can be $\sim \mathcal{O}(10^{-13})$ second) is short compared to the duration of the CCSN explosion ($\sim \mathcal{O}(1)$ second). This means that $\sim \mathcal{O}(10^{13})$ time steps could be needed for solving the system fully explicitly. On the other hand, solving the moment equations fully implicitly requires inverting global band-structured matrices whose sizes depend on the phase-space discretization. Such a global inversion is both expensive and unfriendly to parallelization. To circumvent these challenges, implicit-explicit (IMEX) methods are taken into consideration. By treating the transport terms in the two-

IOP Conf. Series: Journal of Physics: Conf. Series **1225** (2019) 012013 doi:10.1088/1742-6596/1225/1/012013

moment equations explicitly and the collision terms implicitly, IMEX methods are subject only to a time step governed by the explicit transport terms, and the matrices to be inverted are block diagonal. Therefore, IMEX methods require far fewer time steps compared with a fully explicit method, and the computation for each step is easily parallelizable. For the relativistic setting that we have, where the fluid and the neutrinos have comparable propagation speeds, IMEX methods can be efficient.

To model neutrino transport using a two-moment method, two (or at least two) things need to be chosen carefully: an algebraic closure based on Fermi-Dirac statistics for closing the two-moment equations and a convex-invariant, diffusion-accurate IMEX scheme to ensure a physical result. A convex-invariant scheme has the following property: if the solution $u^n \in W$ and W is a convex set, then $u^{n+1} \in W$. Since the neutrino distribution function is bounded ($f \in [0,1]$) by the Pauli exclusion principle, its moments as weighted integrals of a bounded function over the domain $\omega \in \mathbb{S}^2$ are also bounded. We call the moments satisfying the constraints due to Fermi-Dirac statistics *realizable moments* and the realizable $\mathcal{J}$ and $\boldsymbol{\mathcal{H}}$ define a convex set [29]. The algebraic closure should give a realizable $\boldsymbol{\mathcal{K}}$, and the well-posedness of the closure requires realizable $\mathcal{J}$ and $\boldsymbol{\mathcal{H}}$. This explains why an algebraic closure based on Fermi-Dirac statistics is needed. Realizability of $\mathcal{J}$ and $\boldsymbol{\mathcal{H}}$ after each time step requires a convex-invariant IMEX scheme. Since the realizable moments form a convex set, it is possible to construct a realizability-preserving method using a convex-invariant IMEX scheme for two-moment neutrino transport. In addition, the physics of neutrino transport in CCSNe requires the IMEX scheme to be diffusion-accurate.

The study of moment realizability and realizability-preserving methods with diffusion-accurate IMEX schemes motivates this work. Gottlieb et al. [30] showed that standard strong-stability-preserving IMEX schemes cannot have an order higher than first without a restricted time step. One way to obtain the second-order (or higher-order) accuracy is to add some correction steps after the standard step [31, 32]. Unfortunately, the correction steps can deteriorate the accuracy of the IMEX scheme in the diffusion limit or restrict the time step. To keep things simple, we focus on IMEX schemes without correction steps and require them to be high-order (second or higher order) in the streaming limit and diffusion-accurate. We call these IMEX schemes *PD-ARS*.

`thornado` is our toolkit for high-order neutrino-radiation hydrodynamics based on high-order Runge-Kutta Discontinuous Galerkin (RKDG) methods. It is being developed at the University of Tennessee, Knoxville and Oak Ridge National Laboratory. It currently includes solvers for the Euler equations for fluid dynamics and the two-moment approximation of the radiative transfer equation [33]. In this paper, we focus on the transport methods in `thornado` with emphasis on IMEX.

This paper is organized as follows: Section 2 discusses the mathematical model, algebraic closures, and the constraints on the moments and algebraic closures imposed by Fermi-Dirac statistics; Section 3 gives a first-order finite-volume spatial discretization and shows how the spatial discretization preserves constraints in an IMEX step; Section 4 discusses how to use convex combination to construct two PD-ARS schemes, one with second-order accuracy in the streaming limit and the other with third-order accuracy in the same limit; Section 5 presents the results of the numerical tests, which demonstrate the properties of the PD-ARS schemes; Section 6 summarizes the achievements of this paper and discusses future work.

2. Two-Moment Model

2.1. Transport Equations

We consider neutrino transport through a static background in a non-relativistic framework. For simplicity, we only include neutrino–matter interactions due to emission, absorption, and isoenergetic scattering. It is possible to include other interactions, such as neutrino–electron scattering and pair processes, in the two-moment model. However, these interactions result in dense coupling across neutrino energies and across neutrino species, and we have not yet considered the realizability-preserving property of our scheme in this case. We leave this for future work.

After scaling to dimensionless units, the Boltzmann equation can be written as

$$\partial_t f + \boldsymbol{\ell} \cdot \nabla f = \frac{1}{\tau} \mathcal{C}(f), \tag{1}$$

where the distribution function $f = f(\omega, \varepsilon, \boldsymbol{x}, t)$ gives the number of neutrinos propagating in the direction $\omega \in \mathbb{S}^2$, with energy $\varepsilon \in \mathbb{R}^+$, at position $\boldsymbol{x} \in \mathbb{R}^3$ and time $t \in \mathbb{R}^+$. $\boldsymbol{\ell}(\omega) \in \mathbb{R}^3$ is the unit vector parallel to the neutrino three-momentum: $\boldsymbol{p} = \varepsilon\,\boldsymbol{\ell}$. On the right-hand side, τ is a collision time scale. In opaque regions, where neutrinos have frequent interactions with the background, $\tau \ll 1$. In transparent regions, where neutrinos rarely interact and stream freely, $\tau \gg 1$. The collision term, $\mathcal{C}(f)$, which models emission, absorption, and isoenergetic scattering is given by

$$\mathcal{C}(f) = \xi\,\big(\,f_0 - f\,\big) + (1-\xi)\,\Big(\frac{1}{4\pi}\int_{\mathbb{S}^2} f\,d\omega - f\,\Big), \tag{2}$$

where $\xi = \sigma_{\mathrm{A}}/(\sigma_{\mathrm{A}} + \sigma_{\mathrm{S}})$ is the ratio of the absorption opacity σ_{A} to the total opacity. The scattering opacity is σ_{S}. The limit $\xi = 1$, when $\sigma_{\mathrm{S}} = 0$, corresponds to pure absorption, while $\xi = 0$, when $\sigma_{\mathrm{A}} = 0$, corresponds to pure scattering. The equilibrium distribution function for neutrinos is given by the Fermi-Dirac distribution:

$$f_0(\boldsymbol{z}) = \frac{1}{e^{(\varepsilon - \mu(\boldsymbol{x}))/T(\boldsymbol{x})} + 1}, \tag{3}$$

where $\boldsymbol{z} := \{\varepsilon, \boldsymbol{x}\}$, T is the material temperature in energy units and μ is the neutrino chemical potential. Both T and μ depend on the spatial position $\boldsymbol{x}$.

2.2. *Two-Moment Model*

Approximate solutions to the Boltzmann equation, Eq. (1), can be found by solving the two-moment model. To this end, define the angular moments of the distribution function as follows:

$$\{\,\mathcal{J}, \boldsymbol{\mathcal{H}}, \boldsymbol{\mathcal{K}}\,\}(\boldsymbol{z}, t) = \frac{1}{4\pi}\int_{\mathbb{S}^2} f(\omega, \boldsymbol{z}, t)\,\{\,1, \boldsymbol{\ell}, \boldsymbol{\ell}\otimes\boldsymbol{\ell}\,\}\,d\omega. \tag{4}$$

The zeroth moment, $\mathcal{J}$, is referred to as the particle density. The first moment, $\boldsymbol{\mathcal{H}}$, is the particle flux, and the second moment, $\boldsymbol{\mathcal{K}}$, is proportional to the stress tensor. By integrating Eq. (1) over the momentum-space angular dimension we obtain equations for the zeroth and the first moments:

$$\partial_t \boldsymbol{\mathcal{M}} + \nabla \cdot \boldsymbol{\mathcal{F}} = \frac{1}{\tau} \boldsymbol{\mathcal{C}}(\boldsymbol{\mathcal{M}}), \tag{5}$$

with $\boldsymbol{\mathcal{M}} = (\mathcal{J}, \boldsymbol{\mathcal{H}})^T$, $\boldsymbol{\mathcal{F}} = (\boldsymbol{\mathcal{H}}, \boldsymbol{\mathcal{K}})^T$, and

$$\boldsymbol{\mathcal{C}}(\boldsymbol{\mathcal{M}}) = \boldsymbol{\eta} - \boldsymbol{\mathcal{D}}\boldsymbol{\mathcal{M}}, \tag{6}$$

where $\boldsymbol{\eta} = (\xi\,f_0, \boldsymbol{0})^T$, $\boldsymbol{\mathcal{D}} = \mathrm{diag}(\xi, \boldsymbol{I})$, and $\boldsymbol{I}$ is the identity matrix. Hence, the process of solving the Boltzmann equation, Eq. (1), for the neutrino distribution function $f(\omega, \boldsymbol{z}, t)$, is replaced by solving the two-moment equations for the neutrino number density, $\mathcal{J}(\boldsymbol{z}, t)$, and flux, $\boldsymbol{\mathcal{H}}(\boldsymbol{z}, t)$.

2.3. Algebraic Closures

The moment equation for $\boldsymbol{\mathcal{H}}$ involves the higher moment $\boldsymbol{\mathcal{K}}$ and the two-moment model is open. To close the two-moment model, we consider algebraic closures. For the two-moment model, algebraic closures give an approximation to $\boldsymbol{\mathcal{K}}$ using the lower moments:

$$\boldsymbol{\mathcal{K}} = \boldsymbol{k}\mathcal{J}, \tag{7}$$

where $\boldsymbol{k}$ is the Eddington tensor. By assuming that the distribution function is symmetric about a preferred direction $\widehat{\boldsymbol{h}} = \boldsymbol{\mathcal{H}}/|\boldsymbol{\mathcal{H}}|$, Levermore [27] proposed a simple form for the Eddington tensor:

$$\boldsymbol{k} = \frac{1}{2}\big[\,(1-\chi)\,\boldsymbol{I} + (3\chi - 1)\,\widehat{\boldsymbol{h}}\otimes\widehat{\boldsymbol{h}}\,\big], \tag{8}$$

where $\chi = \chi(\mathcal{J}, |\boldsymbol{\mathcal{H}}|)$ is the Eddington factor. Thus, the two-moment model is closed by specifying the scalar χ in terms of $\mathcal{J}$ and $|\boldsymbol{\mathcal{H}}|$.

2.4. Constraints on the Moments

Neutrinos are fermions and obey the Pauli exclusion principle. Because of this, the neutrino distribution function is bounded; i.e. $f \in [0,1]$. As a result, the angular moments $\mathcal{J}$ and $\boldsymbol{\mathcal{H}}$ and the Eddington factor χ satisfy the following bounds [27, 34, 35, 36]:

$$\mathcal{J} \in [0,1], \quad (1-\mathcal{J})\mathcal{J} - |\boldsymbol{\mathcal{H}}| \geq 0, \tag{9}$$

$$\chi_{\min} = \max\big(1 - \frac{2}{3\mathcal{J}}, h^2\big) \leq \chi \leq \min\big(1, \frac{1}{3\mathcal{J}} - \frac{\mathcal{J}}{1-\mathcal{J}}h^2\big) = \chi_{\max}, \tag{10}$$

where $h = |\boldsymbol{\mathcal{H}}|/\mathcal{J}$ is the flux factor. The inequalities in Eqs. (9) and (10) hold for moments as defined in Eq. (4) with $f \in [0,1]$. In a relativistic model (e.g., [37, 38]), the evolved quantities — functions of the moments in Eq. (4) — satisfy different bounds. We are currently exploring extensions of the realizability-preserving scheme presented here to the relativistic case.

The constraints in Eq. (9) define realizable moments $\boldsymbol{\mathcal{M}}$. For fermions, realizable moments can only be constructed from a distribution satisfying the bounds $f \in [0,1]$. Moreover, the set of realizable moments is convex: let $\mathcal{R}$ be the realizability set and $\boldsymbol{\mathcal{M}}_1, \boldsymbol{\mathcal{M}}_2 \in \mathcal{R}$, then $\lambda\boldsymbol{\mathcal{M}}_1 + (1-\lambda)\boldsymbol{\mathcal{M}}_2 \in \mathcal{R}$ for any $\lambda \in [0,1]$ [29]. As we will see later in Section 3, this convexity makes it possible to design a realizability-preserving discretization for solving the two-moment model numerically.

The inequalities in Eq. (10) deserve further attention. They are as important as those in Eq. (9), in maintaining consistency of the two-moment model with respect to Fermi-Dirac statistics. When designing a numerical scheme for the two-moment model that maintains realizable $\boldsymbol{\mathcal{M}}$, it is also necessary for the Eddington factor to satisfy the bounds in Eq. (10).

However, recently reported CCSN simulations using two-moment neutrino transport with algebraic closures have employed Eddington factors that can violate the bounds in Eq. (10). As examples, we consider the Eddington factors discussed in [39], where the suitability of several algebraic closures for two-moment neutrino transport was evaluated. In Fig. 1, we plot the Eddington factor versus the flux factor for two occupancies: $\mathcal{J} = 0.1$ (low occupancy) and $\mathcal{J} = 0.9$ (high occupancy). Of the algebraic closures plotted, few satisfy the bounds on the Eddington factor in Eq. (10). Kershaw [35], Wilson [40, 41], Levermore [27], Minerbo [23], and Janka 2 [42] closures may work fine when the occupancy is low. When the occupancy is high, the Eddington factor due to these closures exceeds the upper bound for Fermi-Dirac statistics. The Eddington factor of Janka 1 [43] may violate both the upper and lower bound on the Eddington factor. Only the closure due to Cernohorsky & Bludman [44] satisfies both the upper and lower bounds. This is not surprising as this is the only closure based on Fermi-Dirac statistics. Although the Levermore and Minerbo closures do not satisfy the bounds in Eq. (10), they are widely used in simulations of neutrino transport in CCSNe and compact binary mergers; e.g., O'Connor & Couch [24], Pan et al. [25], Glas et al. [26], Just et al. [21], and Foucart et al. [45] use the Minerbo closure, while Vartanyan et al. [22], Cabezon et al. [28], Kuroda et al. [19], and Fujibayashi et al. [46] use the Levermore closure. When employing these closures in conditions of high occupancy, a numerical scheme may evolve the moments outside the realizable domain of Fermi-Dirac statistics given in Eq. (9). If this were to happen, the update step may give $\mathcal{J} > 1$. Considering the fact that the collision term contains blocking factors — i.e. $(1-\mathcal{J})\times$ something positive — $\mathcal{J} > 1$ would change the blocking factors' sign, and it would be difficult to predict the impact of the subsequent induced errors on the simulation outcome. Besides, even with the simplified collision term Eq 6, $\mathcal{J} > 1$ could result in information loss (such as loss to the neutrino distribution for the groups that exceed the bounds), closure failure, and other unpredictable consequences. If treatments are developed to map the unrealizable moments into the realizable domain, they should be developed to conserve lepton number, energy, and momentum.

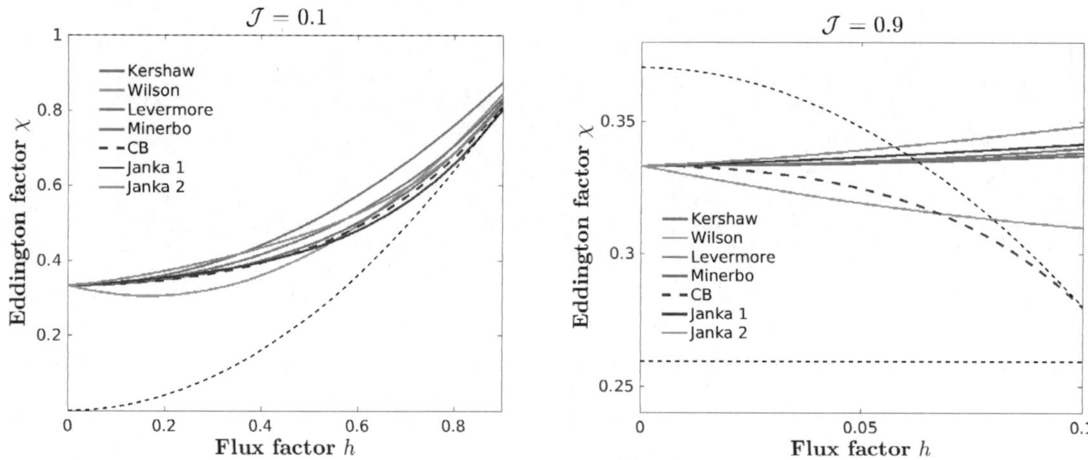

Figure 1. Plot of Eddington factors χ versus flux factor h for different values of $\mathcal{J}$ for various algebraic closures: $\mathcal{J} = 0.1$ (left panel, low occupancy) and $\mathcal{J} = 0.9$ (right panel, high occupancy). In each panel we plot the Eddington factors of two-moment closures due to Kershaw (red), Wilson (yellow), Levermore (green), Minerbo (light blue), Cernohorsky & Bludman (blue), Janka 1 (purple), and Janka 2 (pink) . We also plot $\chi_{\min}$ and $\chi_{\max}$, defined in Eq. (10) (lower and upper dashed black lines, respectively).

3. Spatial Discretization

We discretize the two-moment model with a simple first-order finite volume method to illustrate how the closure affects the realizability-preserving property of the method. By assuming that the moments at time level t^n ($\boldsymbol{\mathcal{M}}^n$) satisfy the bounds in Eq. (9), our goal is to identify sufficient conditions such that the moments at time level t^{n+1} ($\boldsymbol{\mathcal{M}}^{n+1}$) also satisfy the bounds. To simplify, we limit the discussion to one spatial dimension and employ a uniform Cartesian mesh. (The extension to multiple spatial dimensions and high-order discretization using the discontinuous Galerkin method is given in [29].)

We divide the spatial domain D into N uniform cells and denote the i-th cell by $\boldsymbol{K}_i$, with $i = 1, \ldots, N$; i.e.,

$$D = \cup_{i=1}^{N} \boldsymbol{K}_i \quad \text{with} \quad \boldsymbol{K}_i = \{\, x : x \in (x_{i-1/2}, x_{i+1/2})\},$$

and cell width $\Delta x = D/N$. The cell-average of the moments is defined as

$$\boldsymbol{\mathcal{M}}_i = \frac{1}{\Delta x} \int_{\boldsymbol{K}_i} \boldsymbol{\mathcal{M}} dx. \tag{11}$$

Integrating Eq. (5) over each cell $\boldsymbol{K}_i$ gives

$$\frac{d\boldsymbol{\mathcal{M}}_i}{dt} = -\frac{1}{\Delta x}\Big(\widehat{\boldsymbol{\mathcal{F}}}(\boldsymbol{\mathcal{M}}_i, \boldsymbol{\mathcal{M}}_{i+1}) - \widehat{\boldsymbol{\mathcal{F}}}(\boldsymbol{\mathcal{M}}_{i-1}, \boldsymbol{\mathcal{M}}_i)\Big) + \frac{1}{\tau}\mathcal{C}(\boldsymbol{\mathcal{M}}_i), \tag{12}$$

where $\widehat{\boldsymbol{\mathcal{F}}}(\boldsymbol{\mathcal{M}}_a, \boldsymbol{\mathcal{M}}_b)$ is the numerical flux and $\frac{1}{\tau}\mathcal{C}(\boldsymbol{\mathcal{M}}_i)$ is the collision term evaluated with $\boldsymbol{\mathcal{M}}_i$. In this paper we use the global Lax-Friedrichs flux (setting the largest absolute eigenvalue of the flux Jacobian to one):

$$\widehat{\boldsymbol{\mathcal{F}}}_{\text{LF}}(\boldsymbol{\mathcal{M}}_a, \boldsymbol{\mathcal{M}}_b) = \frac{1}{2}\big(\boldsymbol{\mathcal{F}}(\boldsymbol{\mathcal{M}}_a) + \boldsymbol{\mathcal{F}}(\boldsymbol{\mathcal{M}}_b) - (\boldsymbol{\mathcal{M}}_b - \boldsymbol{\mathcal{M}}_a)\big). \tag{13}$$

By treating the transport term explicitly with forward Euler and the collision term implicitly with backward Euler, we have

$$\boldsymbol{\mathcal{M}}_i^{n+1} = \widetilde{\boldsymbol{\mathcal{M}}}_i^n + \frac{\Delta t}{\tau}\mathcal{C}(\boldsymbol{\mathcal{M}}_i^{n+1}), \tag{14}$$

where we have defined

$$\begin{aligned}
\widetilde{\boldsymbol{\mathcal{M}}}_i^n &= \boldsymbol{\mathcal{M}}_i^n - \frac{\Delta t}{\Delta x}\Big(\widehat{\boldsymbol{\mathcal{F}}}_{\text{LF}}(\boldsymbol{\mathcal{M}}_i^n, \boldsymbol{\mathcal{M}}_{i+1}^n) - \widehat{\boldsymbol{\mathcal{F}}}_{\text{LF}}(\boldsymbol{\mathcal{M}}_{i-1}^n, \boldsymbol{\mathcal{M}}_i^n)\Big) \\
&= (1-\beta)\boldsymbol{\mathcal{M}}_i^n + \beta\left[\frac{1}{2}\big(\boldsymbol{\mathcal{M}}_{i+1}^n - \boldsymbol{\mathcal{F}}(\boldsymbol{\mathcal{M}}_{i+1}^n)\big) + \frac{1}{2}\big(\boldsymbol{\mathcal{M}}_{i-1}^n + \boldsymbol{\mathcal{F}}(\boldsymbol{\mathcal{M}}_{i-1}^n)\big)\right], \quad \beta = \frac{\Delta t}{\Delta x}.
\end{aligned} \tag{15}$$

Considering Eqs. (14) and (15) and assuming that $\boldsymbol{\mathcal{M}}_i^n$ is realizable for all i, it can be shown (Lemma 3 in [29]) that $\boldsymbol{\mathcal{M}}_i^{n+1}$ is realizable provided $\frac{\Delta t}{\tau} > 0$. In Eq. (15), if $\beta \in [0,1]$, $\widetilde{\boldsymbol{\mathcal{M}}}_i^n$ is expressed as a convex combination of $\boldsymbol{\mathcal{M}}_i^n$ and the expression in the square brackets on the right-hand side of Eq. (15). It follows that $\widetilde{\boldsymbol{\mathcal{M}}}_i^n$ is realizable if the expression inside the square brackets is realizable. It can be shown (Lemma 2 in [29]) that the expression in square brackets is realizable for a distribution satisfying $f \in [0,1]$. For the two-moment model considered here, realizability depends on the algebraic closure. Specifically, if $\boldsymbol{\mathcal{M}}_i^n$ is realizable and the Eddington factor satisfies the bounds in Eq. (10), then $\boldsymbol{\mathcal{M}}_i^{n+1}$ is realizable provided $\beta \in [0,1]$. Thus, realizability of $\boldsymbol{\mathcal{M}}_i^{n+1}$ requires both a closure based on Fermi-Dirac statistics and a Courant-Friedrichs-Lewy (CFL) condition $\Delta t \leq \Delta x$.

4. Time Integration

Suppose that an algebraic closure based on Fermi-Dirac statistics is used (i.e., the Eddington factor satisfies Eq. (10)). Here we consider the construction of an Implicit-Explicit (IMEX) time integration scheme that maintains the bounds in Eq. (9). The semi-discretization of the two-moment model results in a system of ordinary differential equations of the form

$$\dot{\boldsymbol{u}} = \boldsymbol{\mathcal{T}}(\boldsymbol{u}) + \boldsymbol{\mathcal{Q}}(\boldsymbol{u}), \tag{16}$$

where the solution vector

$$\boldsymbol{u}(t) = (\boldsymbol{\mathcal{M}}_1(t), \ldots, \boldsymbol{\mathcal{M}}_N(t))^T \tag{17}$$

IOP Conf. Series: Journal of Physics: Conf. Series **1225** (2019) 012013 doi:10.1088/1742-6596/1225/1/012013

is the collection of all cell-averaged moments, $\boldsymbol{\mathcal{T}}$ is the transport operator, corresponding to the first term on the right-hand side of Eq. (12), and $\boldsymbol{\mathcal{Q}}$ is the collision operator, corresponding to the second term on the right-hand side of Eq. (12).

Since the set of realizable moments is convex, convex-invariant schemes, which map the initial values into this set, can be used to design realizability-preserving methods for the two-moment model. Ideally, the scheme should also be high-order accurate and work well in the asymptotic diffusion limit (characterized by frequent collisions and long time scales). The following discussion considers the construction of such convex-invariant schemes.

4.1. Standard IMEX Schemes

Treating the transport operator explicitly and the collision operator implicitly, a standard s-stage IMEX scheme takes the following form [47]:

$$\boldsymbol{u}^{(i)} = \boldsymbol{u}^n + \Delta t \sum_{j=1}^{i-1} \tilde{a}_{ij}\, \boldsymbol{\mathcal{T}}(\boldsymbol{u}^{(j)}) + \Delta t \sum_{j=1}^{i} a_{ij}\, \boldsymbol{\mathcal{Q}}(\boldsymbol{u}^{(j)}), \quad i = 1, \ldots, s, \tag{18}$$

$$\boldsymbol{u}^{n+1} = \boldsymbol{u}^n + \Delta t \sum_{i=1}^{s} \tilde{w}_i\, \boldsymbol{\mathcal{T}}(\boldsymbol{u}^{(i)}) + \Delta t \sum_{i=1}^{s} w_i\, \boldsymbol{\mathcal{Q}}(\boldsymbol{u}^{(i)}), \tag{19}$$

where $(\tilde{a}_{ij})$ and (a_{ij}), coefficients of the i-th stage, are elements of matrices $\tilde{A}$ and A, respectively. The matrices $\tilde{A}$ and A are lower triangular. ($\tilde{A}$ is strictly lower triangular so that the transport part is explicit.) The vectors $\tilde{\boldsymbol{w}} = (\tilde{w}_1, \ldots, \tilde{w}_s)^T$ and $\boldsymbol{w} = (w_1, \ldots, w_s)^T$ are the weights in the assembly step in Eq. (19). These coefficients and weights must satisfy certain order conditions for consistency, accuracy, and other properties. For second-order temporal accuracy, the following conditions are required [48]:

$$\sum_{i=1}^{s} \tilde{w}_i = \sum_{i=1}^{s} w_i = 1, \tag{20}$$

and

$$\sum_{i=1}^{s} \tilde{w}_i\, \tilde{c}_i = \sum_{i=1}^{s} \tilde{w}_i\, c_i = \sum_{i=1}^{s} w_i\, \tilde{c}_i = \sum_{i=1}^{s} w_i\, c_i = \frac{1}{2}, \tag{21}$$

where $\tilde{c}_i = \sum_{j=1}^{s} \tilde{a}_{ij}$ and $c_i = \sum_{j=1}^{s} a_{ij}$.

The IMEX scheme is called globally stiffly accurate (GSA) if the coefficients satisfy [49]:

$$a_{si} = w_i \quad \text{and} \quad \tilde{a}_{si} = \tilde{w}_i, \quad \text{for} \quad i = 1, \ldots, s. \tag{22}$$

Then, $\boldsymbol{u}^{n+1} = \boldsymbol{u}^{(s)}$, which is simplifying because the assembly step in Eq. (19) is omitted. IMEX schemes are further classified by the structure of the implicit matrix A. If A is invertible, the IMEX scheme is of type A [47]. If $a_{i1} = 0$ for $i = 1, \ldots, s$, $w_1 = 0$, and the submatrix consisting of the last $s - 1$ rows and columns is invertible, the IMEX scheme is of type ARS [50, 47].

4.2. Convex-Invariant IMEX Schemes

To be convex-invariant, the coefficients and weights defining the IMEX scheme must satisfy additional constraints. Our goal is to find constraints on a_{ij}, $\tilde{a}_{ij}$, $\tilde{w}_i$, and w_i that enable each $\boldsymbol{u}^{(i)}$ in Eq. (18) to be expressed as a convex combination of realizable states. Following Hu et al. [32], the stage values in Eq. (18) can be rewritten as

$$\boldsymbol{u}^{(i)} = \sum_{j=0}^{i-1} c_{ij} \Big[\boldsymbol{u}^{(j)} + \hat{c}_{ij}\, \Delta t\, \boldsymbol{\mathcal{T}}(\boldsymbol{u}^{(j)}) \Big] + a_{ii}\, \Delta t\, \boldsymbol{\mathcal{Q}}(\boldsymbol{u}^{(i)}), \quad i = 1, \ldots, s, \tag{23}$$

IOP Conf. Series: Journal of Physics: Conf. Series **1225** (2019) 012013 doi:10.1088/1742-6596/1225/1/012013

where c_{ij} and $\hat{c}_{ij} \equiv \tilde{c}_{ij}/c_{ij}$ are defined in terms of a_{ij} and $\tilde{a}_{ij}$. For IMEX schemes of type ARS, c_{ij} and $\tilde{c}_{ij}$ are given by [32]

$$\begin{aligned} c_{i0} &= 1 - \sum_{j=2}^{i-1}\sum_{l=j}^{i-1} a_{il} b_{lj}, \qquad c_{ij} = \sum_{l=j}^{i-1} a_{il} b_{lj}, \\ \tilde{c}_{i0} &= \tilde{a}_{i1} + \sum_{j=2}^{i-1} a_{ij}\tilde{b}_{j1}, \qquad \tilde{c}_{ij} = \tilde{a}_{ij} + \sum_{l=j+1}^{i-1} a_{il}\tilde{b}_{lj}, \end{aligned} \tag{24}$$

$$b_{ii} = \frac{1}{a_{ii}}, \quad b_{ij} = -\frac{1}{a_{ii}}\sum_{l=j}^{i-1} a_{il} b_{lj}, \quad \tilde{b}_{ij} = -\frac{1}{a_{ii}}\Big(\tilde{a}_{ij} + \sum_{l=j+1}^{i-1} a_{il}\tilde{b}_{lj}\Big). \tag{25}$$

Note that $c_{i1} = \tilde{c}_{i1} = 0$ in Eq. (24), so that $\sum_{j=0}^{i-1} c_{ij} = 1$.

If the IMEX scheme is GSA, $\boldsymbol{u}^{n+1} = \boldsymbol{u}^{(s)}$. Moreover, if $c_{ij}, \tilde{c}_{ij} \geq 0$ and $a_{ii} > 0$, each stage in Eq. (23) is a convex combination of explicit Euler steps (with time step $\hat{c}_{ij}\Delta t$), followed by an implicit Euler step. Each of the explicit Euler steps has a time step condition that ensures its realizability given by $\hat{c}_{ij}\,\Delta t \leq \Delta x$; the CFL condition of the scheme. Using results proved in [29] and discussed in Section 3, the IMEX scheme is convex-invariant and realizability-preserving for the two-moment model in Section 3 provided

$$\max(\hat{c}_{ij})\,\Delta t \leq \Delta x. \tag{26}$$

(This CFL condition becomes more restrictive with high-order DG spatial discretization [29].)

4.3. Diffusion Accurate, Convex-Invariant IMEX Schemes

Accuracy in the diffusion limit is another important property to consider when an IMEX scheme is applied to the two-moment model. In the diffusion limit, the distribution function is nearly isotropic, so $\boldsymbol{\mathcal{K}} \approx \frac{1}{3}\mathcal{J}\boldsymbol{I}$ and $\boldsymbol{\mathcal{H}} \approx -\frac{1}{3}\tau\nabla\mathcal{J}$, and the two-moment model is approximately governed by (e.g., [51])

$$\partial_t\mathcal{J} + \nabla\cdot\boldsymbol{\mathcal{H}} = 0 \quad \text{and} \quad \boldsymbol{\mathcal{H}} = -\tau\,\nabla\cdot\boldsymbol{\mathcal{K}}. \tag{27}$$

In the context of IMEX schemes, the above relationships imply that the following relations should hold [29]:

$$\boldsymbol{e}_i^T A^{-1}\tilde{A}\,\boldsymbol{e} = 1, \quad i = 1,\ldots,s, \tag{28}$$

where $\boldsymbol{e}_i$ is the ith column of the $s\times s$ identity matrix, $\boldsymbol{e}$ is the vector of ones, and A and $\tilde{A}$ are the matrices of the coefficients $(\tilde{a}_{ij})$ and (a_{ij}). Eq. (28) implies:

$$c_i = \tilde{c}_i, \quad i = 1,\ldots,s. \tag{29}$$

We have proved in [29] that only IMEX schemes of type ARS can be both diffusion accurate and convex-invariant. (Another short proof follows from the fact that IMEX schemes of type A have $\tilde{c}_1 = 0$ while $c_i \neq 0$.)

4.4. PD-ARS IMEX schemes

Unfortunately, coefficients satisfying the order conditions in Eqs. (20)-(21) and the conditions for convex-invariance do not exist for the standard IMEX scheme in Eqs. (18)-(19), unless a small time step is invoked that makes the scheme essentially explicit. To circumvent this problem, correction steps can be introduced after the assembly step in Eq. (19) (e.g., [52, 32]).

However, the correction steps can impose time step constraints for realizability or accuracy in the diffusion limit that ruin the efficiency gains expected from the IMEX scheme. Because of this, we sacrifice overall high-order accuracy, and aim for IMEX schemes that are high-order accurate in the streaming limit, diffusion accurate, and convex-invariant. Combining these requirements we seek GSA IMEX schemes of type ARS with coefficients satisfying the following constraints [32, 29]:

(i) Consistency of the implicit coefficients:

$$\sum_{i=1}^{s} w_i = 1. \tag{30}$$

(ii) High-order accuracy in the streaming limit. For second-order accuracy:

$$\sum_{i=1}^{s} \tilde{w}_i = 1 \quad \text{and} \quad \sum_{i=1}^{s} \tilde{w}_i\, \tilde{c}_i = \frac{1}{2}. \tag{31}$$

For third-order accuracy:

$$\sum_{i=1}^{s} \tilde{w}_i = 1, \quad \sum_{i=1}^{s} \tilde{w}_i\, \tilde{c}_i = \frac{1}{2}, \quad \sum_{i=1}^{s} \tilde{w}_i\, \tilde{c}_i^{\,2} = \frac{1}{3} \quad \text{and} \quad \sum_{i=1}^{s} \tilde{w}_i\, \tilde{a_{ij}} \tilde{c}_j = \frac{1}{6}. \tag{32}$$

(iii) Diffusion accuracy:

$$c_i = \tilde{c}_i, \quad i = 1, \ldots, s. \tag{33}$$

(iv) Convex-invariance:

$$\begin{aligned} &a_{ii} > 0, \quad c_{i0}, \tilde{c}_{i0} \geq 0, \quad \text{for} \quad i = 2, \ldots, s, \\ &\text{and} \quad c_{ij}, \tilde{c}_{ij} \geq 0, \quad \text{for} \quad i = 3, \ldots, s, \quad \text{and} \quad j = 2, \ldots, i-1, \end{aligned} \tag{34}$$

with $\sum_{j=0}^{i-1} c_{ij} = 1$, for $i = 1, \ldots, s$, and $c_{\text{SCH}} := \min_{\substack{i=2,\ldots,s \\ j=0,2,\ldots,i-1}} \frac{1}{\tilde{c}_{ij}} > 0$.

(Note that the greater the c_{SCH}, the larger the time step can be. And $c_{\text{SCH}} \leq 1$.)

(v) Having less than five stages ($s \leq 4$).

(vi) Are globally stiffly accurate: $a_{si} = w_i$ and $\tilde{a}_{si} = \tilde{w}_i$, $\quad i = 1, \ldots, s$.

Fortunately, these IMEX schemes are easy to find. (The constraint in (v) is introduced from efficiency considerations to limit the number of implicit solves.) We call the IMEX schemes satisfying the above conditions PD-ARS (see also Definition 3 in [29]), and we provide two optimal PD-ARS schemes below: PD-ARS2 and PD-ARS3, each limiting to the optimal second-order and third-order SSPRK schemes from [53], respectively.

4.4.1. PD-ARS2 The optimal 3-stage PD-ARS, PD-ARS2, in the standard double Butcher tableau form, with explicit tableau ($\tilde{A}$) on the left and implicit tableau (A) on the right, is given by

$$\begin{array}{c|ccc} 0 & 0 & 0 & 0 \\ 1 & 1 & 0 & 0 \\ 1 & 1/2 & 1/2 & 0 \\ \hline & 1/2 & 1/2 & 0 \end{array} \qquad \begin{array}{c|ccc} 0 & 0 & 0 & 0 \\ 1 & 0 & 1 & 0 \\ 1 & 0 & 1/2 & 1/2 \\ \hline & 0 & 1/2 & 1/2 \end{array} \tag{35}$$

Note its explicit tableau is SSPRK2. For this scheme, only two implicit solves are needed per time step and $c_{\text{SCH}} = 1$, which implies that the time step restriction for preserving moment realizability is only due to the explicit part.

IOP Conf. Series: Journal of Physics: Conf. Series **1225** (2019) 012013 doi:10.1088/1742-6596/1225/1/012013

4.4.2. PD-ARS3 The optimal 4-stage PD-ARS, PD-ARS3, is given in its standard double Butcher tableau form (explicit tableau on the left and implicit tableau on the right) by

$$\begin{array}{c|ccc} & & & \\ 1 & 1 & & \\ 1/2 & 1/4 & 1/4 & \\ 1 & 1/6 & 1/6 & 2/3 \\ \hline & 1/6 & 1/6 & 2/3 \end{array} \qquad \begin{array}{c|cccc} 0 & 0 & 0 & 0 & \\ 1 & 0 & 1 & 0 & \\ 1/2 & 0 & 1/4 & 1/4 & \\ 1 & 0 & 1/6 & 1/6 & 2/3 \\ \hline & 0 & 1/6 & 1/6 & 2/3 \end{array} \tag{36}$$

Its explicit tableau is SSPRK3. This scheme requires three implicit solves per time step, and $c_{\text{SCH}} = 1$. Since PD-ARS3 is not more accurate than PD-ARS2 in collision-dominated regions (see our results in Section 5), it may not offer any practical advantage over PD-ARS2.

5. Numerical Tests

In this section we present numerical results obtained with the PD-ARS schemes in Section 4. The tests in Section 5.1 are designed to compare the accuracy of the schemes in streaming, absorption, and scattering-dominated regimes in one spatial dimension. The test in Section 5.2 demonstrates the convex-invariance of PD-ARS schemes. All the tests in this section were computed with third-order accurate spatial discretization (polynomials of degree $k = 2$) and time step $\Delta t = 0.1 \times \Delta x$, using the DG scheme from [29].

5.1. Accuracy Tests

To compare the accuracy of the IMEX schemes, we applied our PD-ARS schemes and the SSP2332 scheme from Pareschi & Russo [47] to problems with known smooth solutions in streaming, absorption (damping), and scattering-dominated (diffusing) regimes in one spatial dimension. All the tests in this subsection were computed with the maximum entropy closure in the low-occupancy limit (i.e., the Minerbo closure). In the streaming test, the second- and third-order accurate explicit strong-stability-preserving Runge-Kutta methods from [30] (SSPRK2 and SSPRK3, respectively) are also included. To compare the numerical results to analytic solutions, the averaged absolute error or the averaged relative error are computed in the L^1-error norm. We compute the absolute error for the streaming and diffusion tests and the relative error for the damping test. They are averaged over the cell with an equal-weight quadrature for the cell integrals. To examine the convergence, we let the number of elements (N) vary from 8 to 128.

5.1.1. Sine Wave Streaming The sine wave streaming test is designed to test accuracy in the free-steaming regime; i.e. $\sigma_{\text{A}} = \sigma_{\text{S}} = 0$. A periodic domain of unit length is used and the initial condition is $\mathcal{J}_0 = \boldsymbol{\mathcal{H}}_0 = 0.5 + 0.49 \times \sin\big(2\pi\, x\big)$. We evolve the test until the sine wave has completed 10 crossings of the domain. Figure 2 plots the absolute error for the number density versus the number of elements N. We see the errors obtained with SSPRK3 and PD-ARS3 are smallest and decrease as N^{-3}, as expected for a scheme combining third-order accurate time stepping with third-order accurate spatial discretization. For all the other schemes, using second-order accurate explicit time stepping, the error decreases as N^{-2}. Among the second-order schemes, SSP2332 has the smallest error. In the streaming limit, the PD-ARS schemes reduce to SSPRK schemes — PD-ARS2 to SSPRK2 and PD-ARS3 to SSPRK3, respectively. Therefore, the absolute errors of PD-ARS schemes and SSPRK schemes are indistinguishable.

5.1.2. Sine Wave Damping The next test, adapted from [54], is designed for absorption-dominated regimes, with $\sigma_{\text{S}} = 0$ and $f_0 = 0$, which results in exponential damping of the wave amplitude. A periodic domain $D = \{x : x \in [0,1]\}$ and initial condition $\mathcal{J}_0 = \boldsymbol{\mathcal{H}}_0 = 0.5 + 0.49 \times \sin\big(2\pi\, x\big)$ are used. The amplitude of the analytical solution decreases as $e^{-\sigma_{\text{A}} t}$.

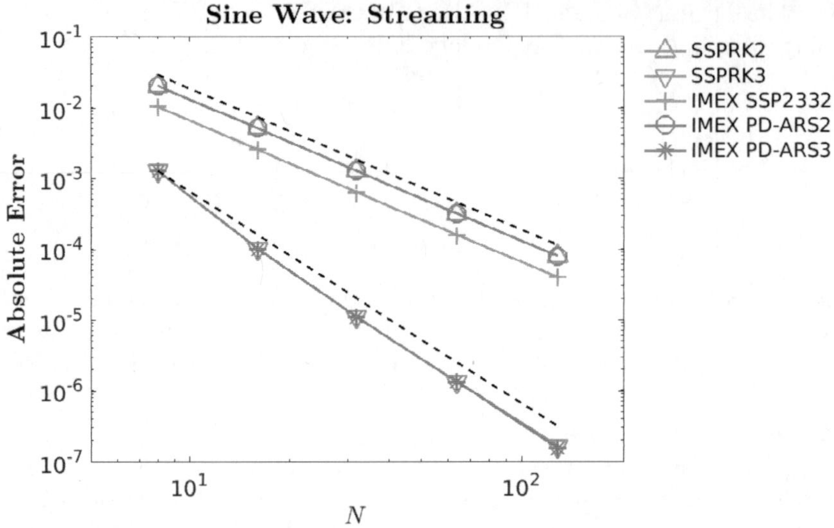

Figure 2. Absolute error versus number of elements, N, for the streaming sine wave test. Results employing various time stepping schemes are compared: SSPRK2 (cyan triangles pointing up), SSPRK3 (cyan triangles pointing down), SSP2332 (green crosses), PD-ARS2 (light red circles) and PD-ARS3 (light red asterisks). Black dashed reference lines are proportional to N^{-2} (top), and N^{-3} (bottom), respectively.

For $\sigma_{\rm A} = 0.1$, 1 and 10 we evolve the test until the initial condition has been damped by a factor e^{-10}. Figure 3 shows convergence results of the test in the relative error. Results for $\sigma_{\rm A} = 0.1$, 1, and 10 are plotted with red, green, and blue lines, respectively. SSP2332 is the most accurate among these schemes for $\sigma_{\rm A} = 1$ and 10. For $\sigma_{\rm A} = 0.1$, PD-ARS2 is the most accurate scheme for $N = 8$ and $N = 16$. We have seen the same behavior for the scheme proposed by McClarren et al. [55] (PC2 in [29]). Since these are special cases for $N = 8$ and $N = 16$, we do not recommend PD-ARS2 instead of SSP2332 for the damping problem. Only SSP2332, a second-order accurate scheme, displays a second-order convergence rate. The PD-ARS schemes are first-order accurate.

5.1.3. Sine Wave Diffusion The final test with known smooth solutions, adopted from [56], is the sine wave diffusion test; i.e. $\sigma_{\rm A} = 0$ and $f_0 = 0$. A periodic domain $D = \{x : x \in [-3, 3]\}$ with initial conditions $\mathcal{J}_0 = 0.5 + 0.49 \times \sin\left(\frac{\pi x}{3}\right)$ and $\mathcal{H}_0 = -\frac{1}{3\sigma_{\rm S}}\frac{\partial \mathcal{J}_0}{\partial x}$ are used. The reference diffusion solution is given by $\mathcal{J} = \mathcal{J}_0 \times \exp\left(-\frac{\pi^2 t}{27\sigma_{\rm S}}\right)$ and $\mathcal{H} = (3\,\sigma_{\rm S})^{-1}\partial_x\mathcal{J}$. We evolve with $\sigma_{\rm S} = 10^2$, 10^3, and 10^4, and adjust the end time so that $t_{\rm end}/\sigma_{\rm S} = 1$, at which time the amplitude of the sine wave has been reduced by a factor $e^{-\pi^2/27} \approx 0.694$ for all values of $\sigma_{\rm S}$. Figure 4 shows the absolute error, obtained using different values of $\sigma_{\rm S}$, for various IMEX schemes at $t = t_{\rm end}$, versus N. Results for $\sigma_{\rm S} = 10^2$, 10^3, and 10^4 are plotted with red, green, and blue lines, respectively. SSP2332 and PD-ARS schemes display third-order accuracy for the number density, $\mathcal{J}$, and second-order accuracy for $\mathcal{H}_x$, and their errors are difficult to distinguish. For $\sigma = 10^2$, the errors in the number density $\mathcal{J}$ do not drop below 10^{-6} because of differences between the two-moment model and the diffusion equation used to obtain the analytic solution. For larger values of the scattering opacity, $\sigma = 10^3$ or 10^4, the two-moment model agrees better with the diffusion model, and we observe convergence over the entire range of N. PD-ARS2 behaves as well as SSP2332 in the diffusion region but requires 33% less implicit solves per time

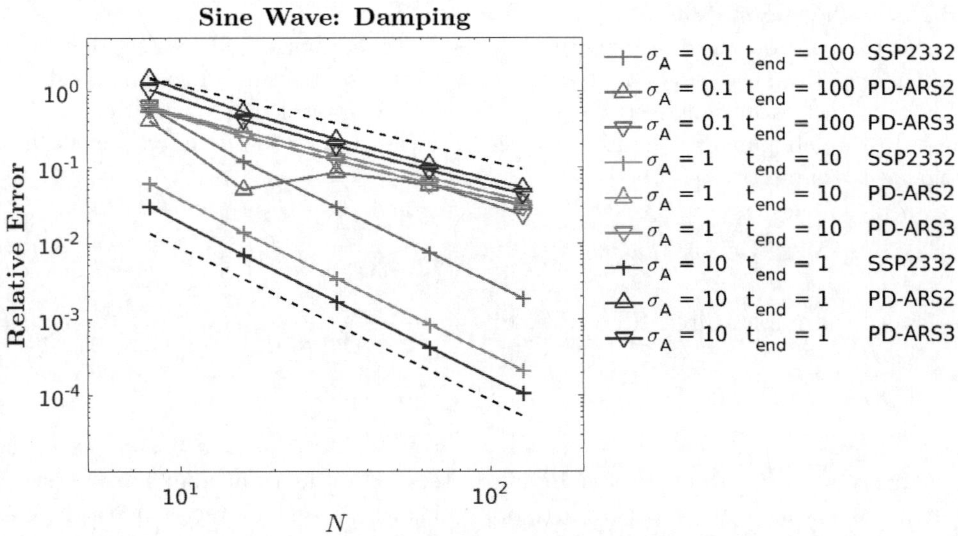

Figure 3. Relative error versus number of elements, N, for the damping sine wave test. Results for different values of the absorption opacity $\sigma_{\rm A}$, employing various IMEX time stepping schemes, are compared. Errors for $\sigma_{\rm A} = 0.1$, 1, and 10 are plotted with red, green, and blue lines, respectively. The IMEX schemes employed are SSP2332 (+), PD-ARS2 (triangles pointing up) and PD-ARS3 (triangles pointing down). Black dashed reference lines are proportional to N^{-1} (top) and N^{-2} (bottom), respectively.

step.

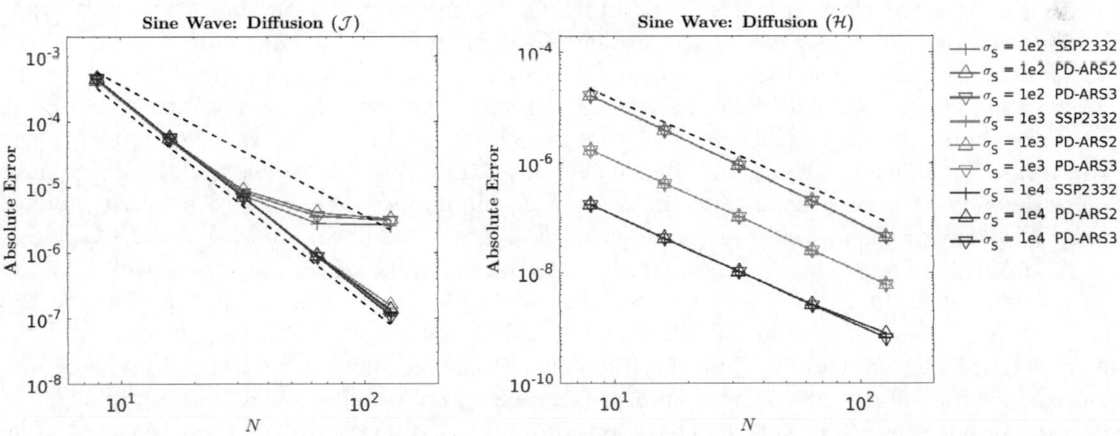

Figure 4. Absolute error for the number density $\mathcal{J}$ (left) and the number flux $\mathcal{H}_x$ (right) versus number of elements for the sine wave diffusion test. Results with different values of the scattering opacity, $\sigma_{\rm S}$, employing different IMEX schemes, are compared. Errors with $\sigma_{\rm S} = 10^2$, 10^3, and 10^4 are plotted with red, green, and blue lines, respectively. The IMEX schemes employed are: SSP2332 (+), PD-ARS2 (triangles pointing up), and PD-ARS3 (triangles pointing down). Black dashed lines in the left plot are reference lines proportional to N^{-2} (top) and N^{-3} (bottom), respectively. Black dashed line in the right plot is a reference line proportional to N^{-2} .

5.2. Neutrino Stationary State Test

In this section we consider a more "realistic" test: two-dimensional multigroup neutrino transport with emission, absorption, and isoenergetic scattering through a stationary background. This test is designed to test the realizability-preserving properties of the PD-ARS schemes. In the left panel in Figure 5 we plot the thermal state of the background, which mimics a collapsed stellar core:

$$\text{Mass Density:} \quad \rho = 4 \times 10^{14} \times \frac{7.5}{(7.5 + (r/5 \text{ km})^4)} \text{ g cm}^{-3}, \tag{37a}$$

$$\text{Temperature:} \quad T = 1.5 \times 10^{11} \times \frac{1}{1 + (r/50 \text{ km})^2} \text{ K}, \tag{37b}$$

$$\text{Electron Fraction:} \quad Y_e = 0.25 \times \left(1 + \frac{1}{1 + (r/50 \text{ km})^{-12}}\right), \tag{37c}$$

where the radius $r = \sqrt{(x^1)^2 + (x^2)^2}$ is in kilometers. In the right panel in Figure 5 we plot the neutrino opacities, computed by interpolation in a pre-calculated opacity table based on [57]. This test is computed using Cartesian coordinates on a two-dimensional domain $D = \{\boldsymbol{x} \in \mathbb{R}^2 : x^1 \in [0, 200] \text{ km}, x^2 \in [0, 200] \text{ km}\}$, using a grid of 128 elements in each direction, 10 energy groups covering 0-300 MeV, reflecting inner boundaries, and outflow outer boundaries. Because we use Cartesian coordinates in two spatial dimensions, this problem has cylindrical geometry, and results in an artificial stratification of the radiation quantities. We initialize the neutrino number density to $\mathcal{J} = 10^{-99}$, the number flux density to $\boldsymbol{\mathcal{H}} = 0$, and evolve until an approximate steady state is reached ($t = 5$ ms). The background is kept fixed during the entire run. For this test, we employed both CB and Minerbo closures. We attempted to run this test with our PD-ARS schemes, SSP2332 from [47], IMEXRKCB2 proposed by Cavaglieri & Bewley [58], and the IMEX PC2 scheme proposed by McClarren et al. [55]. Only the PD-ARS schemes produce realizable moments and are able to evolve to a steady state with either CB or Minerbo closure. SSP2332 and IMEX PC2 failed after a few time steps with either CB or Minerbo closure because of the development of unrealizable moments. Even though IMEXRKCB2 with Minerbo closure can run and reach a steady state in this test, its results are different from that of PD-ARS schemes with CB closure, and there is no guarantee of stability.

Results obtained with the IMEX PD-ARS schemes are plotted in Figure 6 for various times: $t = 0.01$ ms (top panels), 0.35 ms (middle panels), and 5.0 ms (bottom panels). In the left column we plot the solution in the $|\boldsymbol{\mathcal{H}}|\mathcal{J}$-plane. In the middle column we show scatter plots of the number density $\mathcal{J}$ versus radius for select neutrino energies: 5 MeV (red lines), 16 MeV (magenta lines), and 93 MeV (blue lines). In the right column we plot the flux factor $|\boldsymbol{\mathcal{H}}|/\mathcal{J}$ versus radius for the same neutrino energies as in the middle column. In the left panels, each solution point in the domain is marked as a red dot in the $|\boldsymbol{\mathcal{H}}|\mathcal{J}$-plane, and the realizable domain is shown as the light blue region. The figures show that all the states in the simulation with the PD-ARS schemes are realizable. In the middle and right column, we can see how the neutrinos are generated near the core, stream out, and eventually reach an equilibrium distribution over the phase space. The oscillations in the flux factor seen in the right columns are associated with steep gradients in the radiation field as the initial transient propagates through the computational domain. Note that we do not apply any limiters to prevent oscillations in the numerical solution, and these will likely go away when we implement slope limiters; e.g., as described in [59]. The fact that we can still evolve the solution to a steady state speaks to the robustness of the scheme.

6. Summary and Outlook

We have developed IMEX schemes suitable for a two-moment model of neutrino transport that obey Fermi-Dirac statistics. The scheme employs algebraic closure based on Fermi-Dirac

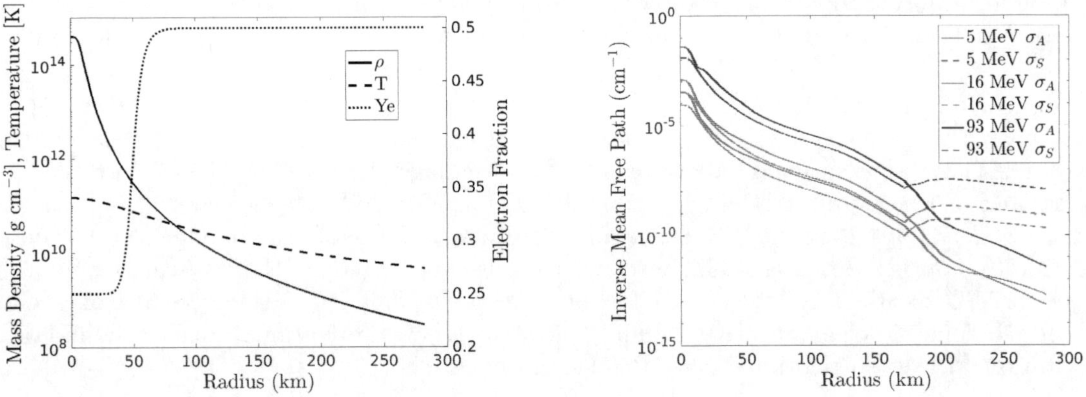

Figure 5. Left panel: thermal state of the background versus radius in the neutrino stationary state test; mass density (solid line), temperature (dashed line), and electron fraction (dotted line). Right panel: corresponding opacities for select neutrino energies; absorptivity ($\sigma_{\rm A}$) and scattering opacity ($\sigma_{\rm S}$).

statistics, high-order discontinuous Galerkin methods for spatial discretization, and convex-invariant time integration to maintain realizability of the moments. Since the realizable domain is convex and its convexity can be inherited by a convex combination, a scheme having convex combinations as its stages can preserve the realizable domain. This encouraged us to construct realizability-preserving time integrators, realizability-preserving IMEX schemes, and a method with a realizability-preserving IMEX time integrator and high-order DG method.

In the applications that motivate this work, the neutrino distribution function can vary from 0 to 1. Hence, we have considered algebraic closures based on Fermi-Dirac statistics for both low and high occupancy. Among the seven algebraic closures we considered – Kershaw [35], Wilson [40, 41], Levermore [27], Minerbo [23], Janka 1 [43], Janka 2 [42], and Cernohorsky & Bludman [44] – only the Cernohorsky & Bludman closure obeys Fermi-Dirac statistics for all occupancies. As a result, we employed the Cernohorsky & Bludman closure for the neutrino stationary state test in Section 5.2. We also ran our code with Minerbo closure, IMEX PC2, IMEX SSP2332 and IMEXRKCB2 schemes, and the results show that only PD-ARS schemes have stability. In addition, closures have impact on the simulation result. As we observed, both using PD-ARS scheme, there are $\sim 30\%$ difference in the neutrino number densities (and relaxation time) between the results obtained with Minerbo closure and that with CB closure. Even though RKCB2 with Minerbo closure luckily survived our test, the results it gave were compromised: they were closer to the results of the PD-ARS scheme with Minerbo closure than to the results of the PD-ARS scheme with CB closure. In what way and to what degree the results are in fact compromised either by the closure or by a particular correction step for unrealizable moments are difficult to determine fully and is left for further study.

Two PD-ARS schemes are proposed. The one with SSPRK2 has second-order accuracy while the other with SSPRK3 has third-order accuracy, and both have the strong-stability preserving property in the streaming limit. Their accuracy was demonstrated on problems with known smooth solutions in streaming, absorption, and scattering-dominated regimes. The neutrino transport test with emission, absorption, and isoenergetic scattering through a stationary background, was designed to test the convex-invariance of our PD-ARS schemes. The neutrino stationary state test shows that a method combining an algebraic closure based on Fermi-Dirac statistics and convex-invariant time integration is promising for robust CCSN simulation.

In this work, we adopted Cartesian coordinates, a linear collision term, and a fixed material background. More realistic problems of scientific interest, such as with energy-exchanging scattering and relativistic effects, are left for future research.

7. Acknowledgment

This research is sponsored, in part, by the Laboratory Directed Research and Development Program of Oak Ridge National Laboratory (ORNL), managed by UT-Battelle, LLC for the U.S. Department of Energy under Contract No. De-AC05-00OR22725. This material is based, in part, on work supported by the U.S. Department of Energy, Office of Science, Office of Advanced Scientific Computing Research. This research was also supported by the Exascale Computing Project (17-SC-20-SC), a collaborative effort of the U.S. Department of Energy Office of Science and the National Nuclear Security Administration, and by the National Science Foundation Gravitational Physics Program (NSF-GP 1806692).

References

[1] Janka H T, Langanke K, Marek A, Martínez-Pinedo G and Müller B 2007 *Physics Reports* **442** 38–74

[2] Mezzacappa A 2005 *Annu. Rev. Nucl. Part. Sci.* **55** 467–515

[3] Janka H T 2012 *Annual Review of Nuclear and Particle Science* **62** 407–451

[4] Burrows A 2013 *Reviews of Modern Physics* **85** 245

[5] Müller B 2016 *Publications of the Astronomical Society of Australia* **33**

[6] Mezzacappa A and Bruenn S W 1993 *The Astrophysical Journal* **405** 637–668

[7] Mezzacappa A and Bruenn S W 1993 *The Astrophysical Journal* **405** 669–684

[8] Mezzacappa A and Bruenn S W 1993 *The Astrophysical Journal* **410** 740–760

[9] Mezzacappa A, Liebendörfer M, Messer O B, Hix W R, Thielemann F K and Bruenn S W 2001 *Physical Review Letters* **86** 1935

[10] Liebendoerfer M, Mezzacappa A, Thielemann F K, Messer O B, Hix W R and Bruenn S W 2001 *Physical Review D* **63** 103004

[11] Liebendörfer M, Messer O B, Mezzacappa A, Bruenn S W, Cardall C Y and Thielemann F K 2004 *The Astrophysical Journal Supplement Series* **150** 263

[12] Livne E, Burrows A, Walder R, Lichtenstadt I and Thompson T A 2004 *The Astrophysical Journal* **609** 277

[13] Liebendoerfer M, Rampp M, Janka H T and Mezzacappa A 2005 *The Astrophysical Journal* **620** 840

[14] Ott C D, Burrows A, Dessart L and Livne E 2008 *The Astrophysical Journal* **685** 1069

[15] Sumiyoshi K and Yamada S 2012 *The Astrophysical Journal Supplement Series* **199** 17

[16] Nagakura H, Sumiyoshi K and Yamada S 2014 *The Astrophysical Journal Supplement Series* **214** 16

[17] Sumiyoshi K, Takiwaki T, Matsufuru H and Yamada S 2014 *The Astrophysical Journal Supplement Series* **216** 5

[18] Nagakura H, Iwakami W, Furusawa S, Okawa H, Harada A, Sumiyoshi K, Yamada S, Matsufuru H and Imakura A 2018 *The Astrophysical Journal* **854** 136

[19] Kuroda T, Takiwaki T and Kotake K 2016 *The Astrophysical Journal Supplement Series* **222** 20

[20] Roberts L F, Ott C D, Haas R, O'Connor E P, Diener P and Schnetter E 2016 *The Astrophysical Journal* **831** 98

[21] Just O, Bollig R, Janka H T, Obergaulinger M, Glas R and Nagataki S 2018 *arXiv preprint arXiv:1805.03953*

[22] Vartanyan D, Burrows A, Radice D, Skinner M A and Dolence J 2018 *Monthly Notices of the Royal Astronomical Society* (*Preprint* arXiv:1809.05106)

[23] Minerbo G N 1978 *JQSRT* **20** 541–545

[24] O'Connor E P and Couch S M 2018 *The Astrophysical Journal* **865** 81

[25] Pan K C, Mattes C, O'Connor E P, Couch S M, Perego A and Arcones A 2018 *arXiv preprint arXiv:1806.10030*

[26] Glas R, Just O, Janka H and Obergaulinger M 2018 *arXiv preprint arXiv:1809.10146*

[27] Levermore C D 1984 *JQSRT* **31** 149–160

[28] Cabezón R M, Pan K C, Liebendörfer M, Kuroda T, Ebinger K, Heinimann O, Thielemann F K and Perego A 2018 *arXiv preprint arXiv:1806.09184*

[29] Chu R, Endeve E, Hauck C D and Mezzacappa A 2018 *JCP submitted* (*Preprint* arXiv:1809.06949)

[30] Gottlieb E, Shu C W and Tadmor E 2001 *SIAM Review* **43** 89–112

[31] Huang J and Shu C W 2017 *Mathematical Models and Methods in Applied Sciences* **27** 549–579

[32] Hu J, Shu R and Zhang X 2018 *SIAM Journal on Numerical Analysis* **56** 942–973

[33] Endeve E, Buffaloe J, Dunham S J, Roberts N, Andrew K, Barker B, Pochik D, Pulsinelli J and Mezzacappa A 2018 *Astronum Conference submitted*

[34] Larecki W and Banach Z 2011 *JQSRT* **112** 2486–2506

[35] Kershaw D 1976 Flux limiting nature's own way — a new method for numerical solution of the transport equation Tech. Rep. UCRL-78378 Lawrence Livermore Laboratory

[36] Shohat J and Tamarkin J 1943 *The Problem of Moments* Mathematical Surveys and Monographs (American Mathematical Society)

[37] Shibata M, Kiuchi K, Sekiguchi Y i and Suwa Y 2011 *Progress of Theoretical Physics* **125** 1255–1287

[38] Cardall C Y, Endeve E and Mezzacappa A 2013 *Physical Review D* **88** 023011

[39] Murchikova E M, Abdikamalov E and Urbatsch T 2017 *MNRAS* **469** 1725–1737

[40] Wilson J R, Couch R, Cochran S, Le Blanc J and Barkat Z 1975 *Annals of the New York Academy of Sciences* **262** 54–64

[41] LeBlanc J M and Wilson J R 1970 *The Astrophysical Journal* **161** 541

[42] Janka H T 1992 *Astronomy and Astrophysics* **256** 452–458

[43] Janka H T 1991 *Technische Universität München* **Ph.D. Thesis** MPA–587

[44] Cernohorsky J and Bludman S A 1994 *Astrophysical Journal* **433** 450 – 455

[45] Foucart F, O'Connor E, Roberts L, Duez M D, Haas R, Kidder L E, Ott C D, Pfeiffer H P, Scheel M A and Szilagyi B 2015 *Physical Review D* **91** 124021

[46] Fujibayashi S, Sekiguchi Y, Kiuchi K and Shibata M 2017 *The Astrophysical Journal* **846** 114

[47] Pareschi L and Russo G 2005 *Journal of Scientific Computing* **25** 129–155

[48] Hairer E 1981 *Numerische Mathematik* **36** 431–445

[49] Dimarco G and Pareschi L 2013 *SIAM Journal on Numerical Analysis* **51** 1064–1087

[50] Ascher U, Ruuth S and Spiteri R 1997 *Applied Numerical Mathematics* **25** 151–167

[51] Jin S and Levermore C 1996 *Journal of Computational Physics* **126** 942–973

[52] Chertock A, Cui S, Kurganov A and Wu T 2015 *SIAM J. Numer. Anal.* **53** 2008–2029

[53] Shu C W and S O 1988 *Journal of Computational Physics* **77** 439–471

[54] Skinner M A and Ostriker E C 2013 *Astrophysical Journal Supplement Series* **206** 21

[55] McClarren R, Evans T, Lowrie R and Densmore J 2008 *Journal of Computational Physics* **227** 7561–7586 ISSN 0021-9991

[56] Radice D, Abdikamalov E, Rezzolla L and Ott C D 2013 *Journal of Computational Physics* **242** 648–669

[57] Bruenn S W 1985 *APJS* **58** 771–841

[58] Cavaglieri D and Bewley T 2015 *Journal of Computational Physics* **286** 172 – 193 ISSN 0021-9991

[59] Cockburn B and Shu C W 1998 *Journal of Computational Physics* **141** 199–224

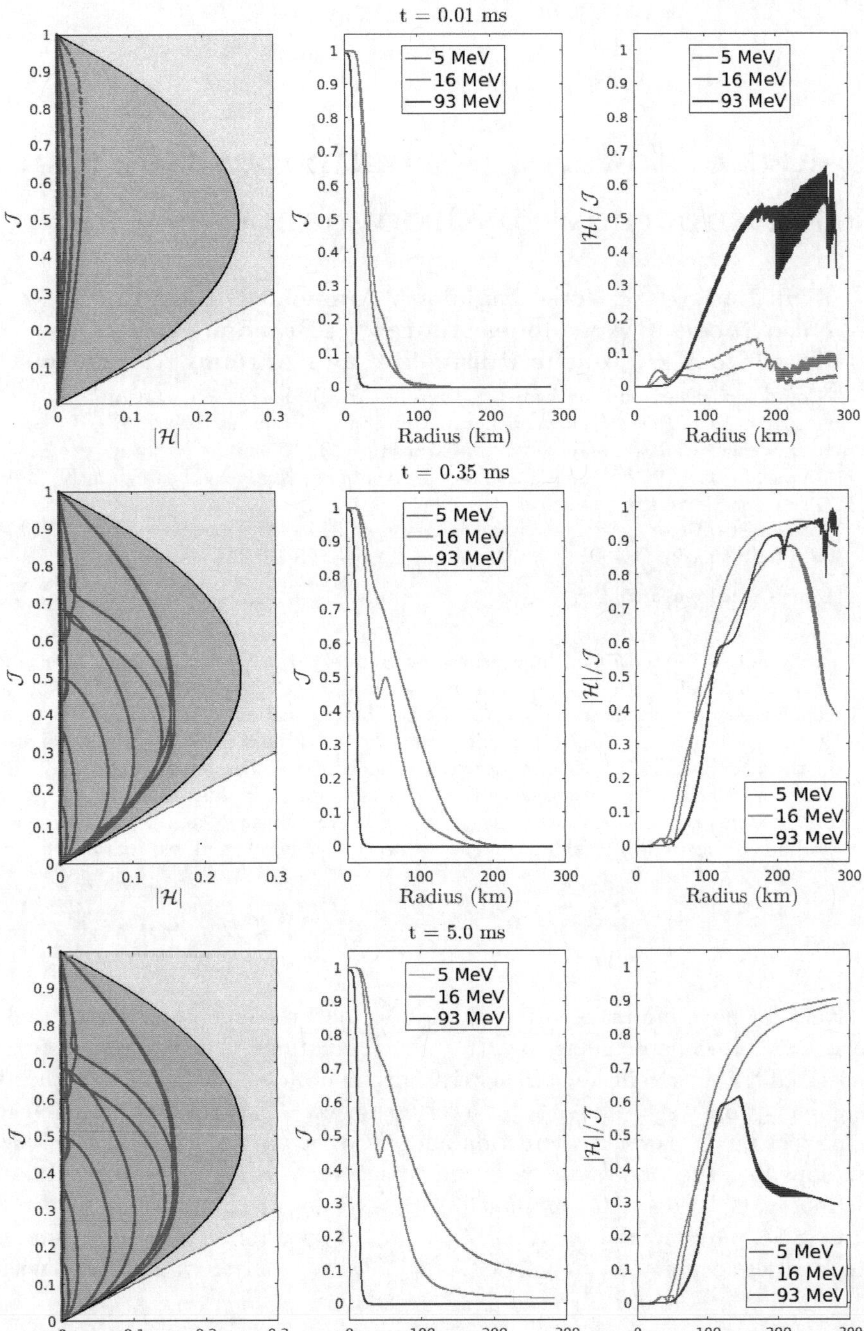

Figure 6. Results from the neutrino stationary state test: moments relative to the realizable domain (left column, the light blue domain is for $f \in [0,1]$, with the black-solid-line as its boundary, while the light red domain is for $f \geq 0$, with the thin-red-solid-line as its boundary), the number density $\mathcal{J}$ versus radius (middle column), and the flux factor $|\mathcal{H}|/\mathcal{J}$ versus radius (right column), at $t = 0.01$ ms, 0.35 ms, and 5.0 ms. For the plots in the left column, each $\boldsymbol{\mathcal{M}} = (\mathcal{J}, \boldsymbol{\mathcal{H}})^T$ state is marked by a red dot, which are all inside the light blue region (the realizable domain for fermions). The results of PD-ARS2 and PD-ARS3 are indistinguishable in these plots.

IOP Conf. Series: Journal of Physics: Conf. Series **1225** (2019) 012014 doi:10.1088/1742-6596/1225/1/012014

thornado-hydro: towards discontinuous Galerkin methods for supernova hydrodynamics*

Eirik Endeve[1,2,6], Jesse Buffaloe[2], Samuel J Dunham[3], Nick Roberts[2], Kristopher Andrew[4,1], Brandon Barker[2], David Pochik[2], Juliana Pulsinelli[5,7] and Anthony Mezzacappa[2,6]

[1]Computer Science and Mathematics Division, Oak Ridge National Laboratory, TN 37831
[2]Department of Physics and Astronomy, University of Tennessee Knoxville, TN 37996
[3]Department of Astronomy, Vanderbilt University, TN 37212
[4]Department of Physics and Astronomy, University of Kentucky, Lexington, KY 40506
[5]Princeton University, NJ 08544
[6]Joint Institute for Computational Sciences, Oak Ridge National Laboratory, TN 37831
[7]Physics Division, Oak Ridge National Laboratory, TN 37831

E-mail: endevee@ornl.gov

Abstract. The **t**oolkit for **h**igh-**or**der **n**eutrino-r**ad**iation hydr**o**dynamics (thornado) is being developed for simulations of core-collapse supernovae (CCSNe) and related problems. Current capabilities in thornado include solvers for the Euler equations — in non-relativistic and special relativistic limits — and the two-moment model of neutrino transport. The spatial discretization in thornado is based on the discontinuous Galerkin (DG) method, which is receiving increased attention from the computational astrophysics community. In this paper, we provide an overview of the numerical methods for the Euler equations in thornado, and present some encouraging preliminary numerical results from a set of basic tests in one and two spatial dimensions.

1. Introduction

Along with solvers for neutrino transport and gravity, the hydrodynamics solver constitutes a major component of core-collapse supernova (CCSN) simulations. The dynamics of the bounce shock and associated fluid instabilities, which drive turbulent flows and help shape the explosion, must be faithfully captured (see, e.g., [23], for a recent review). Essentially, the stellar interior is modeled as a perfect fluid (thermal conduction and viscosity are not explicitly included) using the Euler equations, but the models are extended to accommodate a nuclear equation of state.

The discontinuous Galerkin (DG) method [7, 14] appears as an appealing choice to model fluid flows in CCSNe. DG methods combine elements of spectral and finite volume methods, and achieve high-order accuracy on a compact stencil. Data is only communicated with nearest

* This manuscript has been authored by UT-Battelle, LLC under Contract No. DE-AC05-00OR22725 with the U.S. Department of Energy. The United States Government retains and the publisher, by accepting the article for publication, acknowledges that the United States Government retains a non-exclusive, paid-up, irrevocable, world-wide license to publish or reproduce the published form of this manuscript, or allow others to do so, for United States Government purposes. The Department of Energy will provide public access to these results of federally sponsored research in accordance with the DOE Public Access Plan (http://energy.gov/downloads/doe-public-access-plan).

IOP Conf. Series: Journal of Physics: Conf. Series **1225** (2019) 012014 doi:10.1088/1742-6596/1225/1/012014

neighbors, regardless of the formal order of accuracy, which leads to a high computation to communication ratio, and favorable parallel scalability on heterogeneous architectures has been demonstrated [17]. They can be used in combination with *hp*-adaptivity [26], where, in addition to grid refinement with AMR, the local polynomial degree can be chosen differently, and independently, in different cells. DG methods can easily be applied to problems involving curvilinear coordinates, which is beneficial in numerical relativity [36]. Currently, most hydrodynamics solvers in CCSN codes are based on finite volume methods. Exploring the utility and performance of DG methods to model CCSNe seems like a worthy exercise by itself.

The **t**oolkit for **h**igh-**or**der **n**eutrino-**rad**iation hydr**o**dynamics (`thornado`) is being developed to simulate neutrino-radiation hydrodynamics in CCSNe and related applications in nuclear astrophysics. The spatial discretization of solvers for hyperbolic partial differential equations in `thornado` is based on the DG method, while we foresee using a combination of spectral and continuous finite element methods for solving elliptic equations; e.g., for Newtonian gravity or relativistic gravity employing the conformal flatness approximation [39]. With these approaches, we employ high-order discretization techniques based on a common mathematical framework in all the major model components. Whether the high-order approach will improve accuracy and efficiency of CCSN models remains to be demonstrated. However, in this paper, we provide an initial description and encouraging demonstration of the DG method implemented in `thornado` to solve the Euler equations. We focus on basic tests in one and two spatial dimensions and focus, to some extent, on the implementation of limiters in `thornado`. We note that others (e.g., [25, 28, 43, 10]) have also investigated the DG method in the context of astrophysical applications and reported encouraging results. The DG limiting techniques implemented in `thornado` are similar to those used by [28], but is supplemented with a troubled-cell indicator [12] to prevent excessive limiting (e.g., around smooth extrema), and differs from more recent approaches in the literature (e.g., [15, 9, 43, 35]), which combine the sub-cell resolution of the DG method with high-order finite-volume techniques to improve the performance of the DG limiting procedure. While the results with current limiters are encouraging, these latter methods appear as attractive options for future implementation in `thornado`.

2. The discontinuous Galerkin method

Excellent books and review articles on the DG method are available (see, e.g., [6, 7, 14, 33] and references therein), and we will not go into too much detail here. However, we review some key concepts to introduce notation, and emphasize specific choices for our implementation in `thornado`. To this end, we consider a system of conservation laws with sources of the form

$$\partial_t\big(\sqrt{\gamma}\,\boldsymbol{U}\big)+\sum_{i=1}^{d}\partial_i\Big(\sqrt{\gamma}\,\boldsymbol{F}^i(\boldsymbol{U})\Big)=\sqrt{\gamma}\,\boldsymbol{S}(\boldsymbol{U}), \tag{1}$$

where $\boldsymbol{U}$ is the evolved state vector, $\boldsymbol{F}^i$ are the fluxes, and $\boldsymbol{S}$ is the source vector. We use a formulation of the equations sufficiently general to accommodate curvilinear spatial coordinates encoded in the metric γ_{ij}, giving the squared line element $ds^2=\gamma_{ij}dx^idx^j$, whose determinant is γ. Henceforth, we will assume that the spatial metric is time-independent. To solve Eq. (1), the computational domain $D\subset\mathbb{R}^d$ is divided into a disjoint union $\mathcal{T}$ of open elements $\boldsymbol{K}$, so that $D=\cup_{\boldsymbol{K}\in\mathcal{T}}\boldsymbol{K}$. We require that each element is a box in the logical coordinates; i.e.,

$$\boldsymbol{K}=\{\,\boldsymbol{x}:x^i\in K^i:=(x^i_{\mathrm{L}},x^i_{\mathrm{H}}),\,i=1,\ldots,d\,\}, \tag{2}$$

with the surface elements denoted $\partial\boldsymbol{K}^i=\times_{j\neq i}K^j$. We let $V_{\boldsymbol{K}}$ denote the proper element volume

$$V_{\boldsymbol{K}}=\int_{\boldsymbol{K}}dV,\quad\text{where}\quad dV=\sqrt{\gamma}\prod_{i=1}^{d}dx^i. \tag{3}$$

We also define as a set $\boldsymbol{x} = \{\tilde{\boldsymbol{x}}^i, x^i\}$ and $\Delta x^i = x^i_{\mathrm{H}} - x^i_{\mathrm{L}}$.

We let the approximation space for the DG method, $\mathbb{V}^k$, be constructed from the tensor product of one-dimensional polynomials of maximal degree k. Note that functions in $\mathbb{V}^k$ can be discontinuous across element interfaces. The semi-discrete DG problem is to find $\boldsymbol{U}_h \in \mathbb{V}^k$, which approximates $\boldsymbol{U}$ in Eq. (1), such that for all $v \in \mathbb{V}^k$ and all $\boldsymbol{K} \in \mathcal{T}$

$$\partial_t \int_{\boldsymbol{K}} \boldsymbol{U}_h \, v \, dV + \sum_{i=1}^{d} \int_{\partial \boldsymbol{K}^i} \big(\sqrt{\gamma}\, \widehat{\boldsymbol{F}}^i(\boldsymbol{U}_h)\, v\big|_{x^i_{\mathrm{H}}} - \sqrt{\gamma}\, \widehat{\boldsymbol{F}}^i(\boldsymbol{U}_h)\, v\big|_{x^i_{\mathrm{L}}} \big)\, d\tilde{\boldsymbol{x}}^i$$
$$- \sum_{i=1}^{d} \int_{\boldsymbol{K}} \boldsymbol{F}^i(\boldsymbol{U}_h)\, \partial_i v \, dV = \int_{\boldsymbol{K}} \boldsymbol{S}(\boldsymbol{U}_h)\, v \, dV. \tag{4}$$

In Eq. (4), $\widehat{\boldsymbol{F}}^i(\boldsymbol{U}_h)$ is a numerical flux approximating the flux on the ith surface of $\boldsymbol{K}$. The numerical flux function $\boldsymbol{f}^i$ is evaluated using values from both sides of an element interface; i.e.,

$$\widehat{\boldsymbol{F}}^i(\boldsymbol{U}_h) = \boldsymbol{f}^i(\boldsymbol{U}_h(x^{i,-}, \tilde{\boldsymbol{x}}^i), \boldsymbol{U}_h(x^{i,+}, \tilde{\boldsymbol{x}}^i)), \tag{5}$$

where superscripts $-/+$, e.g. in the arguments of $\boldsymbol{U}_h$, indicate that the function is evaluated to the immediate left/right of x^i. We use the Harten-Lax-van Leer (HLL) flux [13] or the HLLC flux [37, 22] for all the numerical experiments presented in Section 3.

We provide further details on the DG method to arrive at the equations that are actually evolved in `thornado`. We start by introducing some notation, defining the polynomial expansion for $\boldsymbol{U}_h$ and the quadrature rules used to evaluate the integrals in Eq. (4). Then we provide explicit expressions for each of the terms in Eq. (4). For simplicity we consider one spatial dimension ($d = 1$) and drop indices denoting the spatial dimension. In each element K, we use a nodal representation of the conserved variables $\boldsymbol{U}$; i.e.,

$$\boldsymbol{U}(x,t) \approx \boldsymbol{U}_h(x,t) = \sum_{i=1}^{N} \boldsymbol{U}_i(t)\, \ell_i(x), \quad \text{where} \quad \ell_i(\eta) = \prod_{\substack{j=1 \\ j \neq i}}^{N} \frac{\eta - \eta_j}{\eta_i - \eta_j} \tag{6}$$

are Lagrange polynomials defined on the reference element $I = \{\eta : \eta \in (-0.5, 0.5)\}$, and constructed to interpolate the node set $S_N = \{\eta_i\}_{i=1}^{N} \subset I$. The spatial coordinate x and the reference coordinate η are related by the mapping $x(\eta) = x_{\mathrm{L}} + (0.5 + \eta)\,\Delta x$. Then, for any $\eta_j \in S_N$, $\ell_i(\eta_j) = \delta_{ij}$, so that $\boldsymbol{U}_h(x(\eta_j), t) = \boldsymbol{U}_j(t)$. We introduce numerical quadratures to evaluate the integrals in Eq. (4). First we define the M-point quadrature $Q_M : C^0(I) \to \mathbb{R}$ with abscissas $\hat{S}_M = \{\eta_q\}_{q=1}^{M}$ and weights $\{w_q\}_{q=1}^{M}$, normalized such that $\sum_{q=1}^{M} w_q = 1$. The M-point Legendre-Gauss quadrature, which we use, integrates polynomials of degree $\leq 2M - 1$ exactly. Then, if $P_h(x)$ is such a polynomial, we have

$$\frac{1}{\Delta x} \int_K P_h(x)\, dx = \int_I P_h(\eta)\, d\eta = Q_M[P_h] \equiv \sum_{q=1}^{M} w_q\, P_h(\eta_q). \tag{7}$$

Note that the interpolation points S_N and the quadrature points $\hat{S}_M$ generally do not coincide. However, for the sake of computational efficiency we let $M = N$ and $S_N = \hat{S}_N$, which is a spectral-type nodal collocation DG approximation [2]. Inserting Eq. (6) into Eq. (4), letting $v(x) = \ell_k(x)$, and using the quadratures defined above, we obtain

$$\partial_t \int_K \boldsymbol{U}_h \, v \, dV \approx w_k \sqrt{\gamma}_k \, \partial_t \boldsymbol{U}_k \, \Delta x \tag{8}$$

for the time derivative term. For a general metric, the integral in Eq. (8) is approximate since we use the Legendre-Gauss quadrature rule with the nodal points given by the expansion in Eq. (6). However, since $\ell_i(\eta_k) = \delta_{ik}$, this leads to a diagonal mass matrix, and simplifies the implementation. Similarly, we obtain

$$\int_K \boldsymbol{S}(\boldsymbol{U}_h)\, v\, dV \approx w_k \sqrt{\gamma}_k\, \boldsymbol{S}(\boldsymbol{U}_k)\, \Delta x. \tag{9}$$

for the source term. Finally, the volume term (last term on the left-hand side of Eq. (4)) becomes

$$\int_K \boldsymbol{F}(\boldsymbol{U}_h)\, \frac{\partial v}{\partial x}\, dV \approx \sum_{q=1}^{N} w_q\, \sqrt{\gamma}_q\, \boldsymbol{F}(\boldsymbol{U}_q)\, \frac{\partial \ell_k}{\partial \eta}(\eta_q). \tag{10}$$

Using Eqs. (8)-(10) in Eq. (4) results in the semi-discrete form

$$\frac{d\boldsymbol{U}_k}{dt} = -\frac{1}{w_k \sqrt{\gamma}_k \Delta x}\Big\{\Big[\sqrt{\gamma}\widehat{\boldsymbol{F}}\ell_k\big|_{x_{\mathrm{H}}} - \sqrt{\gamma}\widehat{\boldsymbol{F}}\ell_k\big|_{x_{\mathrm{L}}}\Big] - \sum_{q=1}^{N} w_q\, \sqrt{\gamma}_q\, \boldsymbol{F}(\boldsymbol{U}_q)\, \frac{\partial \ell_k}{\partial \eta}(\eta_q)\Big\} + \boldsymbol{S}(\boldsymbol{U}_k). \tag{11}$$

Eq. (11) comprises a system of ordinary differential equations (ODEs), which are integrated in time with an ODE solver. In Section 3 we use the optimal third-order strong stability-preserving Runge-Kutta (SSP-RK3) method from [30]. We assume that the source term $\boldsymbol{S}(\boldsymbol{U}_k)$ is non-stiff so that the equations can be integrated efficiently with explicit methods. In future applications of the DG method to model CCSNe we plan to use a combination of implicit and explicit methods (e.g., IMEX methods [1]) to stably integrate stiff sources due to neutrino-matter interactions.

Limiting of the polynomial representation $\boldsymbol{U}_h$, to reduce oscillations and prevent unphysical states, is a critical step in the DG algorithm. We use the TVD-type slope limiter based on limiting characteristic variables, discussed in [6]. To prevent excessive limiting (e.g., at smooth extrema), we use the troubled-cell indicator (TCI) of [12]:

$$I_{\boldsymbol{K}}(G) = \frac{\sum_j |G_{\boldsymbol{K}} - G_{\boldsymbol{K}}^{(j)}|}{\max_j |G_{\boldsymbol{K}^{(j)}}^{(j)}|}, \tag{12}$$

where $G \in \boldsymbol{G} \subseteq \boldsymbol{U}$. (Here, $\boldsymbol{G}$ consists of the first and last element of $\boldsymbol{U}$.) In Eq. (12), the sum in the numerator and the max in the denominator extend over the neighboring elements sharing a face with the target element $\boldsymbol{K}$. $G_{\boldsymbol{K}}$ is the cell average in $\boldsymbol{K}$, $G_{\boldsymbol{K}}^{(j)}$ is the cell average computed by extrapolating the polynomial representation from the neighbor element $\boldsymbol{K}^{(j)}$ into $\boldsymbol{K}$, and $G_{\boldsymbol{K}^{(j)}}^{(j)}$ is the cell average native to $\boldsymbol{K}^{(j)}$. An element is flagged for limiting if $I_{\boldsymbol{K}}(G) > C_{\mathrm{TCI}}$ for any $G \in \boldsymbol{G}$. To prevent negative mass density and pressure (and superluminal velocity for the relativistic Euler equations), we follow the approaches in [44, 40, 24].

To summarize the full RK-DG algorithm we let $\bar{\boldsymbol{U}}$ denote the global solution vector containing all the unknowns in D and $\bar{\boldsymbol{F}}$ denote the corresponding DG discretization from the right-hand side of Eq. (11), so that the global ODE system can be written as the $d_t\bar{\boldsymbol{U}} = \bar{\boldsymbol{F}}(\bar{\boldsymbol{U}})$. The TVD limiter and the 'positivity' limiter preventing unphysical states are denoted by $\Lambda^{\mathrm{TVD}}\{\}$ and $\Lambda^{\mathrm{Pos}}\{\}$, respectively (e.g., application of the TVD limiter is $\bar{\boldsymbol{U}} := \Lambda^{\mathrm{TVD}}\{\bar{\boldsymbol{U}}\}$). The general s-stage Runge-Kutta time stepping algorithm, which includes application of limiters, can then be written as [7]:

1. Set $\bar{\boldsymbol{U}}^{(0)} = \bar{\boldsymbol{U}}^n$,

2. For $i = 1, \dots, s$ compute:

$$\bar{\boldsymbol{U}}^{(i)} = \Lambda^{\text{Pos}}\Big\{\Lambda^{\text{TVD}}\Big\{\sum_{j=0}^{i-1}\Big(\alpha_{ij}\,\bar{\boldsymbol{U}}^{(j)} + \beta_{ij}\,\Delta t\,\bar{\boldsymbol{F}}(\bar{\boldsymbol{U}}^{(j)})\Big)\Big\}\Big\}, \tag{13}$$

3. Set $\bar{\boldsymbol{U}}^{n+1} = \bar{\boldsymbol{U}}^{(s)}$.

The coefficients α_{ij} and β_{ij} for SSP-RK3 are given in Table 2.1 in [7]. In Eq. (13), for each stage, the TVD limiter is applied first (for elements where $I_{\boldsymbol{K}} > C_{\text{TCI}}$), followed by the positivity limiter. The algorithm is subject to two time step restrictions, the stability condition Δt_{TVD} and the positivity condition Δt_{Pos}, and we take $\Delta t = \min_D(\Delta t_{\text{TVD}}, \Delta t_{\text{Pos}})$. Let $(h/|\lambda|) = \min_{i\in\{1,\dots,d\}}(\Delta x^i/|\lambda^i|)$, where $|\lambda^i|$ is the largest absolute eigenvalue of the flux Jacobian. Then, assuming Cartesian coordinates (e.g., [7, 44]):

$$\Delta t_{\text{TVD}} \leq \frac{1}{d}\frac{(h/|\lambda|)}{(2k+1)} \quad \text{and} \quad \Delta t_{\text{Pos}} \leq \frac{\hat{w}_1 (h/|\lambda|)}{d}, \tag{14}$$

where $\hat{w}_1$ is the first quadrature weight of the Gauss-Lobatto quadrature that integrates the cell average of $\boldsymbol{U}_h$ exactly. For $d = k = 2$, the effective Courant-Friedrichs-Lewy (CFL) factor is 0.1 for Δt_{TVD}, while it is 1/12 for Δt_{Pos}. The positivity limiter checks for unphysical states in all the quadrature points involved in the DG discretization. If any point in an element is flagged as unphysical, the polynomial representation $\boldsymbol{U}_h$ is damped towards its cell average $\boldsymbol{U}_{\boldsymbol{K}}$, which is guaranteed to be physical by the Δt_{Pos} restriction. Note that Δt_{Pos} is a sufficient condition. In practice, the time step can often be set larger than Δt_{Pos}.

3. Preliminary numerical results

In this section we show preliminary numerical results obtained with the DG method implemented in `thornado` to solve the non-relativistic and special relativistic Euler equations. We use the ideal gas equation of state where the pressure p is related to the internal energy density e by $p = (\Gamma - 1)e$, and where Γ is the (constant) adiabatic index. Let the d-dimensional computational domain be $D = \times_{i=1}^{d}[x^i_{\text{m}}, x^i_{\text{M}}]$, where x^i_{m} and x^i_{M} denote the coordinates of the inner and outer boundary in the ith coordinate direction, respectively. To limit the scope of this initial study, we use polynomials of degree $k = 2$ (except in Section 3.1.4, where we also use $k = 1$ and $k = 3$) combined with SSP-RK3 time stepping, and, unless stated otherwise, run all the tests with a CFL factor of $C = 0.1$. We use the HLL Riemann solver when solving the non-relativistic Euler equations and the HLLC Riemann solver when solving the special relativistic Euler equations. The purpose of the tests is to gauge the performance of the DG implementation on a set of benchmarks as an initial measure of its suitability for future CCSN simulations.

3.1. Non-relativistic (NR) hydrodynamics

For the NR Euler equations, the state, flux, and source vectors in Eq. (1) are

$$\boldsymbol{U} = \big(\rho, \rho u_j, E\big)^{\text{T}}, \ \boldsymbol{F}^i = \big(\rho u^i, \Pi^i{}_j, (E+p)u^i\big)^{\text{T}}, \text{ and } \boldsymbol{S} = \big(0, \frac{1}{2}\Pi^{ik}\partial_j\gamma_{ik} - \rho\partial_j\Phi, -\rho u^i\partial_i\Phi\big)^{\text{T}}, \tag{15}$$

where ρ and u^i are the mass density and components of the fluid three-velocity, respectively. The stress tensor is $\Pi^i{}_j = \rho u^i u_j + p\delta^i{}_j$, and $E = e + \frac{1}{2}\rho u_i u^i$ is the fluid (internal plus kinetic) energy density. The sources are due to curvilinear coordinates and Newtonian gravity.

3.1.1. Sod shock tube The first test is the classic Riemann problem due to Sod [34], computed using Cartesian coordinates. The one-dimensional computational domain is $D = [0, 1]$, and a discontinuity is initially located at $x^1 = 0.5$, separating the left and right states

$$\boldsymbol{U}_{\mathrm{L}} = (1.0, \mathbf{0.0}, 2.5)^{\mathrm{T}} \quad \text{and} \quad \boldsymbol{U}_{\mathrm{R}} = (0.125, \mathbf{0.0}, 0.25)^{\mathrm{T}},$$

where the adiabatic index is $\Gamma = 1.4$. The test is run with 100 elements until $t = 0.2$, using $C_{\mathrm{TCI}} = 0.03$. Results are plotted in the left panels of Figure 1. The DG method captures the main features of the solution well without introducing noticeable oscillations near the discontinuities (upper left panel). In the lower left panel we plot locations in the xt-plane of elements flagged for limiting. Limiting is mainly triggered by the contact discontinuity and the shock.

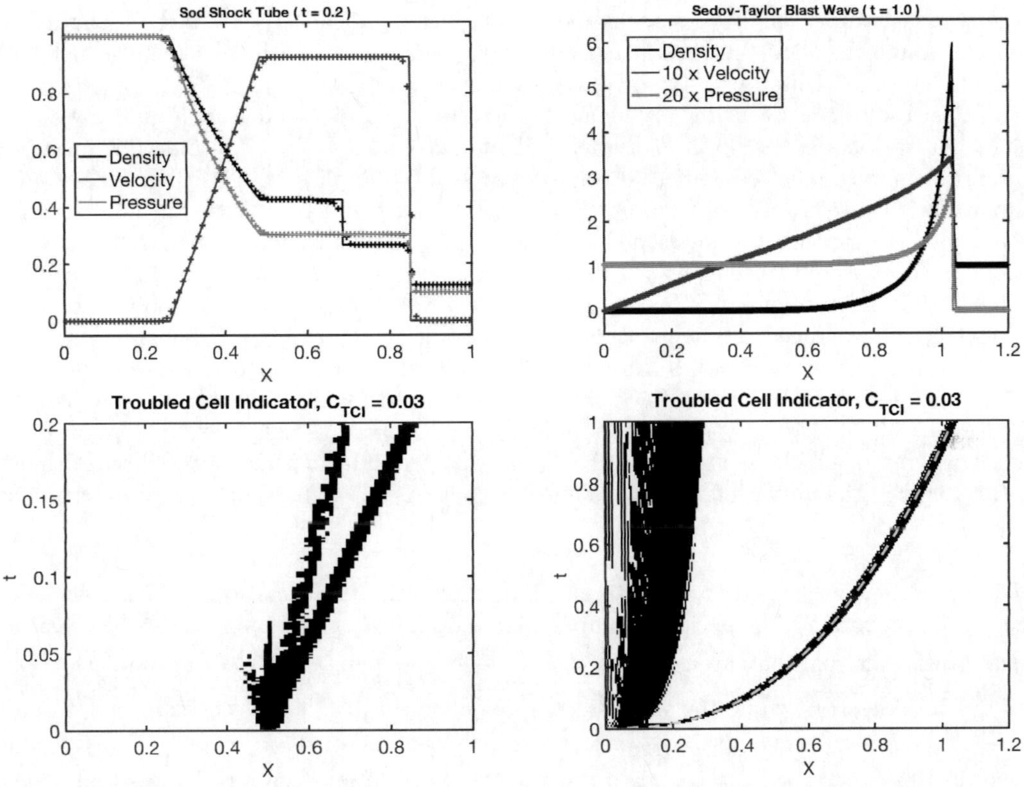

Figure 1. *Upper left panel:* numerical solution of Sod's problem using 100 elements (plusses) compared with the exact solution (solid lines; from [38]). *Upper right panel:* numerical solution of the Sedov-Taylor blast wave using 256 elements (plusses) against the exact solution (solid lines; from [16]). In the lower panels we plot locations in the xt-plane of elements flagged for limiting ($I_{\boldsymbol{K}} > C_{\mathrm{TCI}}$) by the Sod (left) and Sedov (right) problems.

3.1.2. Sedov-Taylor blast wave This test, detailed in [29] (see also §99 in [18]), is computed in spherical polar coordinates with the assumption of spherical symmetry.[2] The computational domain is $D = [0, 1.2]$, and the initial condition consists of a fluid at rest with density $\rho = 1$. The adiabatic index is $\Gamma = 1.4$, and an amount of thermal energy equal to 1 is released in the

[2] In this paper, all tests in spherical polar coordinates are computed with the assumption of spherical symmetry.

innermost element. We use 256 elements and run until $t = 1$, using $C_{\rm TCI} = 0.03$. Again, the DG method captures the characteristic of the exact solution (cf. upper right panel of Figure 1). The maximum density in the shock is somewhat lower for the numerical solution than the exact value (about 5 versus 6). In the lower right panel we map locations of elements flagged for limiting in the xt-plane. The troubled-cell indicator tracks the shock well (the analytic shock trajectory is given by the dashed cyan line), but also flags elements at smaller radii with low density.

3.1.3. Shu-Osher shock tube This test from [31] is suitable for measuring the amount of dissipation in a numerical scheme. The one-dimensional computational domain is $D = [-5, 5]$, and a discontinuity is placed at $x^1 = -4$, separating the states

$$\boldsymbol{U}_{\rm L} = \big(3.857143, 10.14185, 0.0, 0.0, 39.16667\big)^{\rm T} \quad \text{and} \quad \boldsymbol{U}_{\rm R} = \big(1 + 0.2 \times \sin(5x^1), \mathbf{0.0}, 2.5\big)^{\rm T},$$

where the adiabatic index is $\Gamma = 1.4$. The density variations ahead of the Mach=3 shock are compressed through the shock and result in higher-frequency variations downstream, which can be difficult to capture with an excessively dissipative scheme.

We run this test to $t = 1.8$ using 200 elements. In the two upper panels of Figure 2 we plot the density for various values of the troubled-cell indicator threshold $C_{\rm TCI}$; 0.0 (full limiting, green), 0.03 (magenta), 0.3 (red), and 3.0 (blue). Larger $C_{\rm TCI}$ implies less limiting. The results are compared with a reference solution computed with 2048 elements. With full limiting, the scheme is unable to capture the density variations behind the shock (cf. upper right panel in Figure 2). This is expected since the TVD-type limiter clips extrema. The results improve when the troubled-cell indicator is used. When $C_{\rm TCI} = 0.03$, the variations are better resolved, but the amplitudes are much reduced relative to the reference solution. With $C_{\rm TCI} = 3.0$, the variations are well resolved and the amplitudes are comparable to the reference. In the lower panels of Figure 2 we plot locations in the xt-plane of elements flagged for limiting by the troubled-cell indicator: $C_{\rm TCI} = 0.03$ (left), $C_{\rm TCI} = 0.3$ (middle), and $C_{\rm TCI} = 3.0$ (right). With $C_{\rm TCI} = 0.03$, limiting is triggered in a wide region around the shock and around the peaks of the three leftmost density variations in the upper left panel in Figure 2. With $C_{\rm TCI} = 3.0$, limiting is only triggered in the shock.

3.1.4. Isentropic vortex This test from [32] is included to verify the accuracy of the DG method implemented in `thornado`. The periodic computational domain is given by $D = [-5, 5] \times [-5, 5]$. The initial density and pressure are given by $\rho_0 = \big(1 - \frac{(\Gamma-1)\,\beta^2}{(8\pi\Gamma)} \exp\big(1 - r^2\big)\big)^{1/(\Gamma-1)}$ and $P_0 = \rho^\Gamma$, where $r = \sqrt{(x^1)^2 + (x^2)^2}$, while the velocity components are $v_0^1 = 1 - \frac{x^2\beta}{2\pi} \exp\big((1 - r^2)/2\big)$ and $v_0^2 = 1 + \frac{x^1\beta}{2\pi} \exp\big((1 - r^2)/2\big)$. The strength of the vortex is set to $\beta = 5$, and we use $\Gamma = 1.4$. We evolve this test until $t = t_f = 10$, when the vortex has returned to its initial position.

In Figure 3 we plot the density error measured in the infinity norm $L^\infty = \max_D |\rho(\boldsymbol{x}, t_f) - \rho_0(\boldsymbol{x})|$ versus number of cells the x^1-dimension for various configurations. We ran with spatial resolutions of 25×25, 50×50, 100×100, and 200×200. In the first set of runs, we ran without any limiting and varied the polynomial degree k: $k = 1$ (solid red), $k = 2$ (solid black), and $k = 3$ (solid green). We observe the expected order of accuracy for $k = 1$ and $k = 2$ (second- and third-order accuracy, respectively; cf. dotted reference lines). For $k = 3$, the order of accuracy is close to fourth-order for the lower resolutions, but tends to third-order as the resolution increases, presumably due to the use of third-order time stepping. (Note that we ran the $k = 3$ case with a CFL factor of 1/14; cf. Eq. (14).) In the second set of runs, we kept $k = 2$ and ran with limiting turned on. We ran one set with $C_{\rm TCI} = 0.03$ (dashed magenta) and one set with $C_{\rm TCI} = 0.1$ (dashed cyan). When $C_{\rm TCI} = 0.03$, the limiter is activated for the lowest resolution run and significantly reduces the accuracy of the method. For all the other runs, the limiter is not activated, and the results are identical to the runs with the limiter turned off.

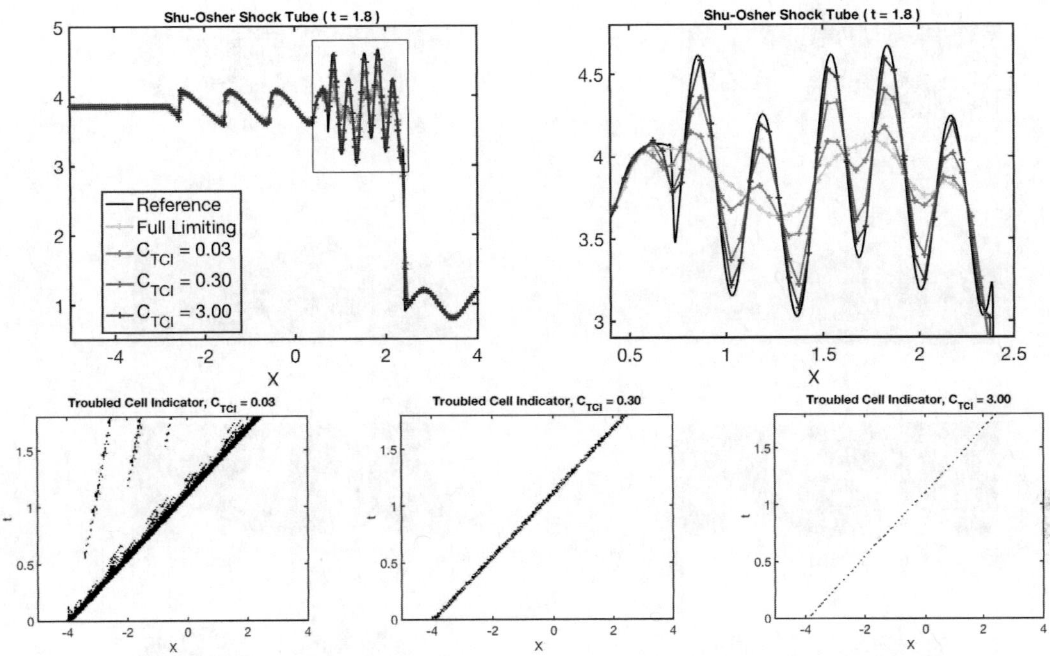

Figure 2. Results for the Shu-Osher shock tube problem computed with 200 elements. The upper panels show the mass density at $t = 1.8$ for various values of the troubled-cell indicator threshold C_{TCI}, compared with a high-resolution (2048 elements) reference solution. In the lower panels we show elements flagged for limiting in the xt-plane for various values of C_{TCI}.

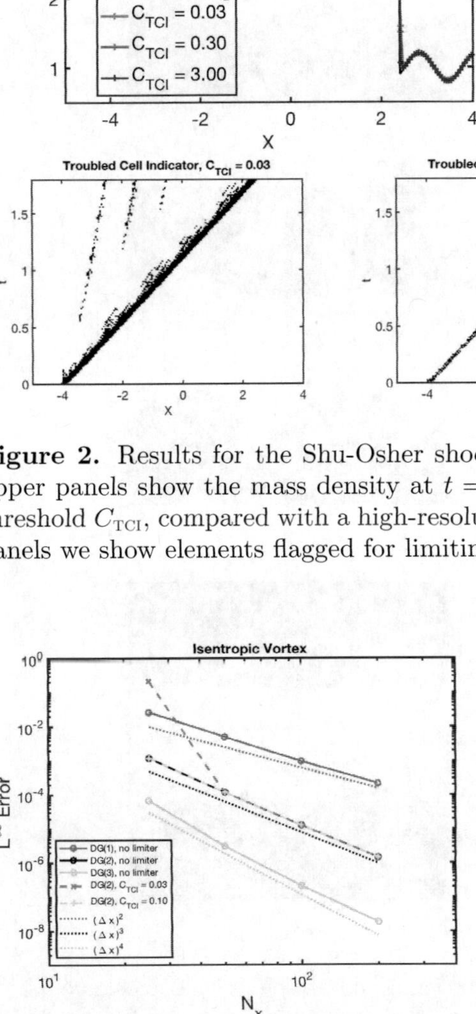

Figure 3. Convergence results for the Isentropic Vortex: L^{∞} error norm versus number of elements in the x^1-dimension.

3.1.5. Kelvin-Helmholtz (KH) instability This hydrodynamic instability is of significant astrophysical relevance — including during the explosion phase of CCSNe — and may occur in the interface separating fluids in relative motion. We compute this problem on the unit square $D = [0, 1] \times [0, 1]$, imposing periodic boundary conditions, and with initial conditions and single-mode perturbations taken from [21]. We use 256^2 elements and evolve until $t = 3.0$, well into the nonlinear regime. Results are displayed in Figure 4. We have computed two models, one with $C_{\mathrm{TCI}} = 0.06$ (left panels) and one without limiting ($C_{\mathrm{TCI}} \to \infty$; right panels).

Early on ($t = 1.5$), the results from the two runs are practically indistinguishable, show no sign of developing secondary billows, and agree visually with results presented in [21]. When $t = 2.4$, the run with $C_{\mathrm{TCI}} = 0.06$ has started to develop secondary billows within the coiled-up interface separating dense (white) and less dense (black) fluids, while the model with no limiting has not. See Figure 5 for the spatial distribution of elements flagged for limiting at $t = 2.4$. When $t = 3.0$, these secondary instabilities appear to have largely destroyed the coil structure for the model with limiting, while the coils remain intact in the model without limiting (although secondary billows have started to form in this

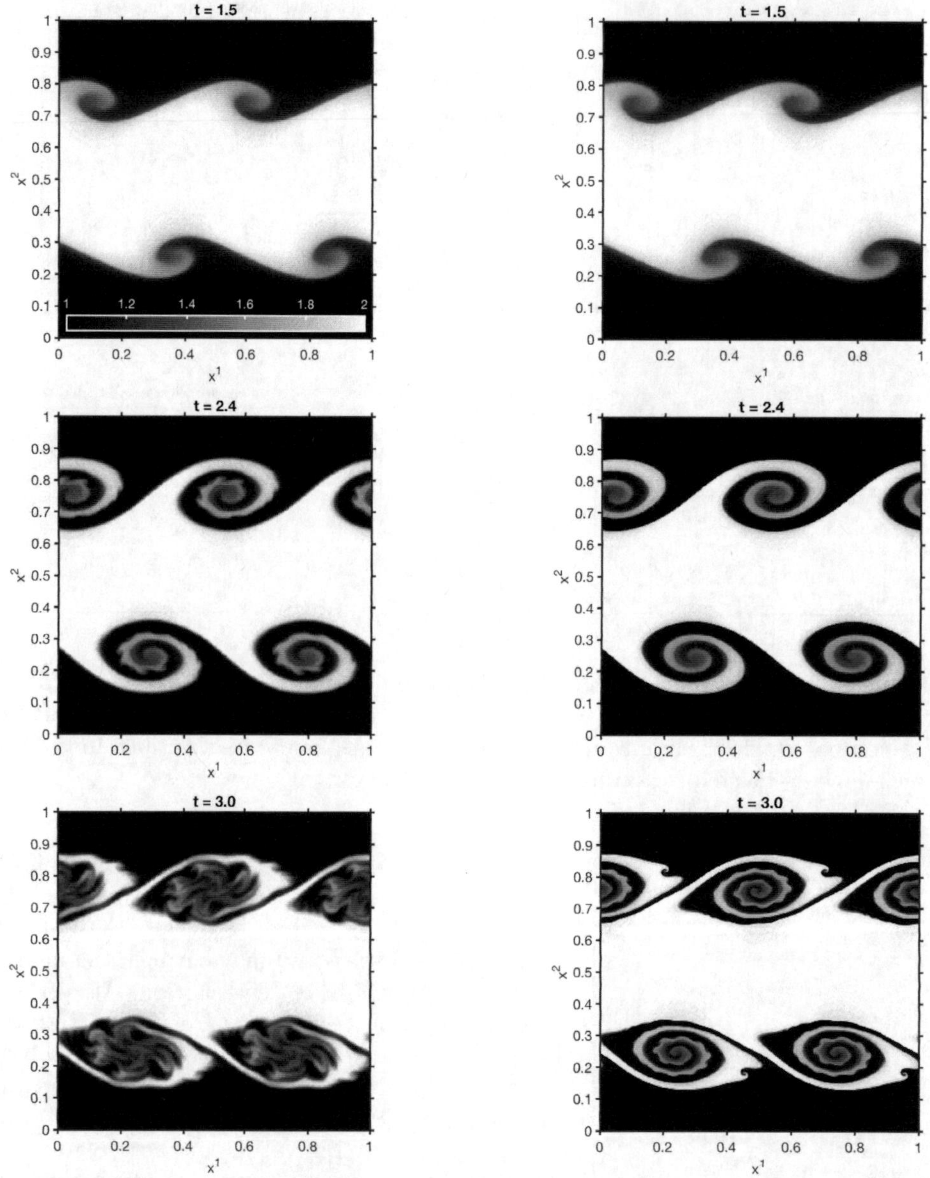

Figure 4. Mass density from the KH instability at various times, computed with 256^2 elements: $t = 1.5$ (top panels), $t = 2.4$ (middle panels), and $t = 3.0$ (bottom panels). Results for $C_{\text{TCI}} = 0.06$ and $C_{\text{TCI}} \to \infty$ (no limiting) are shown in the left and right panels, respectively.

model at this time). As discussed in more detail by [21], the development of the secondary billows may be an artifact of numerical perturbations and diffusion, which is delayed when the resolution is increased. Our results are consistent with this in the sense that the less diffusive model develops secondary billows later.

3.1.6. Liska-Wendroff implosion This test from [19] is computed on a 2D domain $D = [0, 0.3] \times [0, 0.3]$ with reflecting boundary conditions using 256^2 elements. Below the line

$x^2 = 0.15 - x^1$, we initially set $\boldsymbol{U} = (0.125, \mathbf{0.0}, 0.35)^{\mathrm{T}}$, while $\boldsymbol{U} = (1.0, \mathbf{0.0}, 2.5)^{\mathrm{T}}$ elsewhere.

Results for $t = 0.045$ and $t = 2.5$, with $C_{\mathrm{TCI}} = 0.6$, are displayed in Figure 6 (left and right panels, respectively), which agree qualitatively with [19], who compared the results of eight different schemes for this problem. When $t = 0.045$, a shock wave propagates towards the origin of D, while a contact discontinuity is trailing the shock (cf. density contours), and a rarefaction wave is spreading in the opposite directions. When $t = 2.5$, a complex flow pattern with multiple shocks has emerged due to multiple reflections off the boundaries. The initial symmetry about the $x^1 = x^2$ diagonal is maintained, and the results from `thornado` agree qualitatively with the unsplit schemes in [19]. In the right panel of Figure 6, the chatacteristic "jet" has formed and propagates along the diagonal, similar to that displayed by CLAW and WENO in [19].

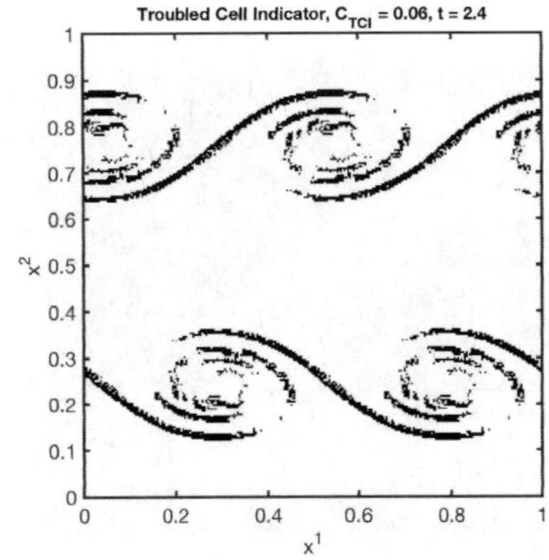

Figure 5. Distribution of elements flagged for limiting in the KH test at $t = 2.4$.

3.1.7. Liska-Wendroff explosion This test, also from [19], is computed on the domain $D = [0, 1.5] \times [0, 1.5]$ with reflecting inner boundaries and outflow outer boundaries using 256^2 elements. Inside the radius $r = \sqrt{(x^1)^2 + (x^2)^2} = 0.4$, we initially set $\boldsymbol{U} = (1.0, \mathbf{0.0}, 2.5)^{\mathrm{T}}$, while $\boldsymbol{U} = (0.125, \mathbf{0.0}, 0.25)^{\mathrm{T}}$ elsewhere. We did not apply any smoothing to the initial discontinuity.

Results for $t = 3.2$, obtained with $C_{\mathrm{TCI}} = 0.1$, are displayed in Figure 7, which agree qualitatively with results in [19]. Initially, this test is similar to the Sod shock tube in cylindrical geometry: a shock and a contact discontinuity propagate in the positive radial direction, while a rarefaction wave propagates toward the origin. Later, the initial shock leaves the computational domain, while an instability (resembling Richtmeyr-Meshkov) develops in the contact discontinuity. A second shock emerges from the origin of the computational domain and runs through the contact discontinuity, which has developed a complex structure (see density contours in Figure 7). The second shock is located around $r = 1.2$ at $t = 3.2$. In the right panel

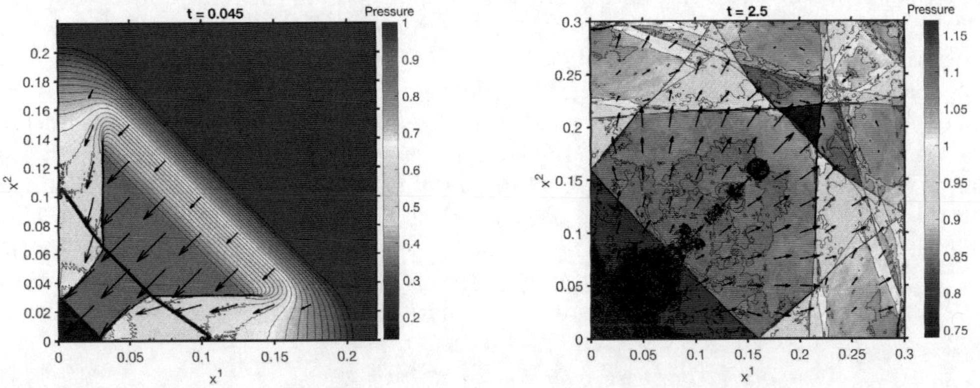

Figure 6. Plots for the Liska-Wendroff implosion problem at $t = 0.045$ (left) and $t = 2.5$ (right), intended to match Figures 4.10 and 4.11 in [19]. The color map shows the pressure distribution, black contours show the mass density, while arrows indicate the velocity.

we plot the distribution of elements flagged for limiting by the troubled-cell indicator at $t = 3.2$. Limiting is mainly focused on elements around the shock and the contact discontinuity.

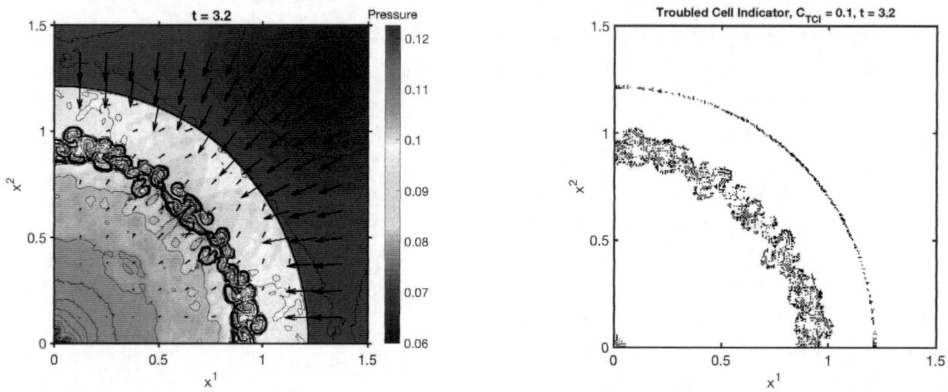

Figure 7. Plots for the Liska-Wendroff Explosion problem at $t = 3.2$. In the left panel, intended to match panels in Figure 4.12 in [19], the color map shows the pressure distribution, black contours show the mass density, while arrows indicate the velocity. In the right panel we plot the distribution of elements flagged for limiting $t = 3.2$.

3.1.8. Standing accretion shock (SAS) This test is of immediate relevance to simulation of CCSNe. The initial conditions, described in detail in [3], have been used by many to study the standing accretion shock instability (SASI) and CCSN explosion dynamics. We follow closely the description in [3] to initialize this test. We use spherical polar coordinates, let $D = [0.2, 2.0]$, and $\Gamma = 4/3$. The gravitational potential is given by the point-mass formula with $GM = 0.5$. A stationary shock is placed at a radius $R_{\rm SH} = 1$. Ahead of the shock, the flow is essentially in free-fall towards the shock with a constant Mach number of 100. The mass accretion rate is held fixed at the outer boundary to $\dot{M} = 4\pi$. Below the shock, $r \in [0.2, 1]$, the settling solution is obtained by solving Bernoulli's equation. Matter flows through the inner boundary, and the density and pressure in boundary elements are set by extrapolating from D assuming $\rho \propto r^{-3}$ and $E \propto r^{-4}$, while the momentum components are held fixed to their initial values.

We use 256 elements and run the tests until $t = 100$ (> 5 dynamical times [3]). Results are shown in Figure 8. In the first test we show results from an unperturbed run to gauge the ability of the DG method to maintain the initial state. In the upper left panel we compare the initial state (dashed lines) with the solution at $t = 100$. Except for a slight shift ($\sim 1\%$) in the position of the shock, the initial and final states are indistinguishable on the plot. Also, there are no oscillations visible in the numerical results. In the upper right panel we show elements flagged for limiting in the rt-plane (we set $C_{\rm TCI} = 0.03$ in this run), which illustrates how limiting is confined to the shock. In the second test we placed a shell with thickness 0.2 and density three times higher than the ambient density ahead of the shock to induce a strong radial perturbation (similar to [3]; see their Figure 4). The initial condition is stable against radial perturbations. In the lower panel in Figure 8 we plot the pressure deviation below the shock, defined as

$$\delta p = (p - \bar{p})/\bar{p}, \quad \text{where} \quad \bar{p} = p_0(r_0/r)^4, \tag{16}$$

and p_0 and r_0 are the pressure and radius at the inner boundary. As the shell falls through the shock, the strong perturbation results in an interesting pattern of waves propagating inside the shocked cavity, and oscillations of the shock position. The configuration eventually settles down

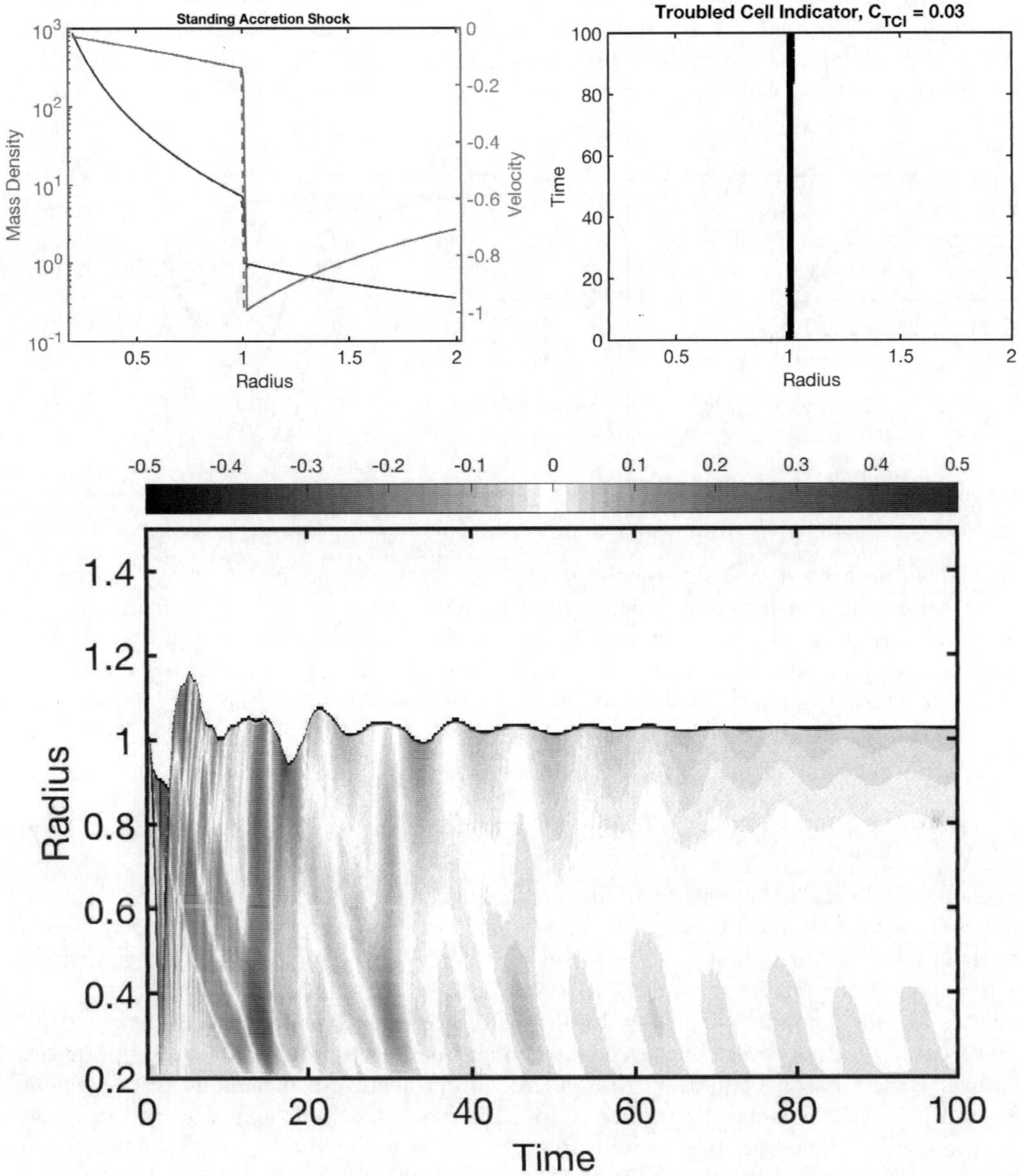

Figure 8. Results from the SAS test using 256 elements. In the upper left panel we plot the mass density (blue) and velocity (red) versus radius for the initial condition (dashed lines) and for $t = 100$ (solid lines) from an unperturbed run. In the upper right panel we show elements flagged for limiting in the rt-plane. In the bottom panel we show results from a perturbed model, where we plot the relative deviation of the pressure below the shock; cf. Eq. (16).

to a configuration that is close to the initial condition (although the position of the shock is slightly larger than that of the initial state.)

3.1.9. Yahil-Lattimer collapse The final test of the NR hydrodynamics in `thornado` involves self-gravity and is due to [41, 42]. It models the self-similar collapse of a polytropic star;

i.e., $p = \kappa\rho^\Gamma$, where κ is the polytropic constant. This test is also of immediate relevance to CCSN simulations. In [41, 42], self-similar solutions to the gravitational collapse problem were constructed for $6/5 \le \Gamma < 4/3$. The solutions smoothly connect an homologously collapsing subsonic inner core (velocity proportional to radius) to a supersonically collapsing outer core in near free-fall ($u \propto r^{-1/2}$). With two dimensional parameters in the model (the gravitational

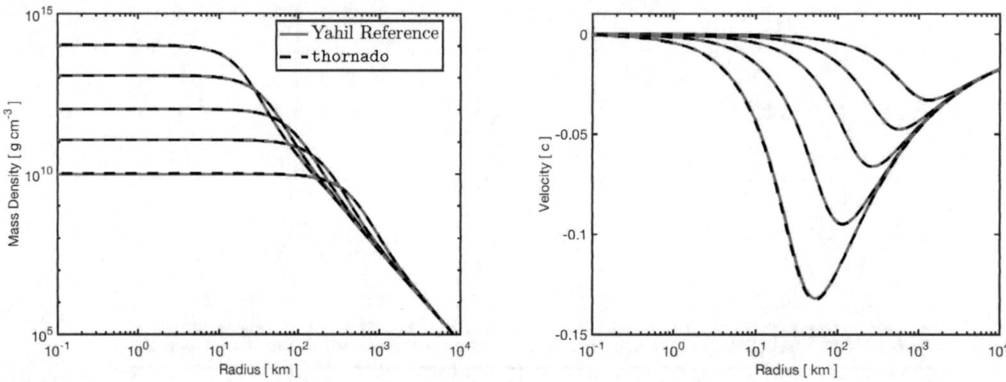

Figure 9. Results from the Yahil-Lattimer collapse test using 256 elements. Results obtained with `thornado` (dashed black) are compared with the reference solution from Yahil [42] (solid red). The mass density (left panel) and velocity (right panel) are plotted versus radius. We compare the solutions at select central densities during collapse, approximately $[10^{10}, 10^{11}, 10^{12}, 10^{13}, 10^{14}]$ g cm^{-3}, which correspond to $(-t) = [51.0, 15.0, 5.0, 1.5, 0.5]$ ms.

constant G and the polytropic constant κ), the dimensionless similarity variable is

$$X = \kappa^{-1/2}\, G^{(\Gamma-1)/2}\, r\, (-t)^{\Gamma-2}, \tag{17}$$

where the origin of time is the moment of infinite central density. All the hydrodynamic variables can be expressed as a function of X, and the time-dependent Euler equations can be recast as a system of ODEs (see [42] for details). To construct a reference to compare with our numerical results we have solved the ODEs given in [42] to obtain these self-similar solutions.

We show results for a model with $\Gamma = 1.30$. We use spherical polar coordinates, and let $D = [0, 1 \times 10^5]$ km, which is covered with 256 elements. The Newtonian gravitational potential is obtained by solving Poisson's equation using a third-order accurate continuous finite element method (e.g., [4]). We use a geometric grid to resolve the mass distribution as the star collapses and the central density increases from about 10^9 g cm^{-3} to about 10^{14} g cm^{-3}. The size of the innermost element is set to 1 km while the size of the last element is about 3×10^3 km. We specify the polytropic constant by setting a reference pressure $p = 6 \times 10^{27}$ erg cm^{-3} for $\rho = 7 \times 10^9$ g cm^{-3} (reasonable values for a massive star in the pre-collapse stage). We also set the collapse time to $(-t) = 150$ ms. Results comparing the mass density and velocity of Yahil's reference solution to those obtained with `thornado` using $C_{\rm TCI} = 0.03$ are plotted in Figure 9. The agreement of the mass density and velocity profiles is excellent throughout collapse.

3.2. Special relativistic (SR) hydrodynamics

Here we show results from solving the SR Euler equations in Cartesian coordinates with `thornado`. In this case the state, flux, and source vectors in Eq. (1) are

$$\boldsymbol{U} = \left(\rho W, \rho h W^2 u_j, \tau\right)^{\rm T}, \;\; \boldsymbol{F}^i = \left(\rho W u^i, \Pi^i{}_j, \rho(hW-1)Wu^i\right)^{\rm T}, \text{ and } \boldsymbol{S} = 0, \tag{18}$$

where W is the Lorentz factor, $h = 1+(e+p)/\rho$ is the specific enthalpy, $\tau = \rho W(hW-1)-p$, and $\Pi^{i}{}_{j} = \rho h W^2 u^i u_j + p\delta^{i}{}_{j}$. To recover primitive from conserved variables we have implemented the procedure for analytic equations of state detailed in Appendix D in [27]. We solve two Riemann problems and compare the numerical results with exact solutions obtained with the solver from [20]. Let the primitive state vector be $\boldsymbol{V} = \left(\rho, \boldsymbol{u}, p\right)^{\mathrm{T}}$. In the first problem (Riemann problem 1), taken from [22], we use $\Gamma = 4/3$ and

$$\boldsymbol{V}_{\mathrm{L}} = \left(1.0, 0.9, 0.0, 0.0, 1.0\right)^{\mathrm{T}} \quad \text{and} \quad \boldsymbol{V}_{\mathrm{R}} = \left(1.0, 0.0, 0.0, 0.0, 10.0\right)^{\mathrm{T}},$$

while in the second problem (Riemann problem 2), also from [22], we use $\Gamma = 5/3$, and let

$$\boldsymbol{V}_{\mathrm{L}} = \left(1.0, \mathbf{0.0}, 10^3\right)^{\mathrm{T}} \quad \text{and} \quad \boldsymbol{V}_{\mathrm{R}} = \left(1.0, \mathbf{0.0}, 10^{-2}\right)^{\mathrm{T}}.$$

In both problems, the computational domain is $D = [0, 1]$, the discontinuity is initially located at $x^1 = 0.5$, and the solutions are integrated to $t = 0.4$. Results are plotted in Figure 10.

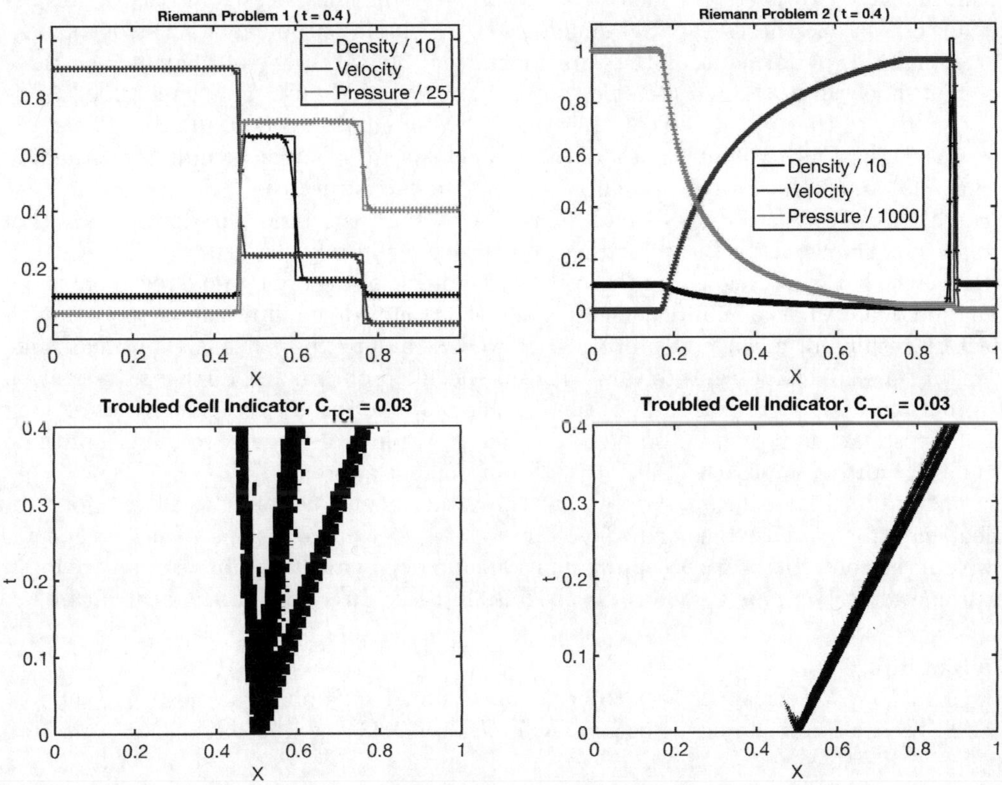

Figure 10. Results from solving the SR Euler equations with `thornado`. Riemann problem 1 (left panels) was solved using 100 elements, while Riemann problem 2 (right panels) was solved using 400 elements. The numerical and exact solutions are plotted with plusses and solid lines, respectively. In the lower panels we plot xt-plane locations of elements flagged for limiting

Again, the main features of the exact solution are captured with the DG method implemented in `thornado`. For Riemann problem 1, we observe some oscillations in the density profile between the leftmost shock and the contact discontinuity located around $x^1 = 0.6$. For Riemann problem 2, the maximum density in the thin density shell between the shock and the contact discontinuity is significantly lower for the numerical solution than the exact solution. This is due to a combination of limited spatial resolution and the action of the slope limiter. We used $C_{\text{TCI}} = 0.03$ in both tests in this section, and we plot locations in the xt-plane of elements flagged for limiting in Figure 10. The troubled-cell indicator tracks the discontinuities well for both Riemann problems. We have found the agreement with the exact solution to improve for larger values of C_{TCI}, but at the expense of somewhat more oscillatory results.

4. Summary and outlook

We have presented preliminary algorithm details and numerical results for solvers of the non-relativistic and special relativistic Euler equations of gas dynamics as implemented in the **t**oolkit for **h**igh-**or**der **n**eutrino-**rad**iation hydr**o**dynamics (`thornado`). The spatial discretization is based on the DG method and the ODEs resulting from this discretization are integrated in time using SSP-RK methods. We employ a spectral-type nodal collocation DG approximation based on Legendre-Gauss points for interpolation and numerical quadrature evaluation [2]. This choice simplifies expressions for the semi-discretized equations, especially when applied to problems involving curvilinear coordinates encoded in a metric (e.g., in numerical relativity). Results from a suite of tests in one and two spatial dimensions — including problems with strong shocks — demonstrate reliable performance of the implementation. The accuracy of `thornado` on three shock tests (Sod shock tube [34], Sedov-Taylor blast wave [29], and Shu-Osher shock tube [31]) is qualitatively similar to that of the well-established CCSN simulation code CHIMERA [5], which is based on the PPM finite volume method [8]. Two tests in spherical symmetry (standing accretion shock [3] and Yahil-Lattimer collapse [41, 42]) demonstrate the DG method's ability to handle conditions relevant to CCSN simulations. We will present a more in-depth analysis of the performance of `thornado` on these (and related) problems in a future study.

The performance of the DG algorithm is sensitive to limiting of the polynomial representation. The combination of a TVD-type limiter (e.g., [6]) and the troubled-cell indicator of [12] seems to give satisfactory results for a range of problems. However, in our experience, the optimal value for the indicator threshold C_{TCI} seems to vary with the specific problem, and further investigation is needed to determine if there exists an optimal value suitable for CCSN simulations. Along these lines of investigation, it would also be interesting to explore and compare the use of the *a posteriori* subcell limiting approach in [9, 11] with our current approach.

Ongoing and planned near-future work within `thornado` include coupling to solvers for two-moment neutrino transport, extensions to accommodate nuclear equations of state, general relativity within the conformal flatness approximation, and deployment within an adaptive mesh refinement framework. We hope to report on progress in these directions in the near future.

Acknowledgments

Eirik Endeve, Anthony Mezzacappa, Nick Roberts, and Samuel J. Dunham acknowledge support from the NSF Gravitational Physics Program (NSF-GP 1505933 and 1806692).

References

[1] Ascher U M, Ruuth S J and Spiteri R J 1997 *Applied Numerical Mathematics* **23** 151
[2] Bassi F, Franchina N, Ghidoni A and Rebay S 2013 *Int. J. Numer. Meth. Fluids* **71** 1322
[3] Blondin J M, Mezzacappa A and DeMarino C 2003 *ApJ* **584** 980
[4] Brenner S C and Scott L R 2008 *The Mathematical Theory of Finite Element Methods* Springer
[5] Bruenn S W *et al Preprint* arXiv:1809.05608
[6] Cockburn B and Shu C-W 1998 *JCP* **141** 199
[7] Cockburn B and Shu C-W 2001 *Journal of Scientific Computing* **16** 173
[8] Colella P and Woodward P 1984 *JCP* **54** 174
[9] Dumbser M, Zanotti O, Loubère R and Diot S 2014 *JCP* **278** 47
[10] Dumbser M, Guercilena F, Köppel S, Rezzolla L and Zanotti O 2018 *Phys. Rev. D* **97** 084053

[11] Fambri F, Dumbser M, Köppel S, Rezzolla L and Zanotti O 2018 *MNRAS* **477** 4543
[12] Fu G and Shu C-W 2017 *JCP* **347** 305
[13] Harten A, Lax P D and Van Leer B 1983 *SIAM Review* **25** 35
[14] Hesthaven J S and Warburton T 2008 *Nodal discontinuous Galerkin methods: Algorithms, analysis and applications* Springer
[15] Huerta A, Casoni E and Peraire J 2012 *Int. J. Numer. Meth. Fluids* **69** 1614
[16] Kamm J R and Timmes F X 2007 *LANL Report, No. LA-UR-07-2849*
[17] Klöckner A, Warburton T, Bridge J and Hesthaven J S 2009 *JCP* **228** 7863
[18] Landau L D and Lifshitz E M 1979 *Fluid Mechanics* Pergamon
[19] Liska R and Wendroff B 2003 *SIAM J. Sci. Comput.* **25** 3
[20] Martí J M and Müller E 2003 *Living Rev. Relativ.* **6** 7
[21] McNally C, Lyra W and Passy J-C 2012 *ApJS* **201** 18
[22] Mignone A and Bodo G 2005 *MNRAS* **364** 126
[23] Müller B 2016 *PASA* **33** 1
[24] Qin T, Shu C-W and Yang Y 2016 *JCP* **315** 323
[25] Radice D and Rezzolla L 2011 *Phys. Rev. D* **84** 024010
[26] Remacle J-F, Flaherty J E and Shephard M S 2003 *SIAM Review* **45** 53
[27] Rezzolla L and Zanotti O 2013 *Relativistic Hydrodynamics* Oxford
[28] Schaal K, Bauer A, Chandrashekar P Rüdiger P, Klingenberg C and Springel V 2015 *MNRAS* **453** 4278
[29] Sedov L I 1959 *Similarity and Dimensional Methods in Mechanics* Academic Press
[30] Shu C-W and Osher S 1988 *JCP* **77** 439
[31] Shu C-W and Osher S 1989 *JCP* **83** 32
[32] Shu C-W 1997 *NASA/CR-97-206253 ICASE Report No. 97-65*
[33] Shu C-W 2016 *JCP* **316** 598
[34] Sod G A 1978 *JCP* **27** 1
[35] Sonntag M and Munz C-D 2017 *Journal of Scientific Computing* **70** 1262
[36] Teukolsky S A 2016 *JCP* **312** 333
[37] Toro E F, Spruce M and Speares W 1994 *Shock Waves* **4** 25
[38] Toro E F 1999 *NUMERICA: A Library of Source Codes for Teaching, Research and Applications*
[39] Wilson J R, Mathews G J and Marronetti P 1996 *Phys. Rev. D* **54** 1317
[40] Wu K and Tang H 2015 *JCP* **298** 539
[41] Yahil A and Lattimer J M 1982 *Supernovae: A Survey of Current Research* Reidel
[42] Yahil A 1983 *ApJ* **265** 1047
[43] Zanotti O, Fambri F and Dumbser M 2015 *MNRAS* **452** 3010
[44] Zhang X and Shu C-W 2010 *JCP* **229** 8918

IOP Conf. Series: Journal of Physics: Conf. Series **1225** (2019) 012015 doi:10.1088/1742-6596/1225/1/012015

Conservation of Gravitational Energy

Tomoyuki Hanawa

Center for Frontier Science, Chiba University, 1-33 Yayoi-cho, Inage-ku, Chiba 263-8522, Japan

E-mail: hanawa@faculty.chiba-u.jp

Abstract. The total energy of an astronomical system is of great interest since the evolution of the system depends on it. Hence we solve the hydrodynamical equations by taking account of the conservation in numerical simulations of astrophysical objects. However, gravitational energy is often taken into account as a source term, and the total energy including gravity is not guaranteed to be conserved. This is partly because it increases the computational cost to solve hydrodynamical equations in the fully conservative form, i.e., those without any source term. This paper shows that the total energy and momentum of a system are fully conserved if the source terms due to gravity are properly taken into account. The error in the total energy is reduced as small as the round-off error. The method is applicable both when the gravitational force is given by the Poisson equation or explicitly as a function. The former case is confirmed by numerical simulations of 2D fragmentation of a self-gravitating gas. This method is applicable not only to the Cartesian uniform grid but an octree grid often used in adaptive mesh refinement.

1. Introduction

Gravitational energy release is often an important source of activity in astronomical objects such as accreting compact stars and progenitors of core-collapse supernovae. Also, young stellar objects are powered by gravitational energy release through accretion and slow contraction. While energy changes its form, the total amount of it is conserved if energy loss from the system through emission is taken into account. If the energy conservation is not reproduced in numerical simulations, they suffer from artificial energy gain or loss and hence overestimate or underestimate the system activity. The total energy conservation is important as well as the total mass conservation.

Despite the importance, the total energy conservation is not guaranteed in most of the current numerical simulations. It is technically difficult to compute the change in the gravitational energy of a self-gravitational system, which is often expressed as

$$E_{\rm g} = \frac{1}{2} \int \rho \phi dV, \tag{1}$$

where ρ and ϕ denote the mass density and gravitational potential, respectively.

The technical difficulty can be reworded as follows. The conservation of total mass is easily realized by solving the equation of continuity in its conservation form. The equation of continuity balances the temporal change in the density with the divergence of the mass flux. If we apply the Gauss's theorem for each numerical cell, the conservation of total mass is guaranteed. On the other hand, gravity appears as a source term in the equation of energy conservation. Thus the conservation of total energy is not guaranteed unless the spatial integral of gravitational energy release over the whole volume is exactly balanced with the temporal change in the total gravitational energy.

We can rewrite the equation of energy conservation into the fully conservative form in which the source term due to the gravity is decomposed into temporal change of energy density and divergence of energy flux[1]. However, we need some additional numerical procedure to solve the fully conservative form since it contains a new variable, the change in the gravitational potential. The fully conservative form has another shortcoming that gravitational energy release is less accurately evaluated in regions of low density. This is because the gravitational energy flux has a large but nearly constant value when the density thereof is small. The numerical divergence of the flux contains a relatively large round-off error especially when the density thereof is low.

In this paper, we propose a new method to evaluate the gravitational energy release. It is based on another form of the equation of energy conservation, in which the gravitational energy is expressed as

$$E_{\rm g} = -\int \frac{\boldsymbol{g}^2}{8\pi G} dV, \tag{2}$$

where G and g denote the gravitational constant and gravity, respectively. Although based on the fully conservative form, the gravitational energy release is expressed as the product of the gravity and mass flux as proposed in some earlier works [2, 3, 4]. As a result, the conservation of total energy including gravity is ensured by a small change in the evaluation of the gravity. As shown in the subsequent section, the gravity is defined on the surface of a numerical cell and the time-averaged value is used for exact balance.

This paper is organized as follows. The equivalence of equations (1) and (2) is proved in §2.1 and a new form of the equation of energy conservation is given in §2.2. The equation of energy conservation is converted into the discretized form in §2.3. Numerical procedure to solve the discretized equations is given in §2.4. Numerical examples are given in §3 and a summary is given in §4.

2. Methods

2.1. Gravitational Energy

First, we prove the equivalence of equations (1) and (2). The gravitational potential is given as the solution of the Poisson equation,

$$\Delta\phi = 4\pi G\rho, \tag{3}$$

and the gravity is defined as the gradient,

$$\boldsymbol{g} = -\boldsymbol{\nabla}\phi. \tag{4}$$

Substituting equation (3) into equation (1) we obtain

$$E_{\rm g} = \frac{1}{8\pi G}\int \phi\Delta\phi dV = -\frac{1}{8\pi G}\int \phi\left(\boldsymbol{\nabla}\cdot\boldsymbol{g}\right) dV = -\int \frac{\boldsymbol{g}\cdot\boldsymbol{g}}{8\pi G} dV + \int_s \frac{\phi}{8\pi G}\boldsymbol{g}\cdot d\boldsymbol{S}, \tag{5}$$

where the second term denotes the surface integral on the boundary in the right-hand side. Thus equations (1) and (2) are equivalent if the surface integral either vanishes or is negligibly small. The condition is realized when the boundary is periodic or placed very far from most of the mass.

The above proof is valid when the gravitational potential is differentiable, i.e, in its integral form. We discuss the condition for which the equivalence of equations (1) and (2) is valid also in the discretized form. In the following, we solve equation (3) in its discretized form,

$$\int_{V_i} 4\pi G\rho dV = -\int_{S_i} \boldsymbol{g} \cdot d\boldsymbol{S}, \tag{6}$$

where V_i and S_i denote the volume and surface of a given numerical cell, respectively. Note that only the gravity normal to cell surface appears in the right-hand side of equation (6). This fact indicates that we should evaluate the gravitational energy density from the normal component of $\boldsymbol{g}$ defined on each cell surface. Equation (5) requires that also the gravitational potential should be defined uniquely on each cell surface. Otherwise, the surface term may not be cancelled on the interface between two adjacent numerical cells. These constraints give us guidelines for constructing the numerical scheme which ensures conservation of total energy.

2.2. Fully Conservative Form of Hydrodynamical Equations

In this subsection, we rewrite all the components of the hydrodynamical equations in the fully conservative form.

The equation of continuity describes conservation of mass and is expressed as

$$\frac{\partial \rho}{\partial t} + \boldsymbol{\nabla} \cdot (\rho \boldsymbol{v}) = 0, \tag{7}$$

where $\boldsymbol{v}$ denote the velocity. Equation (7) is in the fully conservative form, since the source term vanishes. Combined with equation (7), the equation of motion is converted into the equation of momentum conservation,

$$\frac{\partial}{\partial t}(\rho \boldsymbol{v}) + \boldsymbol{\nabla} \cdot (\rho \boldsymbol{v}\boldsymbol{v} + P\boldsymbol{I}) = \rho \boldsymbol{g} = -\rho \boldsymbol{\nabla} \phi, \tag{8}$$

where P and $\boldsymbol{I}$ denotes pressure and the identity matrix, respectively. Equation (8) can be rewritten into the fully conservative form,

$$\frac{\partial}{\partial t}(\rho \boldsymbol{v}) + \boldsymbol{\nabla} \cdot (\rho \boldsymbol{v}\boldsymbol{v} + P\boldsymbol{I} + \boldsymbol{T}_{\rm g}) = 0, \tag{9}$$

$$\boldsymbol{T}_{\rm g} = \frac{\boldsymbol{g} \cdot \boldsymbol{g}}{8\pi G}\boldsymbol{I} - \frac{\boldsymbol{g}\boldsymbol{g}}{4\pi G}, \tag{10}$$

where $\boldsymbol{v}\boldsymbol{v}$ and $\boldsymbol{g}\boldsymbol{g}$ denote the direct product. The equivalence of equations (8) and (9) is confirmed by the algebra,

$$\boldsymbol{\nabla} \cdot \boldsymbol{T}_{\rm g} = -\frac{(\boldsymbol{\nabla} \cdot \boldsymbol{g})}{4\pi G}\boldsymbol{g} + \frac{(\boldsymbol{\nabla} \times \boldsymbol{g}) \times \boldsymbol{g}}{4\pi G} = -\rho \boldsymbol{g}.$$

Remember that two equalities

$$\boldsymbol{\nabla} \cdot \boldsymbol{g} = -4\pi G\rho, \tag{11}$$

$$\boldsymbol{\nabla} \times \boldsymbol{g} = 0, \tag{12}$$

hold for any gravity (see, e.g., [5]).

Similarly, energy conservation is expressed as

$$\frac{\partial}{\partial t}(\rho E) + \boldsymbol{\nabla}(\rho \boldsymbol{v} H) = \rho \boldsymbol{v} \cdot \boldsymbol{g}, \tag{13}$$

where E and H denotes the specific energy and enthalpy, respectively, and defined as

$$E = \frac{\boldsymbol{v}^2}{2} + \varepsilon, \tag{14}$$

$$H = \frac{\boldsymbol{v}^2}{2} + \varepsilon + \frac{P}{\rho}. \tag{15}$$

Here, the symbol, ε, denotes the specific internal energy. Energy gain and loss through radiation are not taken into account in equation (13) for simplicity. We can rewrite the right-hand side of equation (13) as

$$\begin{aligned} \rho \boldsymbol{v} \cdot \boldsymbol{g} &= -\rho \boldsymbol{v} \cdot \boldsymbol{\nabla} \phi \\ &= -\boldsymbol{\nabla} \cdot (\rho \phi \boldsymbol{v}) + \phi \boldsymbol{\nabla} \cdot (\rho \boldsymbol{v}) \\ &= -\boldsymbol{\nabla} \cdot (\rho \phi \boldsymbol{v}) - \phi \frac{\partial \rho}{\partial t} \qquad (16) \\ &= -\boldsymbol{\nabla} \cdot (\rho \phi \boldsymbol{v}) - \frac{\phi}{4\pi G} \frac{\partial}{\partial t} \boldsymbol{\nabla}^2 \phi \\ &= -\boldsymbol{\nabla} \cdot \left[\phi \left(\rho \boldsymbol{v} - \frac{1}{4\pi G} \frac{\partial \boldsymbol{g}}{\partial t} \right) \right] + \frac{\partial}{\partial t} \left(\frac{\boldsymbol{g} \cdot \boldsymbol{g}}{8\pi G} \right), \qquad (17) \end{aligned}$$

by the help of equations (3) and (7). Thus equation (13) is equivalent to

$$\frac{\partial}{\partial t} \left(\rho E - \frac{\boldsymbol{g}^2}{8\pi G} \right) + \boldsymbol{\nabla} \cdot \left[\rho \boldsymbol{v} \left(H + \phi \right) - \frac{\phi}{4\pi G} \frac{\partial \boldsymbol{g}}{\partial t} \right] = 0. \tag{18}$$

As well as equations (7) and (9), equation (18) has no source term. If the flux across the boundary vanishes, the total energy defined as

$$E_{\text{tot}} = \int \left(\rho E - \frac{\boldsymbol{g} \cdot \boldsymbol{g}}{8\pi G} \right) dV. \tag{19}$$

is conserved. This is equal to the total energy defined by

$$E_{\text{tot}} = \int \left(\rho E + \frac{\rho \phi}{2} \right) dV, \tag{20}$$

as shown in the previous section. Thus, a numerical scheme which guarantees total energy conservation is obtained by applying the Gauss's theorem to equation (18).

Remember that Jiang et al.[1] uses another form of the equation of energy conservation,

$$\frac{\partial}{\partial t} \left(\rho E + \frac{\rho \phi}{2} \right) + \boldsymbol{\nabla} \cdot \left(\rho H \boldsymbol{v} + \boldsymbol{F}_{\text{g}} \right) = 0, \tag{21}$$

$$\boldsymbol{F}_{\text{g}} = \frac{1}{8\pi G} \left(\phi \frac{\partial \boldsymbol{g}}{\partial t} - \frac{\partial \phi}{\partial t} \boldsymbol{g} \right) + \rho \boldsymbol{v} \phi. \tag{22}$$

Note that the total energy is defined not by equation (19) but by equation (20) in this form. Accordingly, the energy flux is different from ours and has an additional term.

When the potential is stationary, we can rewrite Equation (16) into

$$\rho \boldsymbol{v} \cdot \boldsymbol{g} = -\boldsymbol{\nabla} \cdot (\rho \phi \boldsymbol{v}) - \frac{\partial}{\partial t} (\rho \phi). \tag{23}$$

Thus the equation of energy conservation is expressed as

$$\frac{\partial}{\partial t} \left[\rho \left(E + \phi \right) \right] + \boldsymbol{\nabla} \cdot \left[\rho \boldsymbol{v} \left(H + \phi \right) \right] = 0, \tag{24}$$

which has no source term. Thus, a numerical scheme which guarantees total energy conservation is obtained by applying the Gauss's theorem to equation (24).

IOP Conf. Series: Journal of Physics: Conf. Series **1225** (2019) 012015 doi:10.1088/1742-6596/1225/1/012015

2.3. Proper Discretization of the Source Terms

Equations (9), (18) and (24) are ideal in the sense that the total energy of a numerical solution remains constant and the error is of the order of round-off. However, they may give a seriously large error in a region where the density is very low while the gravity is large. Both the fluxes of momentum and energy are large while the divergence is very small in such a region. If the fluxes contain relatively small errors, they result in a relatively large error in the divergence.

The above-mentioned problem is solved if we can find a source term which coincides with the numerical derivative of the corresponding flux or energy density. In the following, we seek the discretized forms of equations (8) and (13) which are equivalent to those of equations (9) and (18). The equivalence ensures the conservation of total momentum and that of total energy, since Equations (9) and (18) are in the conservation form without source terms and the total energy and momentum are conserved in the solutions. Such discretized equations should be consistent with the equation of continuity and the Poisson equation, since we use the latter equations to derive equations (9) and (18) from equations (8) and (13).

First, we consider a Cartesian uniform grid in which each numerical cell is a cuboid having widths, Δx, Δy, and Δz, in the x-, y-, and z-directions, respectively. This means that the center of each numerical cell is expressed as

$$(x_i, y_j, z_k) = (i\Delta x, j\Delta y, k\Delta z), \tag{25}$$

where i, j, and k are integers specifying the cuboid. Then equation (6) is expressed as

$$\frac{\phi_{i+1,j,k} - 2\phi_{i,j,k} + \phi_{i-1,j,k}}{\Delta x^2} + \frac{\phi_{i,j+1,k} - 2\phi_{i,j,k} + \phi_{i,j-1,k}}{\Delta y^2} + \frac{\phi_{i,j,k+1} - 2\phi_{i,j,k} + \phi_{i,j,k-1}}{\Delta z^2} = 4\pi G \rho_{i,j,k}, \tag{26}$$

where $\phi_{i,j,k}$ and $\rho_{i,j,k}$ denote the gravitational potential and density at the center of numerical cell numbered (i, j, k). Then the discretized source term should meet the following conditions.

(i) The gravity, $\boldsymbol{g}$, should be evaluated on the numerical cell surface, since both the gravitational acceleration ($\rho \boldsymbol{g}$) and energy release ($\rho \boldsymbol{v} \cdot \boldsymbol{g}$) originate from the divergence of momentum and momentum fluxes, respectively.

(ii) The gravitational energy, $-\boldsymbol{g} \cdot \boldsymbol{g}/(8\pi G)$, should be evaluated from the component normal to the cell surface since the tangential component does not appear in equation (26). The same requirement is derived also from equation (5). The tangential component of the gravity can be derived from equation (11) but need not be computed, although $\boldsymbol{T}_\mathrm{g}$ contains both the normal and tangential components of gravity.

(iii) The gravitational energy release should be evaluated as the product of the normal components of $\rho \boldsymbol{v}$ and $\boldsymbol{g}$. This constraint comes from the fact that the change in the gravitational potential is proportional to that in the density distribution. The change in the density is due to the mass flux across each numerical cell surface. Remember that essentially the same method is used in earlier works [2, 3, 4].

(iv) The gravity should be the average of those at $t = t_0$ and $t = t_0 + \Delta t$, where t_0 and Δt denotes the time, t, at the initial state and the time step, respectively. Otherwise, equation (17) may not hold since the solution may not satisfy,

$$\frac{\partial}{\partial t}\left(\frac{\boldsymbol{g} \cdot \boldsymbol{g}}{8\pi G}\right) = \frac{\boldsymbol{g}}{4\pi G} \cdot \frac{\partial \boldsymbol{g}}{\partial t}. \tag{27}$$

Note that the change in the gravitational energy density can be evaluated as

$$-\frac{[\boldsymbol{g}(t_0 + \Delta t)]^2}{8\pi G} + \frac{[\boldsymbol{g}(t_0)]^2}{8\pi G} = -\left[\frac{\boldsymbol{g}(t_0 + \Delta t) + \boldsymbol{g}(t_0)}{8\pi G}\right] \cdot [\boldsymbol{g}(t_0 + \Delta t) - \boldsymbol{g}(t_0)]. \tag{28}$$

The total momentum is conserved if items (i) and (ii) are fulfilled. The total energy is conserved only when all the items are fulfilled. We need not take account item (iv) into account when the gravitational potential is stationary. Fortunately, these requirements serve as the guidelines to derive a scheme which ensures the conservation of total momentum and energy.

After some algebra, we find that the properly discretized source term is expressed as

$$(\rho g_x)_{i,j,k} = \frac{\rho_{i,j,k}}{2}\left(g_{x,i+1/2,j,k} + g_{x,i-1/2,j,k}\right), \tag{29}$$

$$(\rho g_y)_{i,j,k} = \frac{\rho_{i,j,k}}{2}\left(g_{y,i,j+1/2,k} + g_{y,i,j-1/2,k}\right), \tag{30}$$

$$(\rho g_z)_{i,j,k} = \frac{\rho_{i,j,k}}{2}\left(g_{z,i,j,k+1/2} + g_{z,i,j,k-1/2}\right), \tag{31}$$

$$\begin{aligned}(\rho \boldsymbol{v} \cdot \boldsymbol{g})_{i,j,k} = \frac{1}{2}\Big[&(\rho v_x)_{i+1/2,j,k}\, g_{x,i+1/2,j,k} + (\rho v_x)_{i-1/2,j,k}\, g_{x,i-1/2,j,k} \\ &+ (\rho v_y)_{i,j+1/2,k}\, g_{y,i,j+1/2,k} + (\rho v_y)_{i,j-1/2,k}\, g_{y,i,j-1/2,k} \\ &+ (\rho v_z)_{i,j,k+1/2}\, g_{z,i,j,k+1/2} + (\rho v_z)_{i,j,k-1/2}\, g_{z,i,j,k-1/2}\Big],\end{aligned} \tag{32}$$

$$g_{x,i+1/2,j,k} \equiv \frac{\phi_{i+1,j,k} - \phi_{i,j,k}}{\Delta x}, \tag{33}$$

$$g_{y,i,j+1/2,k} \equiv \frac{\phi_{i,j+1,k} - \phi_{i,j,k}}{\Delta y}, \tag{34}$$

$$g_{z,i,j+1/2,k} \equiv \frac{\phi_{i,j,k+1} - \phi_{i,j,k}}{\Delta z}, \tag{35}$$

in the Cartesian uniform grid. Here all the values should be the time average between $t = t_0$ and $t_0 + \Delta t$. Equations (29) through (32) define the gravity in each cell as the volume average of those defined on the two cell surfaces. Equation (26) can be rewritten as

$$-\frac{g_{x,i+1/2,j,k} - g_{x,i-1/2,j,k}}{\Delta x} - \frac{g_{y,i,j+1/2,k} - g_{x,i,j-1/2,k}}{\Delta y} - \frac{g_{z,i,j,k+1/2} - g_{x,i,j,k-1/2}}{\Delta z} = 4\pi G \rho_{i,j,k}. \tag{36}$$

Equations (29) through (32) give the properly discretized source terms also when the gravitational potential is stationary and given (see, e.g., [4]). We can confirm it by substituting Equations (29) through (35) into the right hand side of Equations (8) and (16).

Next, we consider a non-uniform Cartesian grid in which the separation is expressed as

$$x_{i+1} - x_i = \frac{\Delta x_i + \Delta x_{i+1}}{2} \equiv \Delta x_{i+1/2}, \tag{37}$$

$$y_{j+1} - y_j = \frac{\Delta y_i + \Delta y_{j+1}}{2} \equiv \Delta y_{j+1/2}, \tag{38}$$

$$z_{k+1} - z_k = \frac{\Delta z_k + \Delta z_{k+1}}{2} = \Delta z_{k+1/2}, \tag{39}$$

between the neighboring numerical cells. Here, the symbols, Δx_i, Δy_j, and Δz_k denote the width, depth, and height of the numerical cell numbered (i, j, k). Then equations (33) thorough (35) should be replaced with

$$g_{x,i+1/2,j,k} \equiv \frac{\phi_{i+1,j,k} - \phi_{i,j,k}}{\Delta x_{i+1/2}}, \tag{40}$$

$$g_{y,i,j+1/2,k} \equiv \frac{\phi_{i,j+1,k} - \phi_{i,j,k}}{\Delta y_{j+1/2}}, \tag{41}$$

$$g_{z,i,j+1/2,k} \equiv \frac{\phi_{i,j,k+1} - \phi_{i,j,k}}{\Delta z_{k+1/2}}, \tag{42}$$

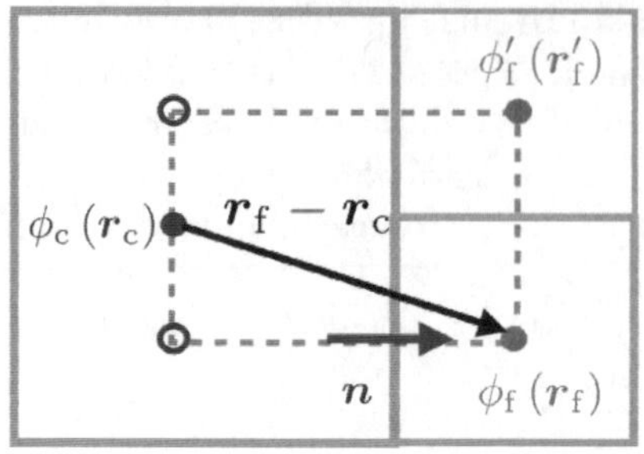

Figure 1. Gravity across the coarse-fine grid boundary in the octree grid. See Equation (43).

However, other formulae for the properly discretized source term remain unchanged.

Almost the same formulae give the properly discretized source term for an octree grid often used in Adaptive Mesh Refinement (AMR). The numerical cells are cubes and some of them are subdivided into 8 cubes recursively in the octree grid. When the neighboring cells have the same size, the same formulae are applied. Only when the neighboring cells have different sizes, i.e., on the coarse-fine grid boundary, we need to modify equations (40) through (42). The normal component of the gravity should be expressed as

$$\boldsymbol{g}_{\rm n} = -\left[\frac{\phi_{\rm f} - \phi_{\rm c}}{(\boldsymbol{r}_{\rm f} - \boldsymbol{r}_{\rm c}) \cdot \boldsymbol{n}}\right] \boldsymbol{n}, \tag{43}$$

where $\boldsymbol{n}$ denotes the unit vector normal to the cell interface. Here the symbols, $\phi_{\rm c}$ and $\boldsymbol{r}_{\rm c}$, denote the potential and position at the coarse cell center, while $\phi_{\rm f}$ and $\boldsymbol{r}_{\rm f}$ do those at the fine cell center (see Figure 1). Any flux across the coarse-fine boundary should be evaluated on the fine grid side. The flux on the coarse grid side should be equated with the sum of those on the fine grid side. Otherwise, the Gauss's theorem, which is essential for a conservation law, does not hold in the discretized equation. This principle applies also to the Poisson equation. Equation (43) implies that the potential at the cell center (the filled circle in Figure 1) is the same as that at the opposing normal to the fine cell (the open circles). In other words, we ignore the gravity tangential to the coarse-fine boundary, i.e., the potential difference between the two open circles. Otherwise, the circulation along the dashed line results in either gain or loss of energy, $\oint \boldsymbol{g} \cdot d\boldsymbol{s} \neq 0$.

The properly discretized source term is likely to be given for other types of numerical grids but it is beyond the scope of this paper.

2.4. Numerical Procedure

As shown in the previous section, our method does not give any constraints on the choice of approximate Riemann solvers. However, the time-averaged gravity should be used to update the velocity and internal energy of the gas. Thus we need to compute the gravity at the advanced stage before completing the update of the velocity and internal energy. The numerical examples to be shown in the next section are obtained by the procedure.

(i) Obtain the gravity, $\boldsymbol{g}(t_0)$, at the initial stage.

(ii) Compute the initial guess for the density, $\rho^*(t_0 + \Delta t)$, at the advanced stage by solving the equation of continuity.

(iii) Compute the initial guess for the gravity, $\boldsymbol{g}^*(t_0 + \Delta t)$ from $\rho^*(t_0 + \Delta t)$.

(iv) Compute the initial guess for the velocity, $\boldsymbol{v}^*(t_0 + \Delta t)$, and internal energy, $\varepsilon^*(t_0 + \Delta t)$, by solving the equation of motion and that of energy conservation.
(v) Improve the density, $\rho(t_0 + \Delta t)$ by using the time-averaged mass flux.
(vi) Obtain the gravity, $\boldsymbol{g}(t_0 + \Delta t)$ from $\rho(t_0 + \Delta t)$.
(vii) Improve the velocity, $\boldsymbol{v}(t_0 + \Delta t)$, and internal energy, $\varepsilon(t_0 + \Delta t)$.
(viii) Go back to step (ii) after replacing $t_0 + \Delta t$ with t_0.

Here, steps (ii) through (iv) are predictor stages while steps (v) through (vii) are corrector ones. This procedure requires to solve Poisson equation twice per cycle since step (i) is performed only once at the initial stage. However, we can omit step (iii), although the total energy is not conserved at the predictor stage then. Our scheme does not increase the number of times solving the Poisson equation.

3. Results

3.1. Jeans Instability in 2D

The first example demonstrates that the total energy is conserved in the 2D fragmentation of initially Jeans unstable gas. The initial condition is constructed by superposing a sinusoidal perturbation on a uniform gas having the initial density, $\rho_0 = 1.0$, and pressure $P_0 = 0.6$. The specific heat ratio is set to be $\gamma = 5/3$. We use the unit system in which the gravitational constant is $G = 1/4\pi$. Thus the Jeans length is $\lambda_J = 2\pi$ at the initial state. The computation box is set to be as large as $(4\sqrt{2}\pi)^2$ and covered with a uniform grid having 32^2 cells. The periodic boundary condition is applied to both the x- and y-directions. The Poisson equation is solved with the Fast Fourier Transform (FFT). The CFL number is set to be 0.4. It took 2400 time steps to follow the evolution of $0 \leq t \leq 107.40$.

Figure 2 shows the fragmentation by a series of snapshots. The top left panel shows the initial stage at $t = 0.00$, while the bottom right does the final one at $t = 107.40$. The color and contours denote the density and entropy, $s = \ln(\gamma P) - \gamma \ln \rho$, respectively, while the arrows denote the velocity. First, the initially uniform gas is fragmented into a sheet as shown in the top right panel. Then the sheet is fragmented into a cylinder.

Figure 3 shows the change in the energy in this model. The left panel denotes the kinetic energy ($E_{\rm kin}$), thermal energy ($E_{\rm th}$), and the gravitational energy ($-E_{\rm g}$) as a function of time. The ordinate is normalized by the initial thermal energy. The gravitational energy decreases by two steps; first at the sheet formation and second at the sheet fragmentation. The right panel denotes the change in the total energy, $E_{\rm tot} = E_{\rm kin} + E_{\rm th} + E_{\rm g}$. The ordinate is normalized by 10^{-13} times of the initial thermal energy, $10^{-13} E_{\rm th}(t = 0)$. The total energy increases little. The small increase is due to the fact that all the computations are done with the double-precision floating-point format.

3.2. Gravitational Collapse on Nested Grid

The second example demonstrates that our method can be applied to a nested grid. The nested grid consists of two levels of uniform grids having 64^3. The coarse grid covers the cube of $|x| \leq 2\pi$, $|y| \leq 2\pi$ and $|z| \leq 2\pi$ while the fine grid does that of $|x| \leq \pi$, $|y| \leq \pi$ and $|z| \leq \pi$ in the Cartesian coordinates. The gravitational potential is fixed to be

$$\phi = -0.6 \left[\cos \frac{x}{2} + \cos \frac{y}{2} + \cos \frac{z}{2} \right] . \tag{44}$$

The reflection boundary is set on the outer boundaries of the coarse grid. The source term is evaluated from the gravity derived from the numerical difference in the potential defined at the cell center.

IOP Conf. Series: Journal of Physics: Conf. Series **1225** (2019) 012015 doi:10.1088/1742-6596/1225/1/012015

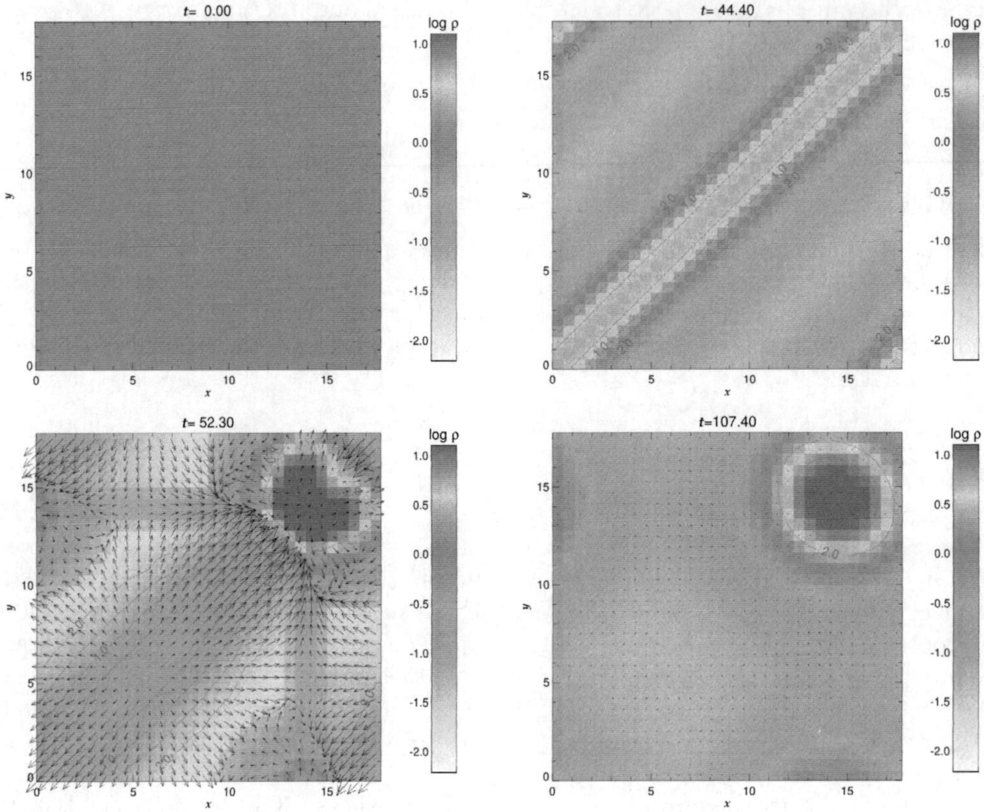

Figure 2. 2D fragmentation of a Jeans unstable gas is shown by a series of snapshot.

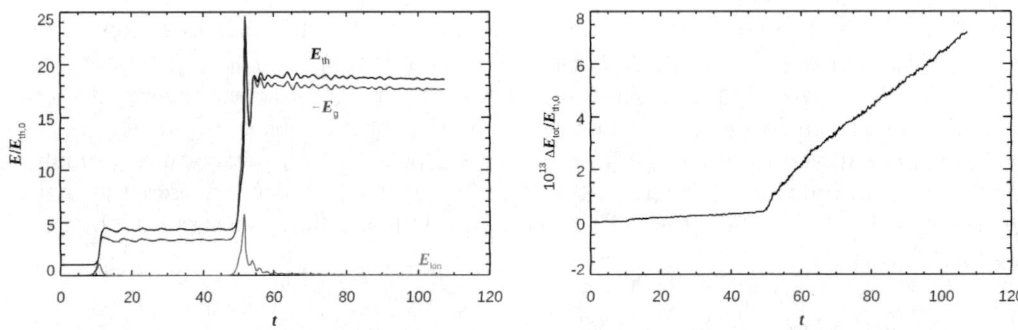

Figure 3. The left panel shows $E_{\rm kin}$, $E_{\rm th}$ and $-E_{\rm g}$ as a function of time, t. The right panel shows the change in the total energy from its initial value as a function of t.

This model follows the collapse of initially uniform gas at rest. The initial density and pressure are set to be $\rho_0 = 1.0$ and $P_0 = 0.6$. The specific heat ratio is set to be $\gamma = 5/3$. The time step is taken to be the same for the coarse and fine grids. The CFL number is set to be 0.25. It took 6400 time steps to follow the evolution of $0 \leq t \leq 48.39$.

Figure 4 is the same as Figure 3 but for the gravitational collapse followed with the nested

IOP Conf. Series: Journal of Physics: Conf. Series **1225** (2019) 012015 doi:10.1088/1742-6596/1225/1/012015

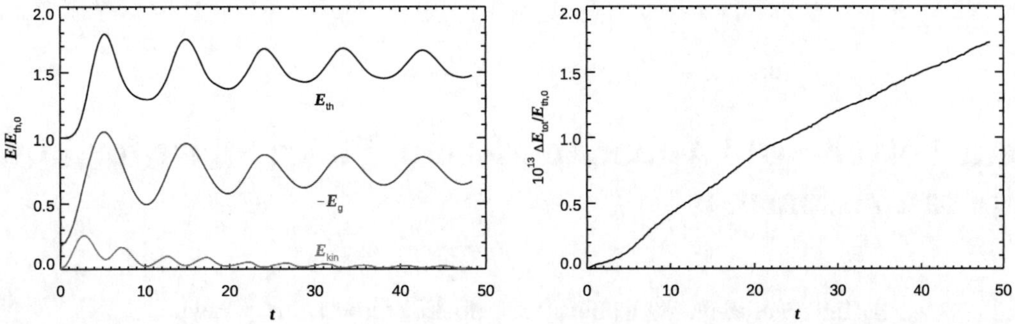

Figure 4. The same as Fig 3 but for the gravitational collapse followed with the nested grid.

grid. The gravitational energy oscillates with a substantial amplitude, while the change in the total energy is of the order of round-off error.

4. Summary

As shown in the previous sections, the total energy including self-gravity can be ensured by a rather simple procedure. The procedure can be applied as far as the conservation form is used for solving both the hydrodynamical and Poisson equations. Poisson equation is solved only once or twice per time step as well as in a conventional scheme of the second-order accuracy in time. The procedure increases computation cost only through increase in the memory to store gravity and mass flux at the predictor step.

The method proposed in this paper is based on equation (18) in which the gravitational energy density is expressed as $-\boldsymbol{g} \cdot \boldsymbol{g}/(8\pi G)$. The method has two key elements. First, the gravity should be evaluated on the cell surface. Second, it should be the average of those at the initial and final stages for each time step [cf. equation (28)]. The former key element implies that the source term has different values inside a numerical cell and the integral over the numerical cell is used in computation. This concept should hold not only for a Cartesian uniform grid but for a variety of numerical grids. The latter means that the gravity should be updated before completing the update of velocity and internal energy. This stands to reason since the gravitational energy is fixed once the density distribution is given. We need to evaluate the gravitational energy at the new time step in advance for ensuring the total energy conservation. However, this point has not been mentioned in literature as far as the author knows.

This paper has outlined the numerical scheme which ensures the conservation of total energy and momentum. Further details will be published elsewhere.

Acknowledgments

This work is supported in part by JSPS KAKENHI Grant Number JP 18H05222.

References

[1] Jiang, Yan-Fei, Belyaev, M., Goodman, J., Stone, J. M. 2013, *New Astronomy*, **19**, 48
[2] Mikami, H., Sato, Y., Matsumoto, T. & Hanawa, T. 2008, *ApJ*, **683**, 357
[3] Springel, V. 2010, *MNRaS*, **401**, 791
[4] Katz, M.P., Zingale, M., Calder, A.C., Swesty, F.D., Almgren, A.S., Zhang, W. 2016, *ApJ*, **819**, 94
[5] Shu, F.H. 1992, The Physics of Astrophysics vol. II Gas Dynamics (University Science Books, Mill Valley) p. 47

IOP Conf. Series: Journal of Physics: Conf. Series **1225** (2019) 012016 doi:10.1088/1742-6596/1225/1/012016

Kinetic Solvers with Adaptive Mesh in Phase Space for Low-Temperature Plasmas

Vladimir Kolobov, [a,b,1] **Robert Arslanbekov** [a] **and Dmitry Levko** [a]

[a] CFD Research Corporation, Huntsville, AL 35806, USA
[b] The University of Alabama in Huntsville, AL 35899, USA

E-mail: vladimir.kolobov@cfdrc.com

Abstract. We describe the implementation of 1d1v and 1d2v Vlasov and Fokker-Planck kinetic solvers with adaptive mesh refinement in phase space (AMPS) and coupling these kinetic solvers to Poisson equation solver for electric fields. We demonstrate that coupling AMPS kinetic and electrostatic solvers can be done efficiently without splitting phase-space transport. We show that Eulerian fluid and kinetic solvers with dynamically adaptive Cartesian mesh can be used for simulations of collisionless plasma expansion into vacuum. The Vlasov-Fokker-Planck solver is demonstrated for the analysis of electron acceleration and scattering as well as the generation of runaway electrons in spatially inhomogeneous electric fields.

1. Introduction

Simulations of low-temperature plasma require coupling kinetic solvers with electromagnetic solvers. With increasing computing power, mesh-based kinetic solvers become competitive to the particle-based (PIC) solvers.[1,2] Advantages of mesh-based kinetic solvers over statistical particle methods for hybrid kinetic-fluid simulations are discussed in Refs. [3,4]. While PIC solvers with Adaptive Mesh Refinement (AMR) are relatively rare,[5,6] using AMR is a proven technique for mesh-based kinetic solvers and electromagnetic solvers, allowing greatly reduce computational cost of both methods. Challenges appear when coupling AMR solvers of different dimensionality. While kinetic solvers operate in phase space, electromagnetic solvers operate in configuration (physical) space.

We have previously shown that splitting velocity and configuration spaces is one way to resolve the coupling challenge for kinetic solvers with adaptive mesh in phase space (AMPS).[7] Here, we further explore this challenge for problems of lower dimensionality. Using symmetry is a well-known method to reduce dimensionality of kinetic solvers. In the present paper, we describe the implementation and application of 1d1v and 1d2v kinetic solvers with AMPS and the coupling of these solvers to 1d Poisson solver for electric field. The choice of coordinate system for kinetic equations depends on the problem type. The best coordinate system in physical space is dictated by the symmetry of the problem. The best coordinate system in velocity space depends on relative importance of collisions and forces. Usually, collision operator is close to diagonal in spherical coordinate system while the force term is nearly diagonal in cylindrical coordinate system.

Collisions of charged particles with neutrals are described using spherical coordinates in velocity space. For transport in configuration (physical) space, we use Cartesian, cylindrical or spherical coordinate systems depending on problem type. Collisions are divided into two types. Boltzmann-type

[1] To whom any correspondence should be addressed.

collisions are associated with a large change of particle momentum – they are described by an integral operator in velocity space. Fokker-Planck-type collisions are associated with small changes of momentum. They are described by Fokker-Planck-type differential operators in velocity space. We solve Vlasov and Fokker-Planck kinetic equation using Finite Volume method with octree Cartesian mesh, without splitting physical and velocity spaces. As a result, mesh adaptation in velocity space triggers mesh adaptation in physical space. This creates some difficulties for calculation of particle density and coupling kinetic solvers to Poisson solver for calculation of electrostatic field.

In this paper, we describe the implementation and application of 1d1v and 1d2v Vlasov and Fokker-Planck kinetic solvers and the coupling of these solvers to the Poisson solver for electric field. We demonstrate that coupling Vlasov and Poisson solvers can be done efficiently without splitting phase-space transport. We also demonstrate that (Eulerian) fluid and kinetic solvers with dynamically adaptive Cartesian mesh can be used for simulations of multi-dimensional problems of collisionless plasma expansion into vacuum.

Collision processes in the Fokker-Planck solver have been implemented using spherical coordinates in velocity space for simulations of electron kinetics in weakly-ionized plasma. We demonstrate these solvers for analysis of electron acceleration by electric fields and scattering by neutral atoms as well as electron impact ionization of atoms and generation of runaway electrons in spatially inhomogeneous electric fields.

2. Coupling Kinetics with Electrostatics

A whole spectrum of mesh-based kinetic solvers has been explored ranging from those without splitting 6d phase space transport[8] to those splitting phase space transport into 1d advection for each dimension.[9] To explore best strategy of coupling AMPS kinetic solvers with electrostatic solvers, we study here 1d1v and 1d2v problems using Basilisk - an open source framework for solving partial differential equations on adaptive Cartesian meshes.[10]

When using adaptive mesh in phase space without splitting velocity and physical spaces, calculations of VDF moments (such as particle density and fluxes) by integration of the VDF over velocity space becomes challenging. The problem of coupling AMPS Vlasov and Poisson solvers has been clearly demonstrated in Ref. [11]. To overcome this challenge, we have developed a methodology illustrated in Figure 1 for the quad-tree grid.

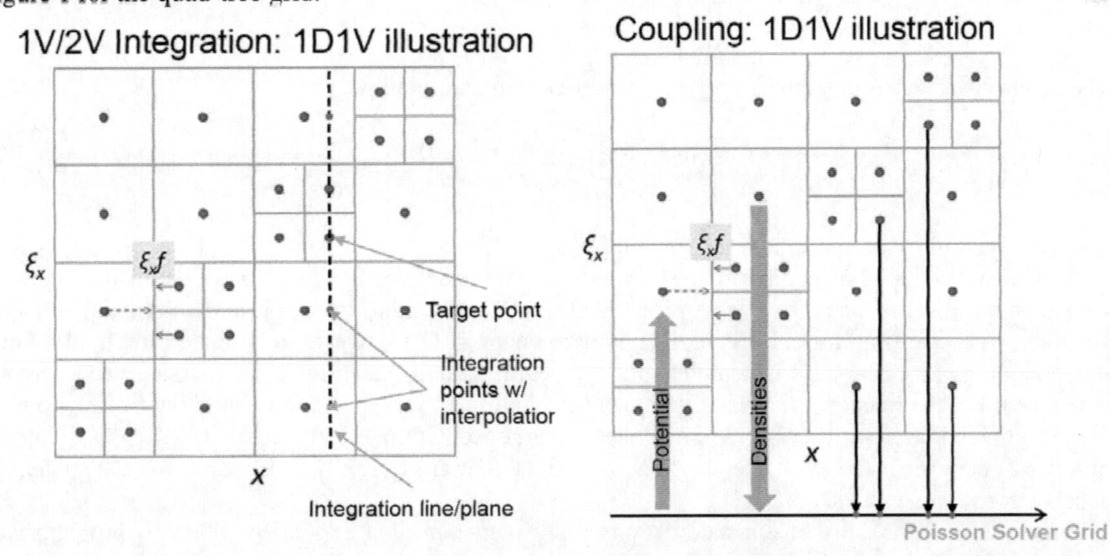

Figure 1. Calculation of velocity space integrals (left) and coupling 1d1v Vlasov and 1d Poisson solvers (right).

IOP Conf. Series: Journal of Physics: Conf. Series **1225** (2019) 012016 doi:10.1088/1742-6596/1225/1/012016

To compute VDF moments (density, mean velocity, temperature, etc.) an integration along an x = const line must be performed. In our method, all cells containing this line are identified, and 2nd order interpolation (involving all neighboring cells) is carried out to compute the VDF required for the integration to proceed.

A one-dimensional Poisson solver is coded inside the 1d1v Vlasov solver. For coupling Vlasov and Poisson solvers, the 1d1v quadtree phase-space grid is projected into 1d Poisson-solver grid using cell centered x-locations. The calculated particle densities are then transferred from the 1d1v grid to the 1d grid. After solving the Poisson equation on the 1d grid, electric potential and electric field are returned back at cell centers and faces of the 1d1v phase-space grid. This method is also used for the 1d2v octree phase-space grid. Below, we demonstrate some results obtained with the coupled solvers.

3. Collisionless Plasma Expansion

Plasma expansion into vacuum is a fundamental problem with important applications in space science, laser ablation, electric propulsion and arc interactions with electrodes. The problem is associated with large variations of plasma density within a computational domain calling for a hybrid fluid-kinetic solver.[12] In this section, we describe our recent advances of the fluid and kinetic solvers for collisionless plasma.

3.1. Fluid Model for Cold Ions

The simplest fluid model of collisionless plasma with cold-ions and Boltzmann electrons comprises of a set of equations for the ion and electron densities, n_i and n_e, ion mean velocity $\vec{u}_i$, and the electrostatic potential, φ:

$$\frac{\partial n_i}{\partial t} + \nabla \cdot (n_i \vec{u}_i) = 0, \tag{1}$$

$$m_i \frac{\partial \vec{u}_i}{\partial t} + m_i (\vec{u}_i \cdot \nabla) \vec{u}_i = -e \nabla \varphi, \tag{2}$$

$$\nabla^2 \varphi = \frac{e}{\epsilon_0} (n_e - n_i), \tag{3}$$

The electron density is described by the Maxwell-Boltzmann relationship:

$$n_e = n_0 \exp\left(\frac{e\varphi}{k_B T_e}\right), \tag{4}$$

with (constant in space and time) electron temperature, T_e.

We have developed a second-order scheme to solve the set of Eqs. (1-4) on dynamically adaptive Cartesian mesh using the approach proposed in Ref. [13]. The non-linear Poisson-Boltzmann equation is solved by converting it to a Poisson–Helmholtz equation. The linear term is treated implicitly while the remaining non-linear term is treated explicitly executing non-linear iterations until a given tolerance is reached. For the linear solver, we use an efficient multigrid solver. A typical number of iterations of the Poisson-Helmholtz solver is 2–5 to achieve convergence within a tolerance of 10^{-8}–10^{-6} and a typical number of non-linear (Newton) iterations is 5–10 for a tolerance of 0.01–0.1, which were sufficient to achieve converged results.

Figure 2 shows results of our calculations for a planar plasma expansion. This 1d problem was studied in Ref. [14] using both Eulerian and Lagrangian approaches with static mesh. Our results show good agreement with the results of [14] for both spatial and temporal profiles. In particular, very sharp ion density peaks are predicted with peak widths of a few λ_D. This was achieved by dynamically adapting grid to the space charge with the maximum refinement levels of 12−13 to resolve the local Debye length.

IOP Conf. Series: Journal of Physics: Conf. Series **1225** (2019) 012016 doi:10.1088/1742-6596/1225/1/012016

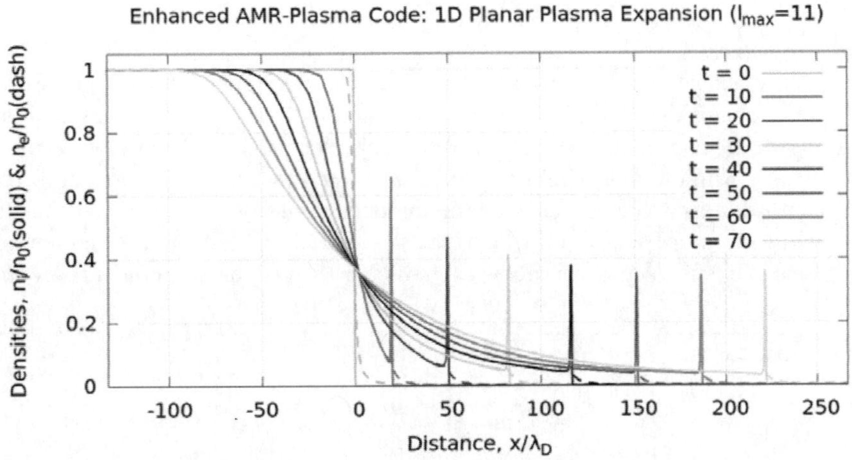

Figure 2. Calculated electron and ion densities for one-dimensional plasma expansion.

Following Ref. [14], we studied effects of the initial ion density profile on plasma dynamics. The initial profile was selected as $n_i(t=0,x) = 0.5 - \arctan(\pi x/l)/\pi$, with the widths of $l = 1$, 5, and 20. According to Ref. [14], the maximum electric field occurs behind the outermost ions, at the point where the electron and ion densities are equal. Positive ions at that position tend to catch up with those moving ahead, producing a peak in the ion density. Our results are in excellent agreement with these theoretical considerations and numerical results.

We have further extended our fluid solver for multi-dimensional problems. Figure 3 shows an example of 3d simulation using an octree Cartesian mesh dynamically adapted to local space charge with the maximum refinement levels of 11-12. The AMR capabilities allow resolving the narrow space-charge zone moving with 3-4 ion-sound speeds while keeping the rest of computational domain (~400 λ_D) coarse for efficient computations and fast convergence rates. The expansion zone remained closely semi-spherical up to several 100's λ_D away from the boundary.

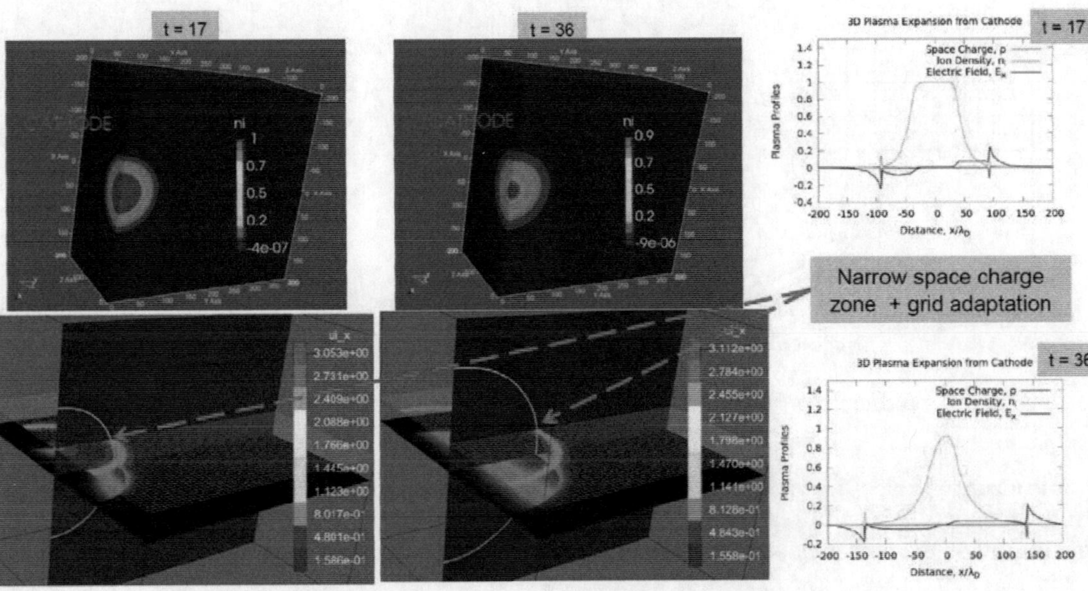

Figure 3. Simulations of 3D plasma expansion from a cathode.

IOP Conf. Series: Journal of Physics: Conf. Series **1225** (2019) 012016 doi:10.1088/1742-6596/1225/1/012016

3.2. Vlasov-Poisson Solver

We have developed a Vlasov solver for ions using quadtree Cartesian mesh in 1d1v phase space. Plasma expansion with the coupled Vlasov-Poisson solver was studied for the planar case. Figure 4 shows spatial distributions of electron and ion densities and the electric field for the case of ion/electron temperature ratio of 0.1, with initial ion density profile as a step function. It is seen that the electric field has a peak at the ion front however no peaks of the ion density are observed. These peaks in the fluid model are artefacts of the numerical treatment of the non-linear term in the ion momentum equation ("wave-breaking" mechanism discussed in Ref [14]). Such a term is absent in the Vlasov formulation.

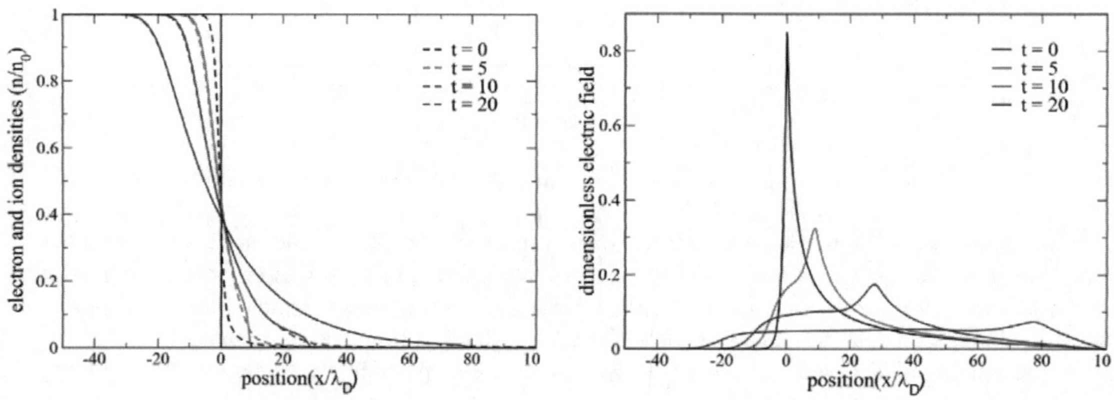

Figure 4. Dynamics of planar plasma expansion with a step initial ion density profile.

For a smooth initial ion density profile with a width about one Debye length, we obtained peaks of the ion density profiles at times > 5 in accordance with the fluid models for smooth initial ion density profiles. However, these peaks disappeared at later times (see Figure 5).

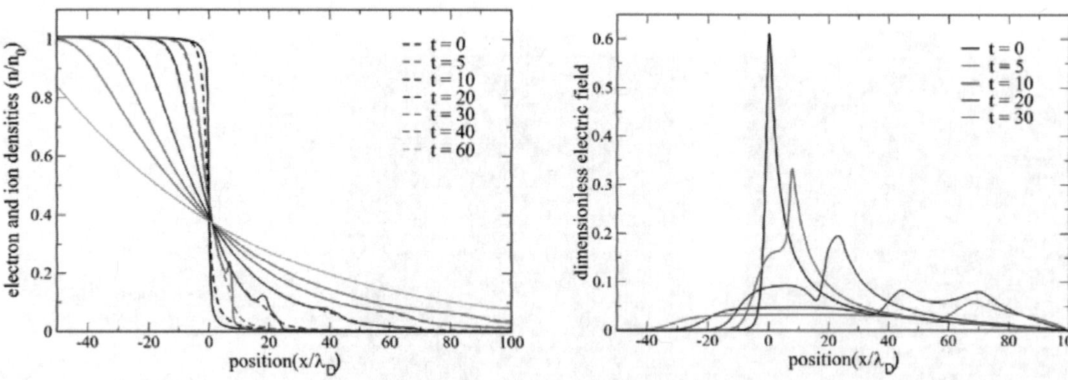

Figure 5. Dynamics of planar plasma expansion with a smooth initial ion density profile.

The mean ion velocities in our simulations reach values of 4–5 in accordance with other (Lagrangian) results (Figure 6). Velocities beyond the ion front depend strongly on the physical model (e.g., fluid vs kinetic) and on the numerical scheme (order of spatial and time accuracy, damping, etc.).

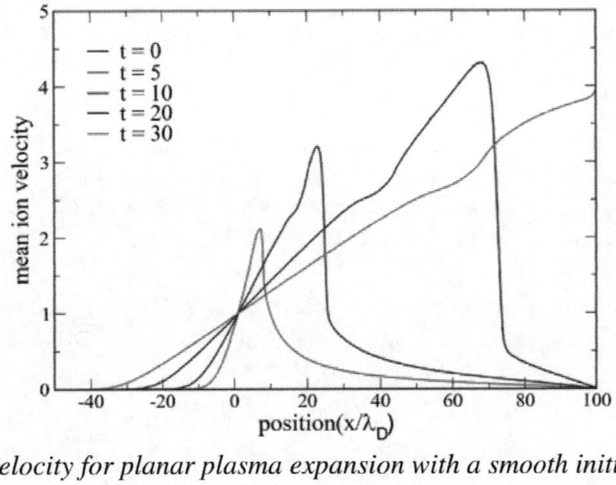

Figure 6. Mean ion velocity for planar plasma expansion with a smooth initial ion density profile.

In conclusion to this section, we can say that the Vlasov-Poisson solver with AMR can be used for simulations of collisionless plasma expansion provided that adequate grids are used in both physical and velocity spaces. The Eulerian fluid code for cold ions and AMR can also be used well to study plasma expansion dynamics in multi-dimensions. We plan to combine these codes into a hybrid kinetic-fluid framework to study collisionless plasma expansion and other problems that can be described by this model.

4. Fokker-Planck Kinetic Solver for Electrons

The Fokker-Planck kinetic equation for the velocity distribution function of charged particles in non-magnetized plasma can be written in the form:

$$\frac{\partial f}{\partial t}+\nabla\cdot(\mathbf{v}f)+\frac{1}{m}\frac{\partial}{\partial \mathbf{v}}\left[\left(F(\mathrm{v})\frac{\mathbf{v}}{\mathrm{v}}-e\mathbf{E}\right)f\right]+\frac{\partial}{\partial \mathrm{v}_i}\left[D_{ij}\frac{\partial f}{\partial \mathrm{v}_i}\right]=I \tag{5}$$

The left side of Eq. (5) describes particle acceleration by the electric force and collision processes with small changes of energy and momentum, which are written in the Fokker-Planck form. The friction force due to energy loss in collisions, $F(\mathrm{v})\mathbf{v}/\mathrm{v}$, is directed towards zero velocity. Small-angle scattering is described by a diffusion tensor:

$$D_{ij}=\nu \mathrm{v}^2\left(\delta_{ij}-\frac{\mathrm{v}_i\mathrm{v}_j}{\mathrm{v}^2}\right) \tag{6}$$

where ν(v) is the collision frequency. Quasi-linear model of wave-particle interactions also results in a Fokker-Planck term, which can be included in the model.

4.1. Spherical Coordinates in Velocity Space

Let us consider a one-dimensional planar problem and use spherical coordinates in velocity space with the axis directed along the electric force. Assuming that the distribution function does not depend on the azimuthal angle, we can rewrite Eq. (5) in the form:

$$\frac{\partial f}{\partial t}+\mathrm{v}\mu\frac{\partial f}{\partial x}-\frac{eE}{m}\left(\mu\frac{\partial f}{\partial \mathrm{v}}+\frac{1-\mu^2}{\mathrm{v}}\frac{\partial f}{\partial \mu}\right)=\frac{1}{\mathrm{v}^2}\frac{\partial}{\partial \mathrm{v}}\left(\mathrm{v}^2Ff\right)+D_\mu\frac{\partial}{\partial \mu}\left(\left(1-\mu^2\right)\frac{\partial f}{\partial \mu}\right)+I \tag{7}$$

where $\mu=\cos\theta$. Now, the diffusion tensor (6) has only one component, $D_\mu(\mathrm{v})$. Introducing $Y=v^2 f$, Eq. (7) can be written in the conservative form:

$$\frac{\partial Y}{\partial t}+\frac{\partial}{\partial x}(UY)+\frac{\partial}{\partial \mathrm{v}}(\mathrm{V}Y)+\frac{\partial}{\partial \mu}[WY]=\frac{\partial}{\partial \mu}\left[D_{\mu}\frac{\partial Y}{\partial \mu}\right]+\mathrm{v}^2 I \tag{8}$$

where

$$U=\mu\mathrm{v}; \qquad V=eE\mu-F; \qquad W=eE\frac{(1-\mu^2)}{m\mathrm{v}}.$$

In the absence of scattering and ionization, a semi-analytical solution of Eq. (8) can be obtained by integration along characteristics:

$$\frac{d\mathrm{v}}{dx}=\frac{eE\mu-F(\mathrm{v})}{m\mathrm{v}\mu}; \qquad \frac{d\mu}{dx}=\frac{eE}{m}\frac{(1-\mu^2)}{\mathrm{v}^2\mu}; \tag{9}$$

The boundary condition $f(x=0)=\Phi(v_0,\mu_0)$ defines the solution $f(x,\mathrm{v},\mu)=\Phi(v_0(\mathrm{v},\mu),\mu_0(\mathrm{v},\mu))$ where μ_0 and v_0 denote the values of μ and v at $x=0$. In the absence of energy losses, Eqs. (9) can be integrated to give:

$$\frac{1-\mu^2}{1-\mu_0^2}=\frac{w_0}{w_0-e\varphi(x)} \tag{10}$$

where $w_0=m\mathrm{v}_0^2/2$ is the electron kinetic energy at $x=0$, and $\varphi(x)=-dE/dx$ is the electrostatic potential. Equation (10) illustrates how rapidly the electron trajectories (characteristics) converge to $\mu=1$ due to electron acceleration by the electric field.

4.2. *Fokker-Planck Solver for Electrons*

The FP equation (8) was solved numerically using adaptive Cartesian mesh in 1d2v phase space (x,v,μ). Figure 7 illustrates the FP solver for the problem of electron acceleration and scattering in a spatially inhomogeneous electric field. It is assumed that the electric field, $E(x)$, decreases linearly at $x<d$, and is zero at $d<x<L$. Electrons are injected at $x=0$ with a Maxwellian distribution and are absorbed at the right boundary, $x=L$. In the ($v_{\parallel}, v_{\perp}$) plane, the VDFs form rings with increasing radius, which are gradually filled by particles (colour) due to scattering.

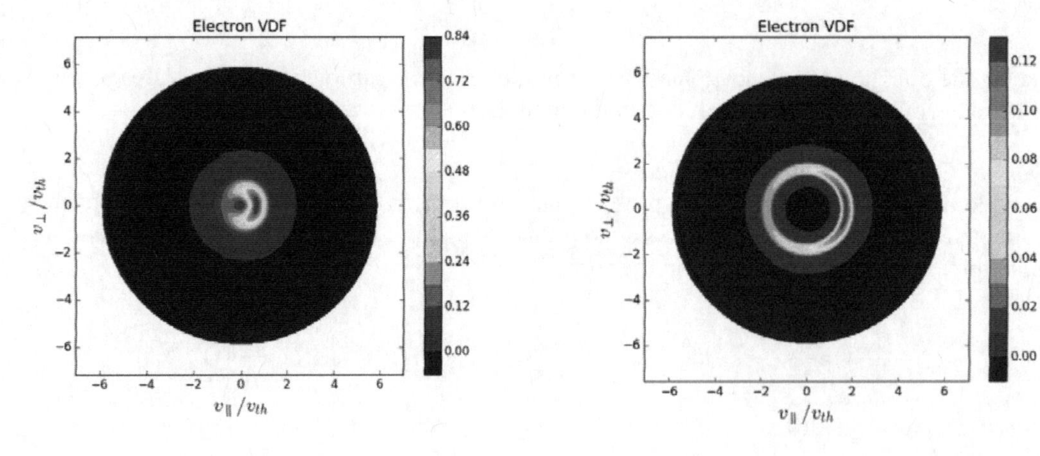

IOP Conf. Series: Journal of Physics: Conf. Series **1225** (2019) 012016 doi:10.1088/1742-6596/1225/1/012016

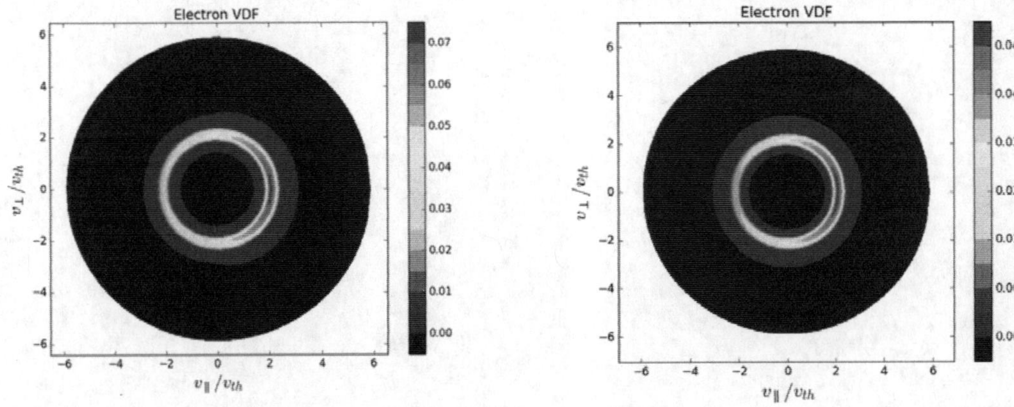

Figure 7. Electron VDF calculated with FP solver at x/L = 0.1, 02, 0.4, and 0.8, for d/L = 0.4.

4.3. Runaway Electrons

Electron runaway in spatially inhomogeneous electric field is a complicated process involving acceleration, energy loss in collisions, scattering and ionization.[15] The energy loss function $F(w)$ has a maximum at about 100 eV, and scattering changes from nearly-isotropic to forward-directed with increasing electron energy. To apply our FP solver for analysis of electron runaway the ionization term was approximated in the simplest form[15]

$$I = \delta(\mathbf{v}) \int \frac{F(v)}{\varepsilon_0} f \, d\mathbf{v} \tag{11}$$

where $\delta(\mathbf{v})$ is a velocity distribution of the secondary electrons, and ε_0 is the cost of ionization. As in the previous section, we consider a linearly decreasing electric field. For easier interpretation of the obtained results, we assume $F(v) = const = F_0$ and $eE_0 / F_0 = 5$, which guaranties the generation of runaway electrons at $eE(x) / F_0 > 1$.

Here, we consider a 1d1v case using the forward-backward scattering model.[16,17] The energy-dependence of the scattering rate is approximated in the form $\exp(\ 0.7v^2)$. Figure 8 illustrates the formation of VDF under effects of acceleration, energy loss, and scattering. Electrons are injected at the left boundary ($x = 0$) with a Maxwellian distribution, move along characteristics (stream lines in the left part of Figure 8), and get absorbed at the anode, at $x = L$. Elastic scattering produces jumps between positive and negative values of v. Due to the rapid decrease of the scattering rate with electron energy there is asymmetry between positive and negative v: slow electrons have near-isotropic VDF, whereas runaway electrons have strongly anisotropic VDF (see the right part of Figure 8). The mesh refinement is based on amplitude of VDF with max-level 8 and min-level 5.

IOP Conf. Series: Journal of Physics: Conf. Series **1225** (2019) 012016 doi:10.1088/1742-6596/1225/1/012016

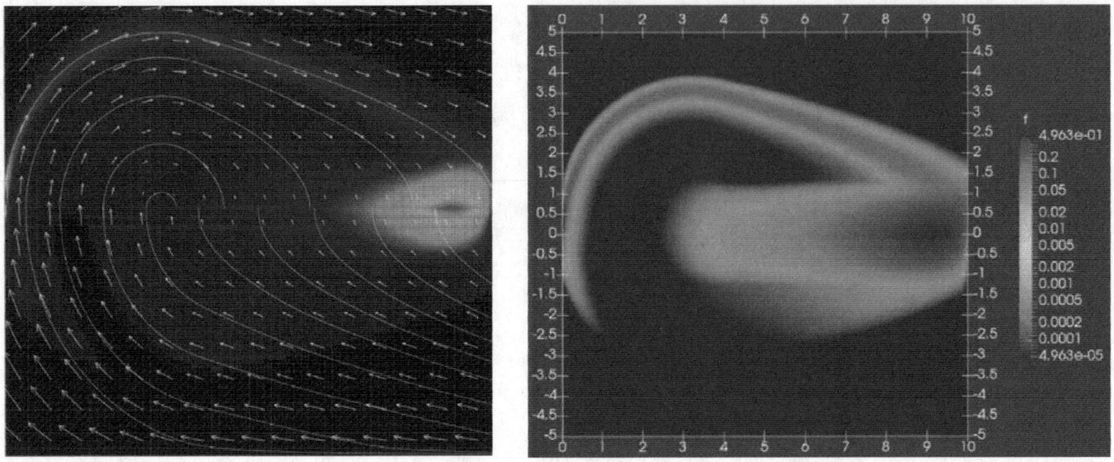

Figure 8. Streamlines (characteristics), adapted computational mesh and calculated VDF in 1d1v phase space (left). The VDF colour map in log scale (on the right). No ionization.

Finally, we include the generation of secondary electrons by electron-impact ionization of the background gas. The velocity distribution of the secondary electrons is assumed in the form $\delta(v) = \sqrt{\xi / \pi} \exp(-\xi v^2)$, which at $\xi \to \infty$ becomes a delta-function considered in the analytical model.15 Figure 9 shows the calculated VDF and spatial distributions of the ionization rate, electron density and temperature. Note the difference in the absolute values of the VDF compared to Figure 8, which is due to the large number of the secondary electrons generated in the present case. The axial distributions of the plasma parameters shown on the right part of Figure 9 correspond to the typical distributions observed in the cathode region of short glow discharges without positive column (see Ref. [15] for details).

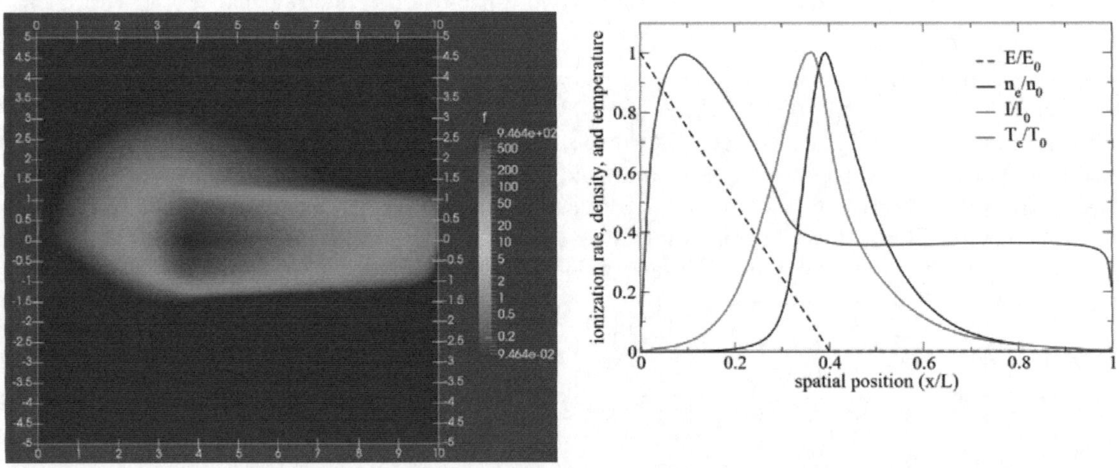

Figure 9. The calculated VDF (colour map in log scale) in the 1d1v phase space (left) and spatial distributions of normalized macro-parameters (right) with ionization.

5. Conclusions

We have described the implementation and application of 1d1v and 1d2v Vlasov and Fokker-Planck kinetic solvers with adaptive mesh in phase space (AMPS). We have demonstrated that coupling kinetic solvers with electrostatic (Poisson) solver can be done efficiently without splitting phase-space transport. We have shown that Eulerian fluid and kinetic solvers with dynamically adaptive Cartesian mesh can be used for simulations of collisionless plasma expansion.

The implemented kinetic solvers have been demonstrated for analysis of electron acceleration, scattering, and generation of runaway electrons in spatially inhomogeneous electric fields. This work opens new opportunities for solving challenging problems of plasma kinetics that can be described by Vlasov-Fokker-Planck equations in 1d2v phase space.

Acknowledgments
This work was partially supported by the DOE SBIR Project DE-SC0015746, by the US Department of Energy Office of Fusion Energy Science Contract DE-SC0001939, and by the NSF EPSCoR project OIA-1655280 "Connecting the Plasma Universe to Plasma Technology in AL: The Science and Technology of Low-Temperature Plasma".

References

[1] S. Zabelok, R. Arslanbekov, and V. Kolobov, Adaptive kinetic-fluid solvers for heterogeneous computing architectures, Journal of Computational Physics 303 (2015) 455
[2] M. Palmroth et al., Vlasov methods in space physics and astrophysics, Living Reviews in Computational Astrophysics, https://doi.org/10.1007/s41115-018-0003-2
[3] V.I. Kolobov, R.R. Arslanbekov, Towards adaptive kinetic-fluid simulations of weakly ionized plasmas, Journal of Computational Physics 231 (2012) 839
[4] https://www.frontiersin.org/research-topics/6432/adaptive-kinetic-fluid-models-for-plasma-simulations-on-modern-computer-systems
[5] K Fujimoto, A new electromagnetic particle-in-cell model with adaptive mesh refinement for high-performance parallel computation, Journal of Computational Physics 230 (2011) 8508
[6] V. Kolobov and R. Arslanbekov, Electrostatic PIC with adaptive Cartesian mesh, Journal of Physics: Conference Series 719 (2016) 012020
[7] R. R. Arslanbekov, V. I. Kolobov, and A. A. Frolova, Kinetic solvers with adaptive mesh in phase space, Phys. Rev. E 88 (2013) 063301
[8] E Deriaz and S Peirani, Six-dimensional adaptive simulation of the Vlasov equations using a hierarchical basis, Multiscale Model. Simul. 16 (2018) 583
[9] S Tanaka et al, Multidimensional Vlasov-Poisson Simulations with High-order Monotonicity and Positivity-preserving Schemes, The Astrophysical Journal 849 (2017) 76
[10] http://basilisk.fr
[11] J. A. F. Hittinger and J. W. Banks, Block-Structured Adaptive Mesh Refinement Algorithms for Vlasov Simulation, J. Comput. Phys. **241**, 118 (2013).
[12] R. Arslanbekov, and V. Kolobov, Adaptive Kinetic-Fluid Models for Expanding Plasmas, Journal of Physics: Conf. Series 1031 (2018) 012018
[13] T. Holgate, and M. Coppins, Numerical implementation of a cold-ion, Boltzmann-electron model for nonplanar plasma-surface interactions, Phys. Plasmas **25** (2018) 043514.
[14] J. E. Allen and M. Perego, On the ion front of a plasma expanding into a vacuum, Phys. Plasmas **21,** 034504 (2014)
[15] V.I. Kolobov and L.D. Tsendin, An analytic model of the cathode region of a short glow discharge in light gases, Phys. Rev. A 46 (1992) 7837
[16] L.P. Babich, K I Bakhov, I M. Kutsyk, et al. Method of simplified kinetic equation for simulating the electron kinetics in dense gases in strong electric fields, Plasma Phys. Rep. 32 (2006) 814.
[17] Y. P. Raizer, *Gas Discharge Physics*, 3rd edition (2009) [in Russian].

IOP Conf. Series: Journal of Physics: Conf. Series **1225** (2019) 012017 doi:10.1088/1742-6596/1225/1/012017

Improved Radiative Transfer in the ANTARES Code

D. Krüger[*,1,2,3], **N. Kostogryz**[2], **D. Fabbian**[1], **F. Kupka**[1]

[1] Institut für Astrophysik, Universität Göttingen, Friedrich-Hund-Platz 1, 37077 Göttingen
[2] Max-Planck-Institut für Sonnensystemforschung, Justus-von-Liebig-Weg 3, 37077 Göttingen
[3] Fakultät für Mathematik, Universität Wien, Oskar-Morgenstern-Platz 1, 1090 Wien

*Contact e-mail: daniel.krueger@uni-goettingen.de

Abstract. One of the most difficult and time-consuming tasks in multi-dimensional stellar radiative-hydrodynamics simulations is the calculation of the radiation field. We present two major improvements we achieved for the ANTARES code in this context. The first one is the use of Bezier splines in combination with the short characteristic method, resulting in a radiative transfer solver which shows more realistic results (less artifacts) and an increased stability compared to the method used so far. The second improvement presented is a radiative transfer solver based on the Eddington approximation. It allows for fast calculation of the radiation field with only a minor loss of precision. Additionally, it shows good numerical properties. These methods for radiative transfer, as newly-implemented within ANTARES, allow improved numerical results in the case of the Bezier method and much faster calculations in the case of the Eddington approximation.

1. Introduction

The ANTARES code was developed to perform numerical simulations of stellar hydrodynamics with realistic microphysics and radiative transfer in 1D, 2D, and 3D [1]. One important focus of these simulations lies on the upper convection zone. A box covering the physical domain to be simulated is thus adopted accordingly and the compressible Navier-Stokes equations are then solved at subsequent time steps on the chosen grid points of the simulation box. Since the stellar plasma and the radiation field exchange energy, the Navier-Stokes equations are coupled with the radiative transfer equation (RTE). The calculation of the radiation field via solution of the RTE usually takes up to $70 - 80\%$ of the whole calculation time in ANTARES.

We consider simulations of convection zones of solar-like stars and white dwarfs with a focus on applications to astero- and helioseismology. Thus, the physical domain simulated is the upper part of the convection zone including the lower part of the photosphere. Calculations covering many sound crossing times to follow a sufficient part of the temporal evolution are needed (e.g. to simulate solar-like p-mode oscillations [2]). Therefore, stable and/or fast computational methods for the radiative transfer are very useful. We present two methods, that are promising in those aspects.

2. Methods for Solving Radiative Transfer in ANTARES

The original radiative transfer scheme in ANTARES uses multiple methods for computation. These shall shortly be described in the following.

The radiation field is found by calculating the formal solution of the radiative transfer equation (RTE), which reads

$$\vec{r} \cdot \nabla I_\nu = \rho\kappa_\nu (S_\nu - I_\nu) \tag{1}$$

with I_ν being the frequency-dependent intensity, ρ being the density, $\vec{r}$ the direction of the ray, S_ν the source function and κ_ν the opacity. We note that dropping velocity- and time-dependent terms in the radiative transfer equation is an approximation which is made in all 3-D radiation hydrodynamics codes used to simulate stellar convection that we are aware of. This is justified by large spatial scales in such simulations where the flow speed is small compared to the speed of light. The validity of this approximation can be derived from the discussion presented in Chap. 6.5 and Chap. 7.2 (Sect. 93) of Mihalas & Mihalas (1984) [3].

We solve the RTE along several rays, thus the equation simplifies to the 1-D case for each ray,

$$\partial_{\tau_\nu} I_\nu = S_\nu - I_\nu, \tag{2}$$

where τ_ν is the optical depth. The dependence on frequency is treated by using the opacity binning method [4, 5]. Usually 4 or 12 bins are used in ANTARES for non-grey atmospheres and the Rosseland mean opacity is used for grey simulations. Additionally, we assume local thermodynamical equilibrium (LTE), i. e. $S_\nu = B_\nu$, where B_ν denotes the Planck function.

For the integration along the rays we use the short characteristic method [6], by fitting a parabola through the points U, P and D of intersection between a given ray and subsequent layers in the simulation box (see Fig. 1) and only using the part of the parabola between U and P. Thus, the solution to eq. 2 reads

$$I_\nu(\tau_{k,\nu}) = I_\nu(\tau_{k-1,\nu})e^{-\delta_{k,\nu}} + \int_{\tau_{k,\nu}}^{\tau_{k-1,\nu}} e^{-(\tau_\nu - \tau_{k,\nu})} S_\nu(\tau) \mathrm{d}\tau_\nu \tag{3}$$

and is numerically solved to get $I_\nu(\tau_k)$, where $\delta_{k,\nu} = \tau_{k,\nu} - \tau_{k-1,\nu}$. The equation to solve then reads

$$I_\nu(\tau_{k,\nu}) = I_\nu(\tau_{k-1,,\nu})e^{-\delta_{k,\nu}} + \alpha_k S_{k-1,\nu} + \beta_k S_{k,\nu} + \gamma_k S_{k+1,\nu}, \tag{4}$$

where α_k, β_k and γ_k are weight functions only depending on the optical depth or in other words on $\delta_{k,\nu}$. The usage of parabolas between k and $k+1$ simplifies the computation but has the drawback that a parabola can overshoot between those two points. This can lead to non-physical results. For horizontal interpolation onto the grid points two equal stencils are used, thus using 4 points in 2-D and 16 points in 3-D [1].

ANTARES uses different methods of angular integration like *Gauss-Radau 3* [7] (GR3) or *Carlson A4* [8], which means that we typically solve Eq. 2 for 10 rays in 2-D and 18 rays in 3-D (in the case of GR3). The direction of each ray differs depending on the method used.

Using the calculated ray-dependent intensities, the mean intensity can easily be derived by summing up over all rays,

$$J_\nu = \frac{1}{4\pi} \int I_\nu \mathrm{d}\omega \approx \sum_i w_i I_\nu(\vec{r}_i), \tag{5}$$

where w_i denotes the corresponding angular weights and $\vec{r}_i$ the ray direction.

The calculation of J_ν gives us an alternative option to calculate the radiative heating Q_{rad}. We can either calculate it as the gradient of the radiative flux,

$$Q_{\mathrm{rad}} = -\nabla \cdot F_{\mathrm{rad}}, \tag{6}$$

or as the difference between J_ν and B_ν,

$$Q_{\mathrm{rad}} = 4\pi\kappa_\nu\rho(J_\nu - B_\nu). \tag{7}$$

In the following section we will drop the index ν but still keeping in mind that the corresponding variables depend on the frequency.

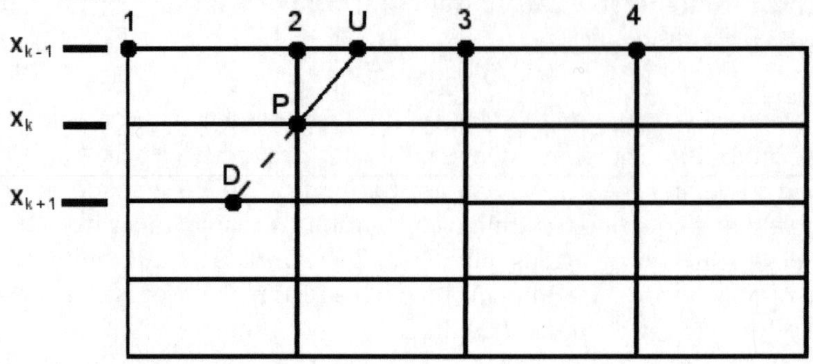

Figure 1. Visual representation of the short characteristic method for a ray entering at the top.

3. Bezier Solver of the Radiative Transfer Equation

The method described above has at least one major problem which is the interpolation of rays. Depending on the method of interpolation the corresponding function can overshoot leading to a larger error for the interpolated values. This overshooting can especially cause the positive definite source function to turn negative leading to unphysical results.

The Bezier method [9] offers a new way to compute the formal solution of Eq. 1, namely controlling the overshooting of the source function. This increases the stability of the calculation of radiative transfer.

To approximate the source function with quadratic Bezier interpolation, the numerical solution of Eq. 3 is written as:

$$I_k = I_{k-1}e^{-\delta_k} + \alpha S_k + \beta S_{k-1} + \gamma C_k, \tag{8}$$

where the coefficients α, β, and γ are calculated by using the expressions from [9] and depend only on optical depth. Note that the coefficients are defined in a different way from the coefficients in Eq. 4. C_k is the Bezier control point that can be expressed using the derivative at the grid points intersecting layer k and $k-1$:

$$\begin{aligned} C_k^0 &= y_k + \frac{x_{k-1} - x_k}{2}\frac{dy_k}{dx}, \\ C_{k-1}^1 &= y_{k-1} - \frac{x_{k-1} - x_k}{2}\frac{dy_{k-1}}{dx}, \\ C_k &= \frac{C_k^0 + C_{k-1}^1}{2}. \end{aligned} \tag{9}$$

For the application to Eq. 8, y corresponds to S and x corresponds to τ. As $\tau_{k-1} - \tau_k < 0$, both C_k^0 and C_{k-1}^1 lie between U and P and thus $C_k \in [U, P]$.

In order to prevent the parabola from overshooting (thus staying between y_k and y_{k+1} with known function values), the harmonic derivatives are calculated at the intersection with the $k-1$ layer and the intersection with the k-th layer.

$$\frac{\mathrm{d}y_k}{\mathrm{d}x} = \frac{d_{\text{front}} \cdot d_{\text{back}}}{\alpha \cdot d_{\text{back}} + (1-\alpha)d_{\text{front}}} \tag{10}$$

where

$$d_{\text{back}} = \frac{y_{k-1} - y_k}{x_{k-1} - x_k}, \tag{11}$$

$$d_{\text{front}} = \frac{y_k - y_{k+1}}{x_k - x_{k+1}}. \tag{12}$$

Overshooting is avoided when the condition $d_{\text{back}} \cdot d_{front} > 0$ is satisfied. When that is not the case, the derivative $\frac{\mathrm{d}y_k}{\mathrm{d}x}$ is set to zero.

As is seen from the definition of the harmonic derivative, the function's values have to be determined at three consecutive intersection points. Due to the usage of the short characteristic method and its specific implementation into ANTARES and given that the RTE is solved locally, we only know the values at three points for every step. These values allow us to define the harmonic derivative for a central point P but not for the previous point U. Therefore, the derivative at U is provided by the slope defined by Eq. 11.

To implement the full Bezier technique an intersection point of the ray with the layer $k-2$ is needed. As this results in a fourth-order method (compared to a second-order method when using only three intersection points), it will lead to a further improvement in the accuracy and precision of the computed models. However, compared to our implementation of the Bezier method (which we call the one-sided Bezier method in the following) this change would decrease the speed of calculations because an additional interpolation at the fourth intersection point is necessary. In comparison to the short characteristic method the one-sided Bezier method usually takes the same amount of calculation time (while being more accurate). The larger number of calculations that need to be done for the one-sided Bezier method are compensated by the higher number of if-statements which only need to be evaluated for the original short characteristic method. These statements take care that variables like the intensity I do not become negative at the boundaries which could happen due to the way of interpolation in the original short characteristic method.

The above integration scheme is described for downward integration, i. e. for a ray from U to P (see Fig. 1). In ANTARES, for each ray the downward and upward integration are performed. For the latter, the points U and D are flipped so we integrate again from U to P and the above formulae do not change.

The Bezier method is also used to calculate the optical depth (with $y = \kappa_\nu$ and x as the height). To solve the RTE, the optical depth at three consecutive intersection points U, P, and D has to be known.

As shown by the results of a comparison for a solar 1-D atmosphere (see Fig. 2), the difference between the methods becomes especially important for the optical depth if the density is very low. This is the case for the photosphere, which is covered by the upper $\sim 20\%$ of our simulated box. The difference found in optical depth results fulfills our expectations since our implementation of the Bezier scheme is a second-order method compared to the trapezoidal rule being a (globally) first-order method. This difference thus becomes especially important in those parts of the atmosphere where the optical depth is strongly nonlinear as a function of opacity.

Fig. 3 shows two snapshots for the radiative flux F_{rad} of a 2-D solar simulation, performed respectively using the Bezier method and the short characteristic method. Weighted parabolas were used for interpolation. We ran the simulation on a 223×171 (vertical $\times$ horizontal) grid with a resolution of 35.08 km in the horizontal direction and 16.5 km in the vertical direction. As time integration method we used a strong-stability-preserving (SSP) method. As boundary conditions we assumed an open top and bottom [10]. We used an adaptive stepwidth ensuring a maximum courant number of $C_{\max} = 0.4$ and thus making sure that the Courant-Friedrichs-Lewy condition

$$C_{\max} = \frac{u \times \Delta t}{\Delta x} \tag{13}$$

IOP Conf. Series: Journal of Physics: Conf. Series **1225** (2019) 012017 doi:10.1088/1742-6596/1225/1/012017

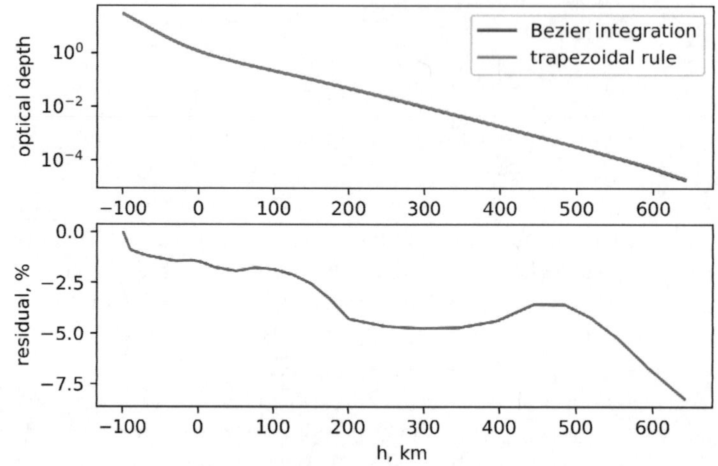

Figure 2. Comparison of values for the optical depth calculated with the Bezier method and the trapezoidal rule for integration. The residual shows that the methods differ mostly towards outer layers (large positive height values), where the density is low. A height value of 0 denotes the solar surface, negative x-values point inside the Sun.

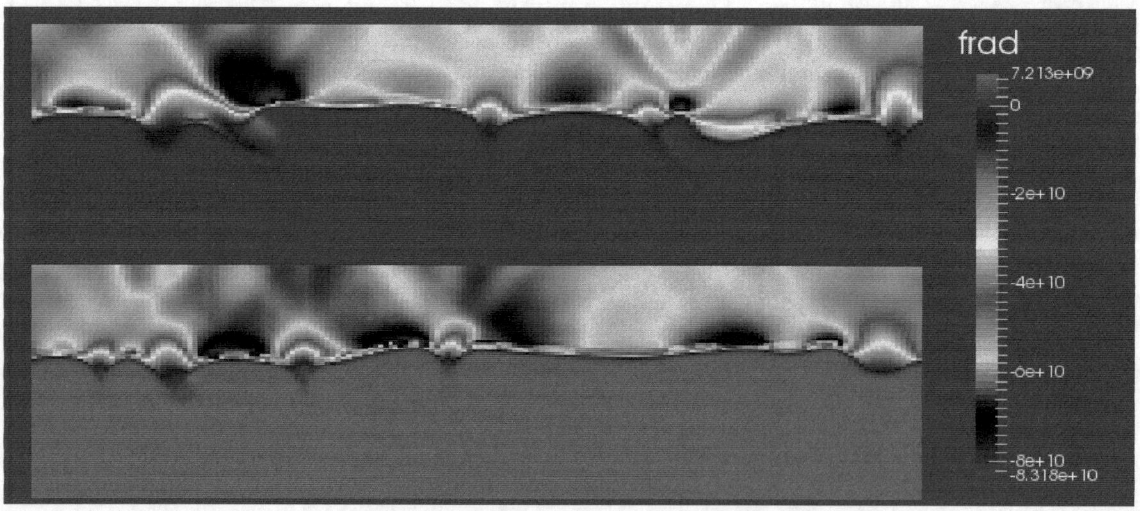

Figure 3. Radiative flux F_{rad} in 2-D ANTARES simulations of the solar atmosphere with the original short characteristic method **(upper image)** and the Bezier method **(lower image)** used for integration. In both cases weighted parabolas were used for interpolation.

holds. Here u is the velocity, Δt the width of the timestep and Δx the resolution of the grid. As one can see from Fig. 3, visible differences between the two simulations lie in the upper part of the box. A comparison of the two images shows that the artifacts along the rays near the areas of orange-yellow F_{rad}-values, generate a cone-like structure which does not represent the real physics. Those artifacts can also partially be seen in the case of the Bezier method but they are weaker. The difference can more easily be seen in a difference plot (Fig. 4) showing $F_{\mathrm{rad,SC}} - F_{\mathrm{rad,Bezier}}$ normalized by the largest value of $|F_{\mathrm{rad,SC}} - F_{\mathrm{rad,Bezier}}|$. As the differences between the two simulations lie in the upper part of the box, the difference plot is cut off and only shows the upper 100 vertical gridpoints. The main differences appear around the superadiabatic peak, which is the same area where the mentioned areas denoted by yellow-orange colours in the plot of F_{rad} show up (see Fig. 3). One can see that the differences are larger along the rays forming a valley-like structure in the difference plot, even though the differences are smaller than in the area around the superadiabatic peak.

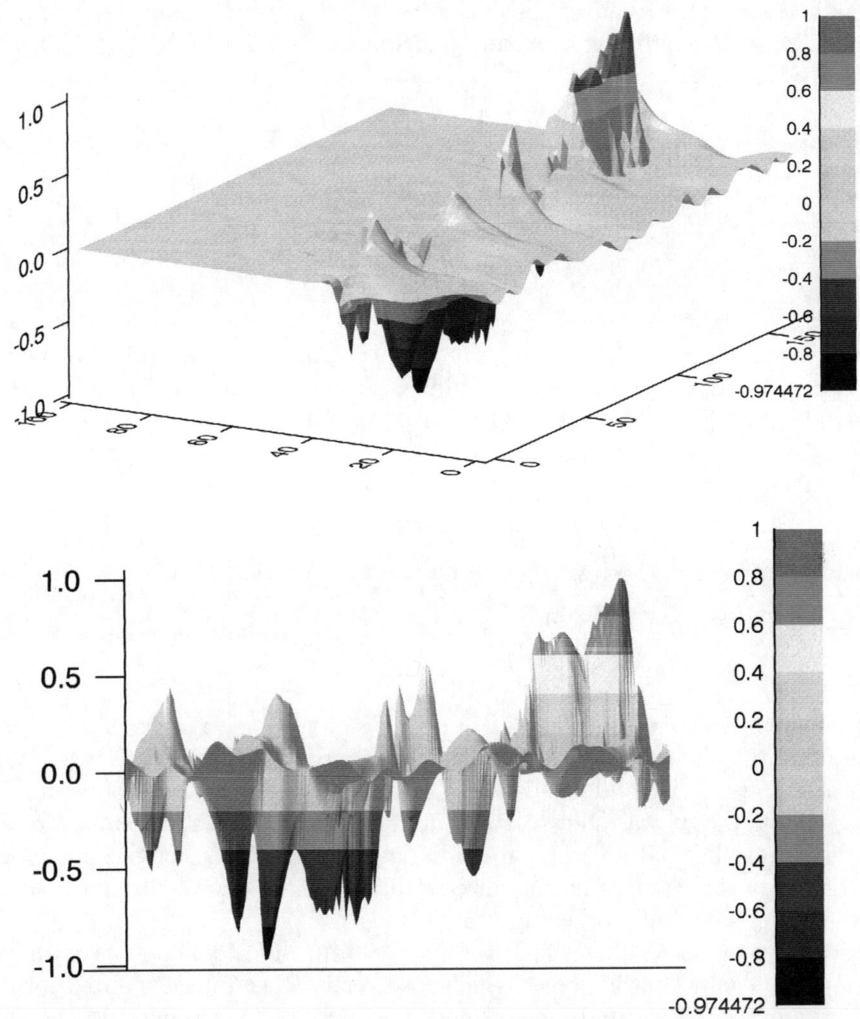

Figure 4. Normalized differences $\frac{F_{\mathrm{rad,SC}} - F_{\mathrm{rad,Bezier}}}{\max(|F_{\mathrm{rad,SC}} - F_{\mathrm{rad,Bezier}}|)}$ between the radiative flux in the original short characteristic method $F_{\mathrm{rad,SC}}$ and the radiative flux calculated with the Bezier method $F_{\mathrm{rad,Bezier}}$ from diagonal viewpoint (upper image) and frontal viewpoint (lower image).

Thus, we expect that implementing the full Bezier method in the future will further improve the simulation results thanks to it being of higher order than our current implementation. In addition to the different order of the methods, the improvement with respect to artifacts is also understandable in terms of the Bezier method being strongly monotone and positive, similarly to the physical problem being treated (see Eq. 1). These results agree with the (semi-)analytical investigations [11, 12, 13, 9] done by other groups.

4. The Eddington Approximation

Often calculation speed is more important than having highly-precise results for radiative transfer, since the additional error due to the faster method is of the order of only a few percent. For this reason, it is of significant interest to use a method that, while not increasing the error too much, is much faster than the standard methods based on ray integration.

Such a method has already been proposed more than fifty years ago [14]. By forming moments, one gets for the radiative transfer equation

$$\nabla \cdot \left(\frac{1}{3\kappa_\nu \rho} \nabla J_\nu \right) - \kappa_\nu \rho J_\nu + \kappa_\nu \rho B_\nu = 0, \tag{14}$$

and the net radiative heating can be expressed as

$$Q_{\text{rad}} = \nabla \cdot \left(\frac{4\pi}{3\kappa_\nu \rho} \nabla J_\nu \right). \tag{15}$$

This has already been used in the treatment of a grey atmosphere by [15], but it can easily be shown that it also holds for non-grey atmospheres as shall be demonstrated in the following. Keeping in mind that LTE holds and using the Eddington approximation we get

$$\int_0^\infty \kappa_\nu J_\nu \mathrm{d}\nu = \int_0^\infty \kappa_\nu B_\nu \mathrm{d}\nu. \tag{16}$$

With the opacity binning method we get the necessary condition

$$\sum_i \kappa_{\nu,i} J_{\nu,i} = \sum_i \kappa_{\nu,i} B_{\nu,i} \tag{17}$$

with the more restrictive condition $J_{\nu,i} = B_{\nu,i}$ holding in the special case of a grey atmosphere. This is a major improvement compared to the Eddington approximation used so far as this allows the inclusion of frequency dependence. Of course, this approximation cannot be used for all kinds of applications. It was shown [15] that the error of this approximation is largest for the part of the atmosphere above the superadiabatic peak. Thus, at the moment we do not recommend the use of this approximation for calculations like element abundance determination from spectral synthesis or limb darkening calculations.

As can easily be seen, Eq. 14 has the form of a generalized Poisson equation and thus can be solved by using a scalar multigrid solver. Such a solver has been implemented into ANTARES and has shown good numerical properties, e.g. weak scaling, meaning that the solution time scales with the number of used processors while keeping the problem size per processor fixed. Therefore, in the non-grey case Eq. 14 is solved for each bin and $\bar{J}$ is then retrieved by summing up over all bins, as opposed to the grey case considered in [15, 14] where $\bar{J}$ can be computed from a single solution of Eq. 14 per time stage (or time step).

The reason for the speed up expected from this approach is the much better scaling of parallelized multigrid solvers when compared to parallelized short characteristic solvers. It has already been shown ([15] for the grey case) that in the non-parallelized case the use of a multigrid solver based on the Eddington approximation is faster than using the most simple ray-integration based solver (10 rays in 3-D). Thus, we expect an even larger speed-up for the parallelized case

IOP Conf. Series: Journal of Physics: Conf. Series **1225** (2019) 012017 doi:10.1088/1742-6596/1225/1/012017

due to the better scaling properties of the multigrid solver. Additionally, we expect the speed-up to be even higher in the non-grey case, as the speed-up holds for each individual bin.

A drawback of this method is that certain quantities like the intensity I_ν are not calculated and thus the method is not usable for all types of simulations and analysis. Another drawback is that the Eddington approximation has a higher relative error than the usual angular integration methods for depths around the superadiabatic gradient, as it is only exact in the limits of very high or very low optical depth [15].

5. Summary

Two new methods for the calculation of radiative transfer were implemented into ANTARES. The Bezier method shows improved results for the calculation of intensity along each ray by significantly reducing visible artifacts. The Eddington method achieves much faster calculations for grey atmospheres and we expect the same for non-grey atmospheres. The two methods now allow us to perform simulations providing physically more realistic results or to drastically reduce the computation time, respectively.

6. Acknowledgment

DK, DF, and FK kindly acknowledge support by the Austrian Science Fund FWF through project P29172.

References

[1] H. J. Muthsam, F. Kupka, B. Löw-Baselli, C. Obertscheider, M. Langer, and P. Lenz. ANTARES - A Numerical Tool for Astrophysical RESearch with applications to solar granulation. *New Astronomy*, 15, July 2010.

[2] F. Kupka, K. Belkacem, R. Samadi, and S. Deheuvels. Studying p-mode damping and the surface effect with hydrodynamical simulations. In K. Zwintz and E. Poretti, editors, *Second BRITE-Constellation Science Conference: Small satellites/big science, Proceedings of the Polish Astronomical Society volume 5, held 22-26 August, 2016 in Innsbruck, Austria. Edited by Konnstanze Zwintz and Ennio Poretti. Polish Astronomical Society, Bartycka 18, 00-716 Warsaw, Poland, pp.222-227*, pages 222–227, September 2017.

[3] D. Mihalas and B. W. Mihalas. *Foundations of radiation hydrodynamics*. 1984.

[4] H.-G. Ludwig, S. Jordan, and M. Steffen. Numerical simulations of convection at the surface of a ZZ Ceti white dwarf. *Astronomy & Astrophysics*, 284:105–117, April 1994.

[5] A. Nordlund. Numerical simulations of the solar granulation. I - Basic equations and methods. *Astronomy & Astrophysics*, 107:1–10, March 1982.

[6] P. Kunasz and L. H. Auer. Short characteristic integration of radiative transfer problems - Formal solution in two-dimensional slabs. *Journal of Quantitative Spectroscopy & Radiative Transfer*, 39:67–79, January 1988.

[7] A. Lowan, N. Davids, and A. Levenson. Table of the zeros of the Legendre polynomials of order 1-16 and the weight coefficients for Gauss' mechanical quadrature formula. *Bull. Amer. Math. Soc.*, 48:734–743, 1942.

[8] B. G. Carlson. In *Methods in Computational Physics 1*. Alder B., Fernbach S. (eds.), 1963.

[9] J. de la Cruz Rodríguez and N. Piskunov. DELO-Bezier Formal Solutions of the Polarized Radiative Transfer Equation. *Astrophysical Journal*, 764:33, 2013.

[10] H. Grimm-Strele, F. Kupka, B. Löw-Baselli, E. Mundprecht, F. Zaussinger, and P. Schiansky. Realistic simulations of stellar surface convection with ANTARES: I. Boundary conditions and model relaxation. *New Astronomy*, 34:278–293, January 2015.

[11] F. Fritsch and R. Carlson. Monotone piecewise cubic interpolation. *SIAM Journal on Numerical Analysis*, 17(2):238–246, 1980.

[12] F. Fritsch and J. Butland. A method for constructing local monotone piecewise cubic interpolants. *SIAM Journal on Scientific and Statistical Computing*, 5(2):300–304, 1984.

[13] L. Auer. Formal Solution: EXPLICIT Answers. In I. Hubeny, D. Mihalas, and K. Werner, editors, *Stellar Atmosphere Modeling*, volume 288 of *Astronomical Society of the Pacific Conference Series*, page 3, January 2003.

[14] W. Unno and E. A. Spiegel. The Eddington Approximation in the Radiative Heat Equation. *Publications of the Astronomical Society of Japan*, 18:85, 1966.

[15] Joel D. Tanner, Sarbani Basu, and Pierre Demarque. Comparing the effect of radiative transfer schemes on convection simulations. *The Astrophysical Journal*, 759(2):120, 2012.

IOP Conf. Series: Journal of Physics: Conf. Series **1225** (2019) 012018 doi:10.1088/1742-6596/1225/1/012018

Test particles in relativistic resistive magnetohydrodynamics

Bart Ripperda[1], Oliver Porth[2] and Rony Keppens[1]

[1] Centre for mathematical Plasma Astrophysics, Department of Mathematics, KU Leuven, Celestijnenlaan 200B, B-3001 Leuven, Belgium

[2] Institut fur Theoretische Physik, Max-von-Laue-Str. 1, D-60438 Frankfurt, Germany

E-mail: `bart.ripperda@kuleuven.be`

Abstract.

The Black Hole Accretion Code (`BHAC`) has recently been extended with the ability to evolve charged test particles according to the Lorentz force within resistive relativistic magnetohydrodynamics simulations. We apply this method to evolve particles in a reconnecting current sheet that forms due to the coalescence of two magnetic flux tubes in 2D Minkowski spacetime. This is the first analysis of charged test particle evolution in resistive relativistic magnetohydrodynamics simulations. The energy distributions of an ensemble of 100.000 electrons are analyzed, as well as the acceleration of particles in the plasmoids that form in the reconnection layer. The effect of the Lundquist number, magnetization, and plasma-β on the particle energy distribution is explored for a range of astrophysically relevant parameters. We find that electrons accelerate to non-thermal energies in the thin current sheets in all cases. We find two separate acceleration regimes: An exponential increase of the Lorentz factor during the island coalescence where the acceleration depends linearly on the resistivity and a nonlinear phase with high variability. These results are relevant for determining energy distributions and acceleration sites obtaining radiation maps in large-scale magnetohydrodynamics simulations of black hole accretion disks and jets.

1. Introduction

Relativistic magnetic reconnection is considered to be the main driving mechanism behind the characteristic energetic flaring activity from accreting compact objects. Instabilities occurring in the accretion disk plasma can produce current sheets where reconnection is triggered. Such reconnection layers are susceptible to the plasmoid instability which breaks the current sheet into a chain of magnetic islands or plasmoids. X-ray and radio emission from supermassive black holes in active galactic nuclei (AGN) and X-ray binaries are attributed to electrons that have gained their energy in reconnection zones in the coronae above the accretion disk. The particles are ejected as energetic plasmoids that are associated with the flaring activity (see e.g. [1] for a review).

Reconnection regions are potentially located nearby the black hole event horizon, such that the plasma flow has to be modelled within the framework of relativistic magnetohydrodynamics. Ideal general relativistic magnetohydrodynamics (GRMHD) simulations are often used to study the accretion disk around compact objects. Such ideal models however, assume that the plasma is infinitely conductive and that resistive dissipation plays no role in the dynamics. For global accretion disk and jet dynamics the Lundquist number $S = v_A L/\eta \approx cL/\eta$ is typically extremely high with the Alfvén speed v_A approximately equal to the speed of light c, L a typical large astrophysical length scale and η, a small but finite resistivity. In such cases, the ideal MHD approximation $S \to \infty$ and $\eta \to 0$ seems adequate. In reconnection regions however, this approximation is inappropriate and the underlying assumption of perfect conductivity is violated such that the Lundquist number is high, but not infinite. For reconnection occurring in ideal GRMHD simulations of accretion disks resistivity is provided by uncontrollable numerical effects due to a finite resolution (e.g. [2]; [3]). A reconnecting plasma layer has a rapidly varying magnetic field $\mathbf{B}$ resulting in a high current density $\mathbf{J}$ that can lead to strong Ohmic dissipation $\sim \eta J^2$ and the formation of a resistive electric field $\mathbf{E} \sim \eta \mathbf{J}$. These time-dependent effects are not captured in ideal MHD descriptions since the electric field is considered a derived variable determined from $\mathbf{E} = -\mathbf{v} \times \mathbf{B}$. In general relativistic resistive magnetohydrodynamics (GRRMHD) instead, the electric field is evolved according to Ampère's law as an independent dynamic variable. Through Ohm's law the effect of a physical and controllable resistivity is taken into account allowing for reconnection and plasmoid formation occuring naturally.

To fully capture all aspects of reconnection resistive MHD is not sufficient either. Reconnection has been shown to be an efficient source of non-thermal particle acceleration explaining the energy emission through episodic flaring from jets and accretion disks ([4]; [5]; [6]; [7]; [8]; [9]; [10]; [11]; [12]; [13]; [14]). MHD describes the fluid properties of the plasma in thermal equilibrium and can by definition not capture the physics of non-thermal particles. The nonlinear interaction between the electromagnetic fields and the particles in the plasma is consistently solved in Particle-in-Cell (PiC) simulations. In such descriptions the plasma is considered to be collisionless and kinetic effects act as an effective resistivity, allowing for reconnection to occur. First efforts are currently being made to numerically simulate general relativistic effects with PiC simulations ([27]; [28]; [29]; [30]; [31] [32]). A fully kinetic description is however prohibitive in large-scale astrophysical simulations of accretion disks and jets. Recent advances to resolve non-thermal physics in accretion disks rely on the combination of large-scale GRMHD simulations, combined with a form of sub-grid particle information. [15], [16], [17] and [18] evolve a thermal electron population alongside the other MHD variables accounting for particle heating. [19] employ a scheme to co-evolve a population of non-thermal electrons to analyze radiation in GRMHD simulations. None of the above works however reproduces the characteristic X-ray flares with hard energy spectra that are typically observed on a daily basis from Sagittarius A* (Sgr A*), the black hole in the Galactic Centre (see e.g. [20]; [21]; [22]; [23]; [24]). [11] and [25] assume a combination of thermal and non-thermal electron distributions, to analyze radiation signatures from GRMHD simulations in postprocessing. Their work indicates that non-thermal electron injection is necessary for modeling the energy spectra and highly variable flaring from Sgr A*. To consistently model non-thermal emission, particle dynamics has to be included in the GRMHD evolution. [2], [13] and [26] have recently overcome this issue by using a special relativistic PiC method to provide non-thermal particle distributions in locally flat slabs of large-scale GRMHD simulations. In these ideal GRMHD simulations, reconnection however occurs due to numerical resistivity instead of a physically motivated resistivity.

Relativistic resistive MHD can describe both the large-scale accretion flow, as well as the reconnection that is induced by instabilities occurring in the accretion disk. However, resistive MHD cannot self-consistently capture particle acceleration either. We combine relativistic resistive MHD simulations of a forming current sheet, with an ensemble of electrons, following the electromagnetic fields from the MHD, according to the relativistic Lorentz force. The test particles evolve simultaneously alongside the MHD but do not exert any feedback on the electromagnetic fields. Resistive electric fields, arising from magnetic reconnection, are efficient particle accelerators ([33]; [34]; [35]; [36]; [37]; [38]), such that non-thermal particle distributions can naturally form in resistive MHD. The test particle assumption made here is valid for a plasma where the non-thermal particles do not dominate the plasma energetics compared to the thermal

IOP Conf. Series: Journal of Physics: Conf. Series **1225** (2019) 012018 doi:10.1088/1742-6596/1225/1/012018

ensemble. In this way, global simulations on astrophysical scales can be carried out without the computational limitations PiC approaches usually suffer from, and particle dynamics can be analyzed within the computationally cheaper test particle approach. However, the assumption that non-thermal particles are not dynamically important becomes invalid in reconnection zones where a substantial fraction of particles typically reach high Lorentz factors. Considering kinetic feedback of these particles to the electromagnetic fields requires a PiC approach. Such a method is however currently impractical due to the large scale-separation between the gyro-motion and the global accretion flow and can therefore only be applied by assuming unrealistic magnetic field strengths for accretion disk scenarios.

We assume a scenario where a turbulent accretion disk has formed around a black hole. In the corona above the accretion disk, a force-free magnetic field is represented by an ensemble of flux loops tied to the disk (see e.g. [39]; [40]; [41]). We focus on two such flux tubes, merging and reconnecting with each other. Since the reconnection dynamics occur in a small region of the corona, we assume a flat slab of spacetime, such that we can ignore the effects of gravity and curved spacetime. The formation of a reconnection sheet in between two merging flux tubes has been investigated by [42] with special relativistic PiC simulations and by [43] with special relativistic resistive magnetohydrodynamics (SRRMHD) simulations. Here, we show a proof-of-principle combination of relativistic resistive MHD and particle methods. We employ the GRRMHD code `BHAC` ([44]; [45]) to study the global scale reconnection dynamics and plasmoid properties of a forming current sheet in between two coalescing Lundquist tubes in 2D Minkowski spacetime. By following an ensemble of 100.000 test particles in MHD fields, we analyze the influence of the Lundquist number, plasma-β and magnetization σ on acceleration sites and particle energy distributions.

2. Numerical setup

Black hole and neutron star magnetospheres and their outflows are considered to be filled with conducting electron-positron plasmas ([46]; [47]; [48]). Therefore we evolve $N_{tot} = 100.000$ electrons with charge q and mass m in the high-resolution SRRMHD simulations of 2D merging Lundquist tubes of [43]. Due to the test particle approximation, positrons can be neglected as their dynamics are equivalent to electron dynamics, with the only difference being their opposite charge. We use a special relativistic Boris scheme to evolve charged test particles evolving in the electromagnetic fields obtained from MHD, according to the Lorentz force [49]. A method to moderate spurious acceleration of test particles, due to a lack of feedback on the electromagnetic fields, was presented in [36] and [37]. For every particle that leaves the current sheet from the outflow region, a thermal particle is inserted at the inflow region of the reconnection zone. Using this approach, electron dynamics and energetics are analyzed in the reconnection region in between the merging flux tubes. We also analyze the effect of the initial spatial distribution of the electrons, and the injection mechanism, which is relevant for future global scale GRRMHD simulations of accretion disks, where particle ensembles are injected in regions showing signs of reconnection. The particle motion is guided by the electromagnetic fields $\mathbf{E}$ and $\mathbf{B}$ as obtained from the MHD evolution, according to the Lorentz force in CGS units with the speed of light set equal to $c = 1$

$$\frac{d\mathbf{u}}{dt} = \frac{q}{mc}\left(\mathbf{E} + \frac{\mathbf{u} \times \mathbf{B}}{\Gamma}\right), \tag{1}$$

where coordinate time t is used as affine parameter, $\Gamma = \sqrt{1+u^2}$ is the particle's Lorentz factor, $\mathbf{u} = \Gamma\mathbf{v}/c$ is the particle four-velocity, divided by the particle's rest mass m and normalized to the speed of light $c = 1$.

Reconnection is triggered in a setup of two coalescing Lundquist tubes ([42]; [43]). The flux tubes are given by the magnetic field

$$\mathbf{B}(r \leq r_j) = \alpha_t c_t J_1(\alpha_t r)\mathbf{e}_\phi + \alpha_t c_t \sqrt{J_0(\alpha_t r)^2 + \frac{C}{(\alpha_t c_t)^2}}\mathbf{e_z}, \tag{2}$$

where J_0 and J_1 are Bessel functions of the zeroth and first kind respectively and the constant $C = 0.01$ is chosen such that the minimum B_z component remains positive, the constant $\alpha_t \approx 3.8317$ is the first root of J_0, and $c_t = 0.262$ ensures that the maximum value of the B_z field is unity, corresponding to the central value in the flux tubes. The solution is terminated at $r = r_j = 1$, corresponding to the first zero of J_1, such that $B_z = B_z(r_j)$ and $B_\phi = 0$ for $r > r_j$. The total current in the flux tube is zero and no surface currents are present in the initial setup. The two flux tubes are just touching each other and the centre positions are set at $\mathbf{x_c} = (\pm r_j, 0, 0)$. The setup is perturbed by a velocity perturbation $\mathbf{v}_{kick} = (\pm 0.1c, 0, 0)$ that pushes the tubes together, with the $\pm$ corresponding to the left and right rope respectively. The kick is applied via the initial electric field as $\mathbf{E} = -\mathbf{v} \times \mathbf{B}$. The dynamical evolution is independent of the kick velocity, such that the choice of $\mathbf{v}_{kick}$ is not restricting the validity of the conclusions ([42]). For the underlying SRRMHD results, as well as the reconnection and plasmoid properties we refer to [43]; For a comparison to PiC results we refer to [42]. The pressure p_0 and density ρ_0 profiles are initially uniform in the force-free equilibrium setup. The values of ρ_0 and p_0 are varied between cases to set the magnetization $\sigma_0 = B_0^2/(\rho_0 h_0)$ and plasma-$\beta_0 = 2p_0/B_0^2$, where $h_0 = 1 + 4p_0/\rho_0$ is the initial enthalpy density where we assume an ideal gas equation of state with adiabatic index $\hat{\gamma} = 4/3$.

We set the typical length to $L = r_j = 10^{10} cm$, such that the whole domain is within $x \in [-3L, 3L]$, $y \in [-3L, 3L]$. In all high-resolution MHD runs of [43] we initialize 100.000 electrons. All particles are initialized from a Maxwellian velocity distribution with temperature $T_0 = 5 \times 10^{-3}\beta_0/\rho_0$, thermal velocity $v_{th} = \sqrt{T_0}$ and thermal Lorentz factor $\Gamma_{th} \approx 1$, in accordance with the MHD temperature as set by $\beta_0 = 2p_0/B^2$ and $\sigma_0 = B^2/(\rho_0 h_0)$. We explore a range of uniform resistivities $\eta = 10^{-2}$, $\eta = 10^{-3}$, $\eta = 10^{-4}$, $\eta = 5 \times 10^{-5}$ and $\eta = 0$, resulting in Lundquist numbers $S = 1/\eta$ of 100, 1000, 10000, 20000 and ∞, where [43] found a critical Lundquist number of $S_c \geq 20000$ for plasmoid formation to occur in SRRMHD. We also explore a non-uniform resistivity model $\eta(r, t) = \eta_0(1 + \Delta_{ei}^2 J)$ [51], depending on the current density magnitude J and the asymptotically small parameter Δ_{ei}^2 that is varied between cases. Two cases with varying plasma-β_0 of 0.1 and 0.5 are considered and three cases with varying magnetization σ_0 of 0.9, 1.0, and 3.3 are explored by varying the pressure p_0 and the density ρ_0 in a regime that is particularly relevant for reconnection in black hole accretion disks [52]. The electromagnetic fields are obtained from SRRMHD simulations, except for run Gi, where the setup is evolved with ideal SRMHD (i.e. $\eta = 0$). The particles are either uniformly initialized in the region of the magnetic flux tubes, $x \in [-2L, 2L]$, $y \in [-1L, 1L]$, or, in runs indicated by an "s", particles are initialized in a smaller box around the current sheet with $x \in [-0.1L, 0.1L]$, $y \in [-1L, 1L]$. In the latter case, any particle that leaves this box is replaced by a thermal particle and inserted at a random y-position $\in [-1L, 1L]$ at $x = \pm 0.05L$. In this way only the particles that accelerate in the formed current sheet are taken into account and bulk acceleration in the initial flux ropes is neglected. The thermal part of the distribution is in this case determined by the injected particles; However, we are mainly interested in the particles that accelerate to non-thermal energies in the reconnection layer.

All runs are summarised in Table 1. The first four parameters, the perturbation velocity $v_{kick,x}$, background plasma-β_0, background magnetization σ_0 and base resistivity η_0 are properties of the SRRMHD simulations. A non-zero Δ_{ei}-parameter indicates a nonuniform resistivity run. The maximum value of the resistivity η_{max}, taken in the whole simulation box and over the whole simulation time, in such cases is given in the last column.

Table 1: The simulated cases with characteristic parameters: kick velocity $v_{kick,x}$; plasma-$\beta_0 = 2p_0/B_0^2$; magnetization $\sigma_0 = B_0^2/(\rho_0 h_0)$; base resistivity η_0; nonuniform resistivity factor Δ_{ei}^2; and maximum resistivity in the domain over the total time evolution η_{max}.

Run	$v_{kick,x}$	β_0	σ_0	η_0	Δ_{ei}^2	η_{max}
A	0.1c	0.1	3.3	5×10^{-5}	0	5×10^{-5}
B	0.1c	0.1	3.3	1×10^{-4}	1	404.7×10^{-4}
C	0.1c	0.1	3.3	1×10^{-4}	0.1	264.6×10^{-4}
D	0.1c	0.1	3.3	1×10^{-4}	0.01	6.7×10^{-4}
E	0.1c	0.1	3.3	1×10^{-4}	0.001	1.3×10^{-4}
F	0.1c	0.1	3.3	1×10^{-4}	0	1×10^{-4}
G	0	0.1	3.3	1×10^{-4}	0	1×10^{-4}
Gi	0	0.1	3.3	0	0	0
H	0.1c	0.1	3.3	1×10^{-3}	0	1×10^{-3}
I	0.1c	0.1	3.3	1×10^{-2}	0	1×10^{-2}
J	0.1c	0.5	0.9	5×10^{-5}	0	5×10^{-5}
K	0.1c	0.5	1.0	5×10^{-5}	0	5×10^{-5}

3. Plasmoid formation and particle acceleration in flux merging events

In Figure 1, the spatial particle distribution is shown for runs G (left-hand column, $v_{kick,x} = 0$, $\eta = 10^{-4}$, $\beta_0 = 0.1$, $\sigma_0 = 3.3$), F (middle column, $v_{kick,x} = 0.1$, $\eta = 10^{-4}$, $\beta_0 = 0.1$, $\sigma_0 = 3.3$) and A (right-hand column, $v_{kick,x} = 0.1$, $\eta = 5\times10^{-5}$, $\beta_0 = 0.1$, $\sigma_0 = 3.3$) at times $t = 5t_c$, $t = 10t_c$, $t = 18t_c$, $t = 24t_c$. The particles are colored by their Lorentz factor Γ, as a proxy for their energy $\mathcal{E} = \Gamma mc^2$. Particles accelerate at early times, in the original flux ropes. They are then drawn towards the current sheet and continue to accelerate in the current sheet in runs F and A. The particles are expelled in the reconnection outflow regions at the top and the bottom of the reconnection zone, after they have moved through the current sheet. They then start moving around the flux ropes, guided by the resistive electric field, potentially re-entering the current sheet. In run A ($\eta = 5\times10^{-5}$), one can observe plasmoids (at $t_c = 18$ and $t_c = 24$), consisting of energetic particles in the current sheet. These plasmoids are expelled in the reconnection outflow regions at the top and bottom of the current sheet. In unperturbed run G (left-hand column), the flux tubes do not merge and no current sheet forms. Contrary to the PiC results in [42], here we observe particle heating in the flux ropes. This is caused by the relatively high resistivity $\eta = 10^{-4}$, that was chosen to be close to the critical Lundquist number $S_c = 1/\eta \approx 20.000$; where plasmoids form for $S > S_c$. The particles accelerate to moderate Lorentz factors of $\sim \mathcal{O}(10)$ inside the flux tubes due to the formation of a resistive electric field $\sim \eta\mathbf{J}$. In ideal SRMHD run Gi (that is not shown here), this acceleration does not occur, due to the ideal approximation $\eta = 0$. Ideal SRMHD simulations however unavoidably show signs of numerical resistivity, resulting in reconnection. Compared to SRRMHD this effect is uncontrolled, since there is no handle on the resistivity.

3.1. Effects of Lundquist number on the energy distribution

Figure 2 shows the distributions for $\Gamma = (1 + u_x^2 + u_y^2 + u_z^2)^{1/2}$, as taken in a box surrounding the current sheet $x \in [-0.05L, 0.05L]$, $y \in [-1L, 1L]$ for run G (left-hand panels, $v_{kick,x} = 0$, $\eta = 10^{-4}$, $\beta_0 = 0.1$, $\sigma_0 = 3.33$), run F (middle panels, $v_{kick,x} = \pm0.1c$, $\eta = 10^{-4}$, $\beta_0 = 0.1$, $\sigma_0 = 3.33$) and run A (right-hand panels, $v_{kick,x} = \pm0.1c$, $\eta = 5\times10^{-5}$, $\beta_0 = 0.1$, $\sigma_0 = 3.33$). We confirmed the results do not change for a slightly smaller or larger box. In the top panels, particles are initialized in the whole simulation box, including the flux tubes $x \in [-2L, 2L]$, $y \in [-1L, 1L]$. In the bottom panels, particles are only initialized in the reconnection layer at $x \in [-0.05L, 0.05L]$, $y \in [-1L, 1L]$, such that the distribution is not affected by particles entering the current sheet after being heated in the flux tubes.

In the unperturbed run G (top left-hand panel), a non-thermal tail, with $\Gamma_{max} \approx 10$ forms at later times ($t \gtrsim 18t_c$, green and red lines) by particles that are accelerated in the flux tubes

IOP Conf. Series: Journal of Physics: Conf. Series **1225** (2019) 012018 doi:10.1088/1742-6596/1225/1/012018

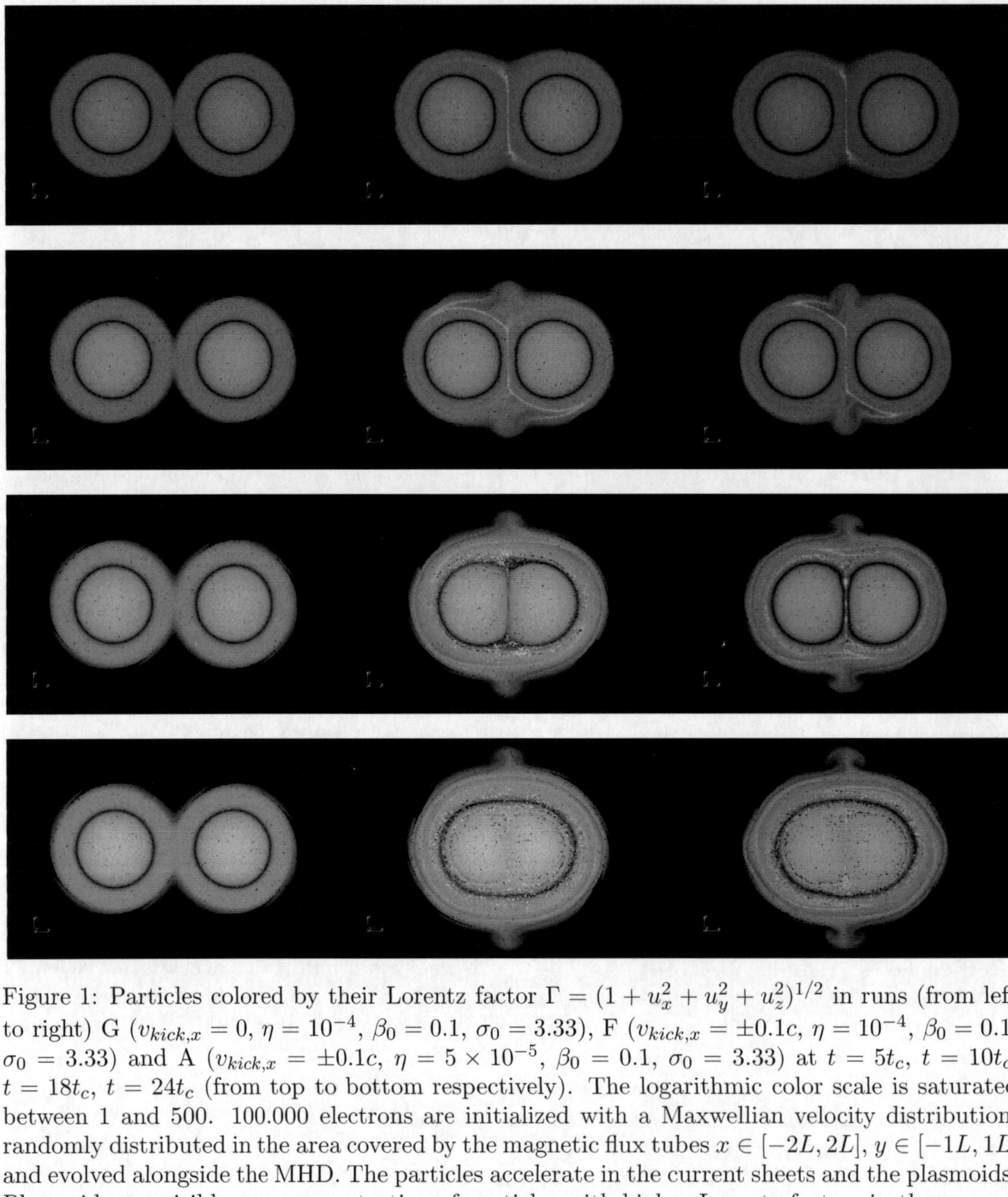

Figure 1: Particles colored by their Lorentz factor $\Gamma = (1 + u_x^2 + u_y^2 + u_z^2)^{1/2}$ in runs (from left to right) G ($v_{kick,x} = 0$, $\eta = 10^{-4}$, $\beta_0 = 0.1$, $\sigma_0 = 3.33$), F ($v_{kick,x} = \pm 0.1c$, $\eta = 10^{-4}$, $\beta_0 = 0.1$, $\sigma_0 = 3.33$) and A ($v_{kick,x} = \pm 0.1c$, $\eta = 5 \times 10^{-5}$, $\beta_0 = 0.1$, $\sigma_0 = 3.33$) at $t = 5t_c$, $t = 10t_c$, $t = 18t_c$, $t = 24t_c$ (from top to bottom respectively). The logarithmic color scale is saturated between 1 and 500. 100.000 electrons are initialized with a Maxwellian velocity distribution, randomly distributed in the area covered by the magnetic flux tubes $x \in [-2L, 2L]$, $y \in [-1L, 1L]$ and evolved alongside the MHD. The particles accelerate in the current sheets and the plasmoids. Plasmoids are visible as a concentration of particles with higher Lorentz factors in the current sheet in between the flux tubes, for case A (right-hand column) at $t = 18t_c$.

and enter the box where the distribution is taken. If the particles are initialized in the box $x \in [-0.05L, 0.05L]$, $y \in [-1L, 1L]$ in run Gs, no non-thermal tail forms and particles accelerate only mildly to $\Gamma_{max} \approx 3$ due to the resistive electric field $\mathbf{E} \sim \eta\mathbf{J}$ present in between the flux tubes. Note that in these runs, no current sheet has formed.

In cases F (top middle panel) and A (top right-hand panel) a non-thermal tail forms in the current sheet from $t \gtrsim 5t_c$ onwards. The maximum Lorentz factor is proportional to the resistivity $\Gamma_{max} \approx 10^3$ in case F ($\eta = 10^{-4}$) and $\Gamma_{max} \approx 5 \times 10^2$ in case A ($\eta = 5 \times 10^{-5}$). In both cases, a "bump" in the energy distribution can be observed at $\Gamma \approx 10^2$ at times $t = 18t_c$ (green lines) and $t = 24t_c$ (red lines). This is caused by electrons entering the current sheet after they have been heated in the flux tubes, due to a resistive electric field $\mathbf{E} \sim \eta\mathbf{J}$. This bulk heating in the large magnetic flux tubes is similar to the effect observed in Figure 1. In PiC simulations, such effects are not observed, due to an intrinsically low resistivity in the flux tubes and ambient plasma [42].

For run Gi (not shown here), in ideal SRMHD ($\eta = 0$), no non-thermal tail forms at all and particles do no accelerate in the flux tubes, confirming that a too large resistivity in the ambient plasma is the cause for the bulk heating. More realistic, non-uniform resistivity models, with a low resistivity in the ambient and a high resistivity in the current sheet, can potentially solve this heating effect in test particles simulations.

initializing particles in the current sheet box at $x \in [-0.05L, 0.05L]$, $y \in [-1L, 1L]$ only, reduces the heating effect and results in a power-law distribution for cases Fs (bottom middle panel) and As (bottom right-hand panel). Here, the particles leaving the current sheet are destroyed, and new thermal particles are injected from the reconnection inflow region at random vertical position $y \in [-1L, 1L]$ and at $x \pm 0.05L$. The maximum Lorentz factor remains similar for both cases as in the top panels, and is still proportional to the resistivity, but the "bump" at $\Gamma \approx 10^2$ disappears.

In runs A and As (right-hand panels), the plasmoid instability is activated from $t_c \approx 18$ onwards for $\eta = 5 \times 10^{-5}$, and a change in the distribution is observable in the green and red lines, compared to run F (middle panels) for $\eta = 10^{-4}$. It is however unclear if this can be attributed to the particle acceleration in the plasmoids, or due to the larger current density in the current sheet due to the plasmoid instability.

Run Fs has been confirmed to give equivalent distributions with 10.000 electrons and in [37] it is shown that for test particles, the total number of particles has little effect on the results, since there is no feedback of the particles on the electromagnetic fields and hence no requirement for a certain number of particles per cell.

3.2. Effects of nonuniform resistivity on the energy distribution

Figure 3 shows the Lorentz factor distributions for, from left to right, case Cs (nonuniform resistivity $\eta = 10^{-4}(1 + 0.1J)$, $\beta_0 = 0.1$, $\sigma_0 = 3.3$), case Js (uniform resistivity $\eta = 5 \times 10^{-5}$, $\beta_0 = 0.5$, $\sigma_0 = 0.9$) and case Ks (uniform resistivity $\eta = 5 \times 10^{-5}$, $\beta_0 = 0.5$, $\sigma_0 = 1.0$). Distributions are again taken in the box around the current sheet at $x \in [-0.05L, 0.05L]$, $y \in [-1L, 1L]$ and particles are initialized in the same box and destroyed once they leave this box through the reconnection outflows.

Run Cs (left-hand panel), with nonuniform resistivity $\eta(r, t) = \eta_0(1 + \Delta_{ei}^2 J)$, with $\eta_0 = 10^{-4}$ and $\Delta_{ei}^2 = 0.1$, should be compared to run Fs with uniform $\eta = 10^{-4}$ (and hence $\Delta_{ei}^2 = 0$) in the bottom middle panel of Figure 1. The nonuniform resistivity in case Cs is close to $\eta = 10^{-4}$ in the ambient plasma and the flux tubes, where the current density remains small. In the current sheet, the resistivity is strongly enhanced due to the large current density to reach a maximum of $\eta_{max} \approx 264.6 \times 10^{-4}$[43]. Due to the strongly enhanced resistivity the MHD evolution is faster, the current sheet broadens and the resistive electric field increases such that particles accelerate at earlier times $t = 10t_c$ (cyan line) to reach a Lorentz factor $\Gamma_{max} \approx 5 \times 10^3$, approximately 20

IOP Conf. Series: Journal of Physics: Conf. Series **1225** (2019) 012018 doi:10.1088/1742-6596/1225/1/012018

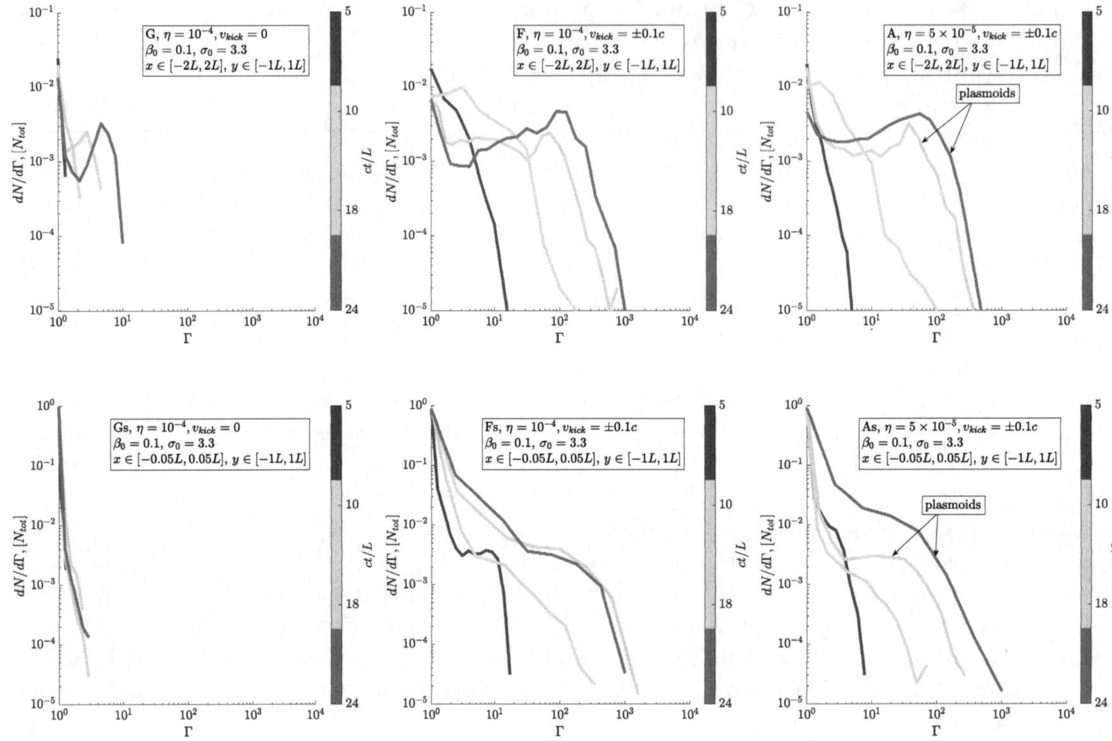

Figure 2: Normalised electron energy distributions ($\mathcal{E}/(mc^2) = \Gamma$) for runs G (top left-hand panel), F (top middle panel), A (top right-hand panel), Gs (bottom left-hand panel), Fs (bottom middle panel) and As (bottom right-hand panel) at times $t = 5t_c$, $t = 10t_c$, $t = 18t_c$ and $t = 24t_c$ as indicated by the color bar. The distributions are taken in a box $x \in [-0.05, 0.05]$, $y \in [-1, 1]$ around the current sheet. In the bottom panels, for runs with indicator "s", particles are initialized in the current sheet and particles leaving the reconnection zone are destroyed. This results in a power-law distribution without the "bump" in energy at $\Gamma \approx 10^2$ as in the top panels where the particles are initialized in the whole simulation box. The heating in the flux tubes causing the bump is attributed to the large resistivity in these regions.

times larger than in case Fs. At later times $t \gtrsim 18t_c$, the current density dissipates due to the enhanced resistivity, the resistive electric field $\mathbf{E} \sim \eta \mathbf{J}$ decreases and particles decelerate again (green and red lines).

3.3. Effects of magnetization on the energy distribution

The middle panel of Figure 3 shows the effect of an increased plasma-$\beta_0 = 0.5$ in run Js (uniform $\eta = 5 \times 10^{-5}$), compared to run As ($\beta_0 = 0.1$, $\eta = 5 \times 10^{-5}$) in the bottom right-hand panel of Figure 1. The increase in β_0 for run Js results in an effectively decreased magnetization $\sigma_0 = B^2/(\rho_0 h_0) = B^2/(\rho_0 + 2B^2\beta_0) = 0.9$, compared to $\sigma_0 = 3.3$ for run As. From $t \gtrsim 10t_c$ onwards, a non-thermal distributions forms in the current sheet, that grows to a maximum Lorentz factor $\Gamma_{max} \approx 4 \times 10^2$, approximately 2.5 times smaller than in run As. In run Js, no plasmoids are formed in the current sheet, due to the smaller magnetization than in run As.

In run Ks, in the right-hand panel of 3, the magnetization is increased to $\sigma_0 = 1.0$ compared

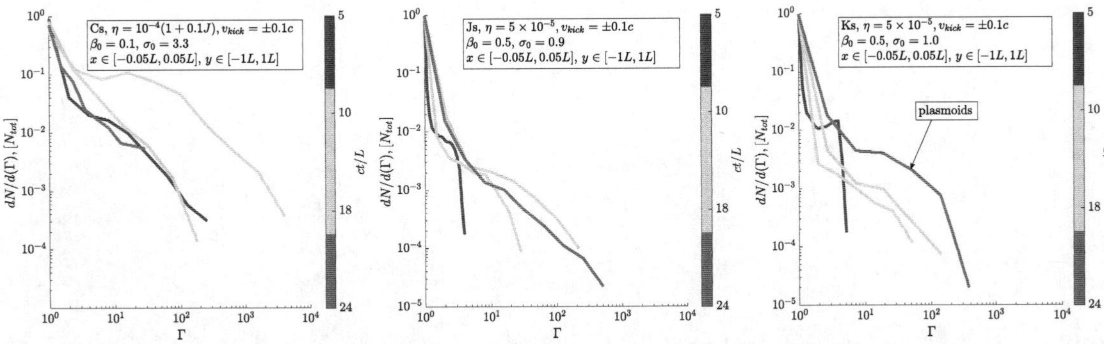

Figure 3: Normalised electron energy distributions ($\mathcal{E}/(mc^2) = \Gamma$) for runs Cs (left-hand panel), Js (middle panel) and Ks (right-hand panel) at times $t = 5t_c$, $t = 10t_c$, $t = 18t_c$ and $t = 24t_c$ as indicated by the color bar. The distributions are taken in a box $x \in [-0.05, 0.05]$, $y \in [-1, 1]$ around the current sheet. Particles are initialized in the current sheet and particles leaving the reconnection zone are destroyed. The left-hand panel shows the effect of a nonuniform resistivity $\eta(r,t) = 10^{-4}(1 + 0.1J)$ in run Cs, compared to uniform resistivity $\eta = 10^{-4}$ run Fs (bottom middle panel in Figure 2). The middle panel shows the effect of an increased $\beta_0 = 0.5$ and decreased $\sigma_0 = 0.9$ for uniform resistivity $\eta = 5 \times 10^{-5}$ in run Js, compared to run As ($\beta_0 = 0.1$, $\sigma_0 = 3.3$, bottom right-hand panel in Figure 2). The right-hand panel shows the effect of plasmoids in run Ks, due to a slightly enlarged $\sigma_0 = 1.0$, compared to run Js.

to run Js (middle panel), for $\beta_0 = 0.5$ and uniform resistivity $\eta = 5 \times 10^{-5}$. Due to the slightly enhanced magnetization, plasmoids are forming now in the current sheet from $t = 20t_c$ onwards [43]. Comparing red line at $t = 24t_c$ in the middle panel (Js, $\sigma_0 = 0.9$, no plasmoid formation) to the red line in the right-hand panel (Ks, $\sigma_0 = 1.0$, plasmoid formation), we observe that the distribution has a steeper power-law for case Ks. The maximum Lorentz factor $\Gamma_{max} \approx 4 \times 10^2$ is equivalent to run Js, where no plasmoids have formed.

3.4. Lorentz factor evolution

In Figure 4, we show the temporal evolution of the maximum Lorentz factor Γ_{max} in all runs, taken over the whole ensemble of particles (i.e. not just in the current sheet). A distinction is made between runs with particles in the whole domain (including the initial flux ropes) and runs where particles that end up outside of the current sheet are neglected and destroyed (indicated by index "s"). In the left-hand panel, all runs with $\beta_0 = 0.1$, $\sigma_0 = 3.3$ and uniform resistivity, are compared, with $\eta = 5 \times 10^{-5}$ for run A and As, $\eta = 10^{-4}$ for runs F and Fs, $\eta = 10^{-3}$ for run Hs, and $\eta = 10^{-2}$ for run Is. Runs G and Gs, with $\eta = 10^{-4}$ but without an initial perturbation (i.e. $\mathbf{v}_{kick} = 0$) are also shown, as well as run Gis that is evolved with the ideal SRMHD module in `BHAC` (i.e. $\eta = 0$).

All resistive runs show an initial growth of the particle Lorentz factor, that is comparable to the Alfvénic growth of the electric field [43]. The acceleration increases for larger resistivity. In runs Hs (dashed black line) and Is (dashed magenta line), the resistivity is so large that the forming current sheet rapidly diffuses, resulting in a decrease in the maximum Lorentz factor at $t_c \approx 10$ and $t_c \approx 20$, respectively. For runs F (solid green line), Fs (dashed green line), A (solid brown line) and As (dashed brown line), there is a subsequent secondary acceleration phase with larger variability after $t_c \approx 5$. These particles are accelerated by the resistive electric field in the current sheet. Runs F and Fs show that Γ_{max} is about two times larger than for

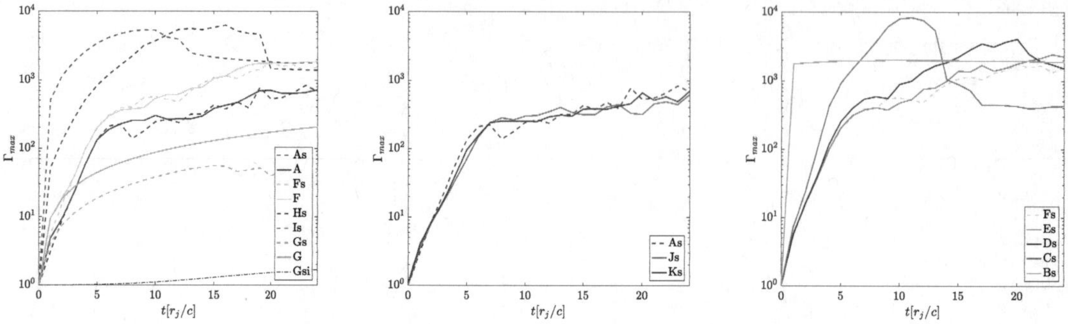

Figure 4: Temporal evolution of Γ_{max} where $max(\Gamma)$ is taken in the whole domain, over all particles. Runs A, As, F, Fs, Hs, Is, G, Gs and Gsi are shown in the left-hand panel, where runs As and A have resistivity $\eta = 5 \times 10^{-5}$, F and Fs have $\eta = 10^{-4}$, Hs has $\eta = 10^{-3}$, and Is has $\eta = 10^{-2}$. Runs G and Gs remain unperturbed (i.e. $\mathbf{v}_{kick} = 0$) and have $\eta = 10^{-4}$ and run Gsi is both unperturbed and evolved in ideal SRMHD (i.e. $\eta = 0$ by definition). An "s" in the index (all dashed lines in the left-hand panel) indicates that in these runs particles are removed when they leave the current sheet at $x \in [-0.1L, 0.1L]$, $y \in [-1L, 1L]$ and in their place a thermal particle is injected at random vertical position $y \in [-1L, 1L]$ and at $x \pm 0.05L$. In the middle panel, all runs with $\eta = 5 \times 10^{-5}$ are shown, where run As has $\beta_0 = 0.1$, $\sigma_0 = 3.3$, run Js has $\beta_0 = 0.5$, $\sigma_0 = 0.9$ and run Ks has $\beta_0 = 0.5$, $\sigma_0 = 1.0$. In the right-hand panel all runs with nonuniform resistivity $\eta = 10^{-4}(1 + \Delta_{ei}^2 J)$ are shown, with $\Delta_{ei}^2 = 0$ for run Fs, $\Delta_{ei}^2 = 0.001$ for run Es, $\Delta_{ei}^2 = 0.01$ for run Ds, $\Delta_{ei}^2 = 0.1$ for run Cs and $\Delta_{ei}^2 = 1$ for run Bs.

runs As and A, in accordance with the factor two difference in resistivity. In runs As and A the plasmoid instability is triggered, which however shows no observable effect on the maximum Lorentz factor. Destroying and injecting particles induces a larger variability for runs As and Fs in the nonlinear regime at $t_c \gtrsim 5$ compared to runs A and F. The maximum Lorentz factor does however not differ much, showing that the particles attain the highest energies inside the current sheet. This confirms the finding that particles accelerate to medium Lorentz factors in the initial flux ropes.

Unperturbed runs G (solid yellow line) and Gs (dashed yellow line) show only an initial acceleration phase that is due to a resistive electric field that forms in the flux ropes. Current sheets do not form in these runs and therefore the acceleration is slower, and Γ_{max} is smaller than for the perturbed runs. Here, the effect of destroying particles that leave the current sheet, and injecting thermal particles, is visible in the maximum Lorentz factor that is reached for run Gs, that is approximately four times smaller than for run G at the final time. This difference quantifies the effect of the bulk heating of particles in the flux tubes. To confirm that this is a resistive effect, a simulation is conducted in the ideal SRMHD module of `BHAC`, such that $\eta = 0$ by definition and $\mathbf{E} = -\mathbf{u} \times \mathbf{B}/\Gamma$, with $\mathbf{u}$ and Γ the fluid momentum and Lorentz factor. In the ideal case Gsi, the Lorentz factor is bounded to $\Gamma_{max} \lesssim 1.7$ on the time scales considered, since no resistive electric field forms. There is no exponential growth phase and particles only mildly accelerate in the unmoving flux ropes due to the ideal electric field $\mathbf{E} = -\mathbf{u} \times \mathbf{B}/\Gamma$. This confirms that even in unperturbed runs, the resistive electric field is the main acceleration mechanism.

In the middle panel of Figure 4, the effects of σ_0 and plasma-β_0 are quantified for runs with uniform resistivity $\eta = 5 \times 10^{-5}$. The larger magnetization of run Bs ($\beta_0 = 0.1$, $\sigma_0 = 3.3$), compared to run Js ($\beta_0 = 0.5$, $\sigma_0 = 0.9$) results in a slightly faster initial acceleration phase.

Increasing the magnetization from $\sigma_0 = 0.9$ for run Js, to $\sigma_0 = 1.0$ for run Ks, results in a minor increase in the slope of Γ_{max}, however still yields a less fast acceleration than for run Bs with lower plasma-β. Runs Js and Ks suggest that a smaller σ limits maximum energy of the particles, in accordance with the PiC findings of [42].

In the right-hand panel of Figure 4, the evolution of Γ_{max} for all nonuniform resistivity runs are shown, and compared to run Fs with uniform resistivity $\eta = 10^{-4}$. For run Bs, with $\Delta^2_{ei} = 1$, the maximum Lorentz factor Γ_{max} grows fast, but then reaches an approximately constant maximum after $t_c \approx 1$. This corresponds to the point where the current sheet breaks-up and rapidly diffuses due to the strongly enhanced resistivity. In run Cs ($\Delta^2_{ei} = 0.1$), the largest Γ_{max} is reached at $t_c \approx 10$, after which the acceleration decays due to the diffusion of the current sheet. In runs Ds ($\Delta^2_{ei} = 0.01$) and Es ($\Delta^2_{ei} = 0.001$), the resistivity is only mildly enhanced, and the maximum Lorentz factor follows the evolution of run Fs with uniform resistivity. The maximum Lorentz factor scales with the maximum of the nonuniform resistivity here, and only in the far nonlinear phase ($t_c \gtrsim 18$) a decrease of Γ_{max} is observable for run Ds. This confirms again that the resistive electric field $\mathbf{E} \sim \eta\mathbf{J}$ is the main acceleration mechanism.

4. Conclusions

We evolve charged test particles in electromagnetic fields obtained from SRRMHD simulations of 2D merging Lundquist tubes for a range of Lundquist numbers, plasma-β and magnetizations. We find high-energy power-law tails forming due to acceleration in the current sheet in cases with $\eta \leq 10^{-4}$. The maximum Lorentz factor of the particle ensemble is proportional to the resistive electric field (and hence proportional to the resistivity). After the initial acceleration in the forming current sheet, a secondary acceleration regime is reached with a larger variability of the Lorentz factor.

We find that electrons accelerate to medium Lorentz factors in the flux tubes and to high Lorentz factors in the forming current sheet in between the flux tubes. The latter is in agreement with PiC results of [42], but the bulk heating in the original flux tubes is not. By conducting an ideal SRMHD simulation (i.e. $\eta = 0$) we have shown that the bulk acceleration in the current sheet is mainly due to the too large resistivity in the ambient plasma and in the two flux tubes.

With a realistic spatiotemporally dependent resistivity and a very low base resistivity, microscopic effects like plasmoid formation and particle acceleration can be restricted to small parts of the domain where a large current density is observed. By prescribing a very small background resistivity $\eta \ll 10^{-5}$ and a current-dependent resistivity below the plasmoid threshold, $\eta < 5 \times 10^{-5}$ in future applications, we expect plasmoid formation and particle acceleration in reconnection regions, and a thermal particle distribution in the ambient and in the flux tubes.

In this work particles are evolved using the special relativistic Boris scheme, neglecting curved spacetime. In accreting black hole systems, the ejection of plasmoids and the associated flaring might occur close to the black hole event horizon. The special relativistic particle integrator used here is incapable of taking the effect of gravity into account. In combination with the capabilities documented in [30] and [50] for particle integrators in curved spacetime, we prepare for future applications of accretion physics and relativistic jet dynamics, and the reconnection physics and particle acceleration associated with them.

Acknowledgements

This research was supported by project GOA/2015-014 (2014-2018 KU Leuven). OP is supported by the ERC synergy grant 'BlackHoleCam: Imaging the Event Horizon of Black Holes' (Grant No. 610058). The computational resources and services used in this work were provided by the VSC (Flemish Supercomputer Center), funded by the Research Foundation

Flanders (FWO) and the Flemish Government - department EWI. BR would like to thank Lorenzo Sironi for useful discussions.

References

[1] Kagan D, Sironi L, Cerutti B and Giannios D 2014 *Space Science Reviews* **191** 1-4
[2] Ball D, Özel F, Psaltis, D, Chan C K and Sironi L 2018 *ApJ* **2** 853
[3] Kadowaki L H S, de Gouveia Dal Pino E M and Stone J M 2018 *ArXiv preprint* arXiv:1803.08557v1
[4] Giannios D 2013 *MNRAS* **1** 431
[5] Sironi L and Spitkovsky A 2014 *ApJL* **1** 783
[6] Melzani M, Walder R, Folini D, Winisdoerffer C and Favre J M 2014 *A&A* **A111** 570
[7] Li X, Guo F, Li H and Li G 2015 *ApJL* **2** 811
[8] Sironi L, Giannios D and Petropoulou M 2016 *MNRAS* **1** 462
[9] Nalewajko K, Zrake, J, Yuan Y, East W E and Blandford R D 2016 *ApJ* **2** 828
[10] Petropoulou M, Giannios D and Sironi L 2016 *MNRAS* **3** 462
[11] Ball D, Özal F, Psaltis, D and Chan C K 2016 *ApJ* **77** 826
[12] Werner G R and Uzdensky D A 2017 *ApJL* **2** 842
[13] Rowan M E, Sironi L and Narayan R 2017 *ApJ* **1** 850
[14] Werner G R, Uzdensky D A, Begelman M C, Cerutti B and Nalewajko K 2018 *MNRAS* **4** 473
[15] Ressler S M, Tchekhovskoy A, Quataert E, Chandra M and Gammie C F 2015 *MNRAS* **2** 454
[16] Ressler S M, Tchekhovskoy A, Quataert E and Gammie C F 2017 *MNRAS* **3** 467
[17] Sadowski A, Wielgus M, Narayan R, Abarca D, McKinney J C and Chael C 2016 *MNRAS* **1** 466
[18] Ryan B R, Ressler S M, Dolence J C, Tchekhovskoy A, Gammie G and Quataert E 2017 *ApJL* **2** 844
[19] Chael A, Narayan R, and Sadowski A 2017 *MNRAS* **2** 470
[20] Baganoff F K. Bautz M W, Brandt W N, Chartas G, Feigelson E D, Garmire G P, Maeda Y, Morris M, Ricker G R, Townsley L K and Walter F 2001 *Nature* **413** 45-48
[21] Genzel R, Schödel R, Ott T, Eckart A, Alexer T, Lacombe F, Rouan D and Aschenbach B 2003 *Nature* **425** 934-37
[22] Eckart A, Baganoff F K, Schödel R, Morris M, Genzel R, Bower G C, Marrone D, Moran J M, Viehmann T, Bautz M W, Brt W N, Garmire G P, Ott T, Trippe S, Ricker G R, Straubmeier C, Roberts D A, Yusef-Zadeh F, Zhao J H and Rao R 2006 *A&A*, **1** 455
[23] Neilsen J, Nowak M A, Gammie C, Dexter J, Markoff S, Haggard D, Nayakshin S, Wang Q D, Grosso N, Porquet D, Tomsick J A, Degenaar N, Fragile P C, Houck J C, Wijns R, Miller J M, and Baganoff F K 2013 **ApJ** **1** 774
[24] Brinkerink C D, Falcke H, Law C J, Barkats D, Bower G C, Brunthaler A, Gammie C, Violette Impellizzeri C M, Markoff S, Menten K M, Mościbrodzka M, Peck A, Rushton A P, Schaaf R and Wright M 2015 *A&A* **A41** 576
[25] Davelaar J, Mościbrodzka M, Bronzwaer T and Falcke H 2018 *A&A* **A34** 612
[26] Chael A, Rowan M, Narayan R, Johnson M and Sironi L 2018 *MNRAS* **4** 478
[27] Watson M and Nishikawa K I 2010 *CPC* **10** 181
[28] Philippov A A, Cerutti B, Tchekhovskoy A and Spitkovsky A 2015 *ApJL* **2** 815
[29] Levinson A and Cerutti B 2018 *submitted to A&A ArXiv preprint* arXiv:1803.04427
[30] Bacchini F, Ripperda B, Chen A Y and Sironi L 2018 *submitted to International Astronomical Union Proceedings Series* **342**
[31] Bacchini F, Ripperda B, Porth O and Sironi L 2018 *submitted to ApJS*
[32] Parfrey K, Philippov A and Cerutti B 2018 *submitted to PRL*
[33] Rosdahl K J and Galsgaard K 2009 *A&A* **73** 551
[34] Gordovskyy M, Browning P K, Kontar E P and Bian N H 2014 *A&A* **72** 561
[35] Pinto R F, Gordovskyy M, Browning P K and Vilmer M 2016 *A&A* **159** 585
[36] Ripperda B, Porth O, Xia C and Keppens R 2017 *MNRAS* **3** 467
[37] Ripperda B, Porth O, Xia C and Keppens R 2017 *MNRAS* **3** 471
[38] Akramov T and Baty H 2017 *Phys. of Plasmas* **8** 24
[39] Uzdensky D A and Goodman J 2008 *ApJ* **1** 682
[40] Goodman J and Uzdensky D A 2008 *ApJ* **1** 688
[41] Li Y-P, Yuan F and Wang Q D 2017 *MNRAS* **3** 468
[42] Lyutikov M, Sironi L, Komissarov S and Porth O 2017 *J. Plasma Phys.* **6** 83
[43] Ripperda B, Porth O, Sironi L and Keppens R 2018 *submitted to MNRAS*
[44] Porth O, Olivares H, Mizuno Y, Younsi Z, Rezzolla L, Moscibrodzka M, Falcke H and Kramer M 2017 *Computational Astrophysics and Cosmology* **1** 4

IOP Conf. Series: Journal of Physics: Conf. Series **1225** (2019) 012018 doi:10.1088/1742-6596/1225/1/012018

[45] Olivares H R, Porth O and Mizuno Y 2018 *J. Phys.: Conf. Ser,* **1031** 012998
[46] Blandford R D and Znajek R L 1977 *MNRAS* **3** 179
[47] Komissarov S S 2004 *MNRAS* **2** 350
[48] Arons J 2012 *Space Science Reviews* **173** 341-67
[49] Ripperda B, Bacchini F, Teunissen J, Xia C, Porth O, Sironi L, Lapenta G and Keppens R 2018 *ApJS* **1** 235
[50] Bacchini F, Ripperda B, Chen A Y and Sironi L 2018 *ApJS* **1** 237
[51] Lingam M, Hirvijoki E, Pfefferlé D, Comisso L and Bhattacharjee A 2018 *Phys. of Plasmas* **24** 042120
[52] Ball D, Özel F, Psaltis D, Chan C K and Sironi L 2018 *ApJ* **2** 853

IOP Conf. Series: Journal of Physics: Conf. Series **1225** (2019) 012019 doi:10.1088/1742-6596/1225/1/012019

The Kelvin-Helmholtz instability and smoothed particle hydrodynamics

Terrence S Tricco

Canadian Institute for Theoretical Astrophysics, University of Toronto, 60 St. George St., Toronto, ON, M5S 3H8, Canada

E-mail: ttricco@cita.utoronto.ca

Abstract. There has been interest in recent years to assess the ability of astrophysical hydrodynamics codes to correctly model the Kelvin-Helmholtz instability. Smoothed particle hydrodynamics (SPH), in particular, has received significant attention, though there has yet to be a clear demonstration that SPH yields converged solutions that are in agreement with other methods. We have performed SPH simulations of the Kelvin-Helmholtz instability using the test problem put forward by Lecoanet et al (2016). We demonstrate that the SPH solutions converge to the reference solution in both the linear and non-linear regimes. Quantitative convergence in the strongly non-linear regime is achieved by using a physical Navier-Stokes viscosity and thermal conductivity. We conclude that standard SPH with an artificial viscosity can correctly capture the Kelvin-Helmholtz instability.

1. Introduction

The goal of numerical simulation is to recreate real physics. Some numerical methods may yield a more accurate solution, but the true test of any numerical method is whether it produces solutions that are convergent upon a 'true' physical answer with increased computational resources.

This work studies the convergence properties of SPH (smoothed particle hydrodynamics; [1, 2]) on the Kelvin-Helmholtz instability. The Kelvin-Helmholtz instability occurs when two fluids move past each other, causing exponential growth of velocity orthogonal to the flow. This develops into vortices which curl back upon themselves – the hallmark of the instability. As the instability becomes non-linear, secondary instabilities form leading to the two fluids becoming well-mixed along the interface. The Kelvin-Helmholtz instability is inherently a mixing instability.

The ability of SPH, a Lagrangian particle-based method for solving the equations of hydrodynamics [1, 2], to correctly model the Kelvin-Helmholtz instability has been questioned in recent years [3, 4, 5, 6, 7, 8]. Agertz et al [9] concluded that there exists "fundamental differences between SPH and grid methods" in their ability to resolve mixing instabilities on the basis of SPH calculations that showed no growth of the Kelvin-Helmholtz instability.

The abysmal results by Agertz et al [9] were a consequence of their initial conditions which contained discontinuities in density, velocity and internal energy, combined with a lack of numerical treat for contact discontinuities. Discontinuities require special care numerically. For SPH, velocity discontinuities are treated by an artificial viscosity, and density discontinuities are handled naturally by the density summation since it makes no assumption about differentiability

IOP Conf. Series: Journal of Physics: Conf. Series **1225** (2019) 012019 doi:10.1088/1742-6596/1225/1/012019

[10]. No treatment for the initial contact discontinuity was used by Agertz et al [9]. It has been demonstrated that including an artificial thermal conductivity is effective in capturing the Kelvin-Helmholtz instability and promoting mixing [10, 11].

Discontinuous initial conditions preclude formal convergence studies of the Kelvin-Helmholtz instability, however, even if the discontinuities are treated numerically. All wavenumber modes are unstable to growth when the interface is discontinuous [12], thus numerical noise may seed secondary instabilities that were not present in the initial conditions. Increasing the resolution of the calculation can alter the solution obtained since this permits access to higher wavenumber modes. Assertions have been made that some methods are of higher quality than others since they produce more detailed structure [13, 8], however, it is inappropriate to attribute structure generated from numerical noise as corresponding to resolved substructure [14, 15, 16].

In this work, we study the convergence properties of SPH on the Kelvin-Helmholtz instability. We use the smooth, well-posed initial conditions of Lecoanet et al [16] to obtain a converged solution, avoiding the issues inherent to discontinuous initial conditions. The calculations utilise a Navier-Stokes viscosity and physical thermal conductivity, thereby permitting the calculations to convergence in resolution to a particular solution in the strongly non-linear regime. Lecoanet et al [16] obtained converged solutions for this problem in both the linear and strongly non-linear regime using the finite-volume grid code ATHENA [17] and the pseudo-spectral code DEDALUS.

2. Physical equations and initial conditions

We solve the set of equations given by

$$\frac{\mathrm{d}\rho}{\mathrm{d}t} = -\rho \nabla \cdot \boldsymbol{v}, \tag{1}$$

$$\frac{\mathrm{d}\boldsymbol{v}}{\mathrm{d}t} = -\frac{\nabla P}{\rho} - \frac{1}{\rho}\nabla \cdot \boldsymbol{\Pi}, \tag{2}$$

$$\frac{\mathrm{d}u}{\mathrm{d}t} = -\frac{P}{\rho}\nabla \cdot \boldsymbol{v} + \nabla \cdot (\boldsymbol{v} \cdot \boldsymbol{\Pi}) + \frac{1}{\rho}\nabla \cdot (\chi \rho \nabla T), \tag{3}$$

$$\frac{\mathrm{d}c}{\mathrm{d}t} = \frac{1}{\rho}\nabla \cdot (\nu_{\mathrm{c}} \rho \nabla c), \tag{4}$$

where ρ is the density, $\boldsymbol{v}$ is the velocity, P is the pressure, $\boldsymbol{\Pi}$ is the Navier-Stokes stress tensor, u is the internal energy, T is the temperature, and χ is the thermal diffusivity. A passive scalar, c, which we call 'colour', is used to quantify the mixing of the two fluid regimes. An ideal equation of state is used, $P = \rho T$, with ratio of specific heats $\gamma = 5/3$. The temperature is related to the internal energy according to $T = (\gamma - 1)u$. The material derivative is $\mathrm{d}/\mathrm{d}t \equiv \partial/\partial t + \boldsymbol{v} \cdot \nabla$. The Navier-Stokes stress tensor is given by

$$\Pi^{ij} = \nu\rho \left(\frac{\partial v^i}{\partial x^j} + \frac{\partial v^j}{\partial x^i} - \frac{2}{3}\frac{\partial v^k}{\partial x^k}\delta^{ij} \right), \tag{5}$$

with shear viscosity ν. The colour is passively advected with the flow, but includes a 'physical' diffusion term analogous to thermal conductivity. The evolution of the Kelvin-Helmholtz instability in the non-linear regime is strongly sensitive to the dissipation, both numerical and physical. In order to enforce one particular solution, the dissipation is made resolution independent through the physical dissipation terms.

The initial conditions are two-dimensional and given by

$$\rho = 1, \tag{6}$$

$$v_x = v_0 \left[\tanh\left(\frac{y - y_1}{a}\right) - \tanh\left(\frac{y - y_2}{a}\right) - 1 \right], \tag{7}$$

$$v_y = A\sin(2\pi x) \left[\exp\left(-\frac{(y - y_1)^2}{\sigma^2}\right) + \exp\left(-\frac{(y - y_2)^2}{\sigma^2}\right) \right], \tag{8}$$

$$P = 10, \tag{9}$$

$$c = \frac{1}{2} \left[\tanh\left(\frac{y - y_1}{a}\right) - \tanh\left(\frac{y - y_2}{a}\right) + 2 \right], \tag{10}$$

where $a = 0.05$, $\sigma = 0.2$, $A = 0.01$, $v_0 = 1$, $y_1 = 0.5$ and $y_2 = 1.5$. The calculations are performed in a periodic box of size $x \in [0, L]$ and $y \in [0, 2L]$ with $L = 1$. The Reynolds number used is Re=10^5, defined according to

$$\mathrm{Re} = \frac{L\Delta v}{\nu}, \tag{11}$$

where $\Delta v = 2v_0$. This yields $\nu = 2 \times 10^{-5}$, and for these calculations $\nu = \chi = \nu_c$.

3. Numerical method

We use SPH to solve equations 1–4. We use a standard formulation of SPH where the density is computed through summation over neighbouring particles, and the smoothing length and density are self-consistently obtained through iteration [18, 19]. Artificial viscosity is used with the Morris & Monaghan limiter [20]. The calculations use the septic (M8) spline, a high-order kernel from the same family as the cubic and quintic B-splines [21].

The Navier-Stokes viscosity is solved using a two-first derivatives implementation, similar to [22, 23, 24]. The Navier-Stokes stress tensor (equation 5) is computed first, then the corresponding accelerations and heating are computed using the result. This implementation exactly conserves energy and momentum. The thermal conductivity and colour diffusion are computed directly using a second derivative in the manner of Brookshaw [25] and Cleary & Monaghan [26]. Total energy and colour are preserved with these schemes.

Calculations are performed using 256×592, 512×1184, 1024×2364 and 2048×4728 particles arranged on triangular lattices, labelled as $n_x = 256$, 512, 1024 and 2048, respectively. The $n_x = 2048$ calculation required $\sim 70\,000$ cpu-hours of computational time.

4. Results

The goal is to obtain convergence of SPH calculations on the Kelvin-Helmholtz instability test of Lecoanet et al [16]. In absence of an analytic solution, convergence is measured by comparing SPH results to the D2048 solution obtained using the pseudo-spectral code DEDALUS at a resolution of $n_x = 2048$ [16].

Figure 1 shows the colour field at $t = 2$, 4, 6 and 8 for the $n_x = 2048$ SPH calculations alongside the D2048 reference solution. The structures in both solutions are remarkably similar at all times. In each case, the seeded mode forms a large singular curl ($t = 2$), which continues to wind ($t = 4$), leading to the two fluids becoming well mixed along the interface at late times ($t = 6$ and 8). One minor difference is the degree of winding of the inner tip of the curl, which is not as tight for the SPH calculations compared to the reference solution. Several of the spurs at in the $t \geq 6$ snapshots are also of slightly different lengths. Despite these small differences, the SPH solution is in close proximity to the reference solution.

Figure 2 shows the growth of the amplitude of the seeded mode in the linear regime, obtained in a manner according to McNally, Lyra & Passy [15]. The measured growth rate of the mode is

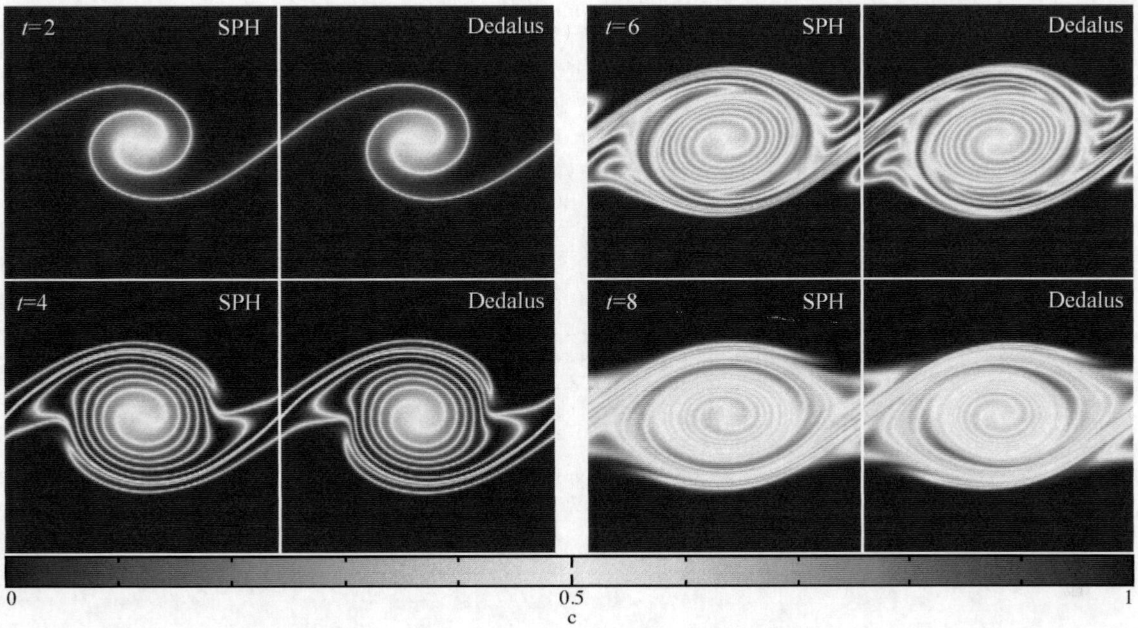

Figure 1. The colour field at $t = 2$, 4, 6 and 8 for the $n_x = 2048$ SPH results with the reference D2048 solution as computed using the DEDALUS code. The SPH results reproduce the reference solution at all times, though with some minor differences in the substructure in the strongly non-linear regime ($t \geq 6$).

converged to $\propto \exp(\pi t)$. It is difficult, however, to obtain an analytic estimate of the growth rate for this problem. An incompressible fluid with discontinuous interfaces should have a growth rate that is $\propto \exp(2\pi t)$ [12]. A smoothed velocity interface will reduce the growth rate by approximately 20% [27]. It is reasonable to expect that the growth rate for this problem should be even slower than that. These calculations are for a compressible fluid, not incompressible, and additionally smooth the initial velocity perturbation used to seed the instability. Thus, while no analytic estimate of the growth rate of this mode is available, the results are at least consistent with the estimates available acting as upper bounds. The growth rate is converged even for our lowest resolution of $n_x = 256$ particles.

The degree is mixing is quantified by an entropic quantity. Defining

$$s = -c\ln(c) \tag{12}$$

as the colour entropy, the total colour entropy is calculated by the volume integral

$$S = \int \rho s \mathrm{d}V, \tag{13}$$

which may computed in SPH by the summation $\sum_a m_a s_a$. The total colour entropy only increases when $\nu_c > 0$. Figure 3 shows the total colour entropy as a function of time, with black circles the values of the reference solution at $t = 2$, 4, 6 and 8. The shape of the total colour entropy curve matches the reference solution (see also [16]), and is converging towards the reference data points as the resolution improves.

Formal convergence of our calculations is assessed through the $\mathcal{L}_2$ error. The SPH particles for each calculation are interpolated to a 2048×4096 grid and compared grid-cell to grid-cell

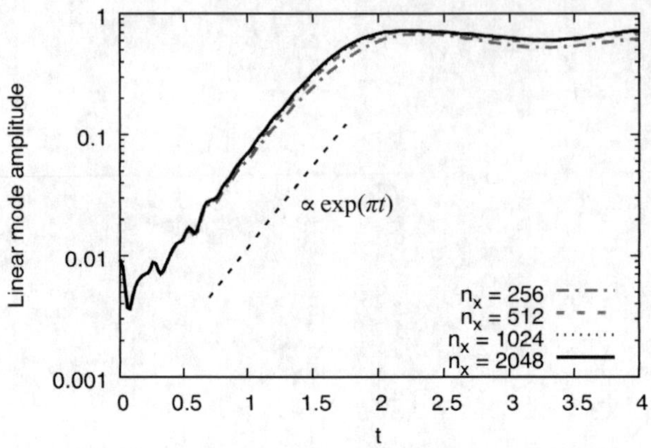

Figure 2. Growth of the seeded mode of the Kelvin-Helmholtz instability. The mode amplitude growth rate is converged to $\propto \exp(\pi t)$ for all resolutions tested.

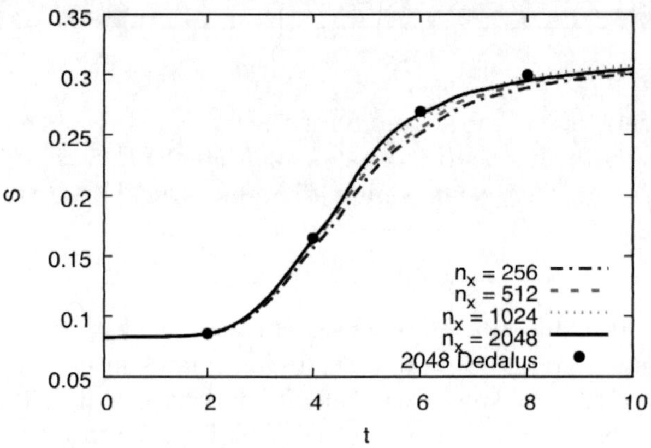

Figure 3. Total colour entropy as a function of time. Black circles are values from the reference solution. The SPH calculations reproduce the magnitude and shape of the total colour entropy from the reference solution.

with the reference solution. The error is computed according to

$$\mathcal{L}_2 = \left[\sum_a (c_a^{\mathrm{SPH}} - c_a^{\mathrm{D2048}})^2 \mathrm{d}V \right]^{1/2}, \tag{14}$$

where $\mathrm{d}V = 2048^{-2}$ is the volume of each grid-cell, and c^{SPH} and c^{D2048} are the colour fields of the gridded SPH data and reference solution, respectively.

Figure 4 shows the $\mathcal{L}_2$ error of the SPH calculations with respect to the reference solution at $t = 2$, 4, 6 and 8. Table 1 lists the measured convergence rates fit to $\propto n_{\mathrm{x}}^{-\Gamma}$. The errors in the SPH calculations may be contrasted to Athena calculations in Figure 2 of [16]. The maximum errors of the SPH calculations are of order 10^{-1}, whereas the A1024 and A2048 calculations have errors on the order of $\sim 10^{-3}$ and $\sim 10^{-4}$, respectively, for $t > 4$.

The error reduces linearly with respect to resolution at $t = 2$, in line with the expected rate of convergence from the artificial dissipation. Even for the highest resolution calculation, the dissipation of kinetic energy from the artificial dissipation remains comparable to the dissipation

IOP Conf. Series: Journal of Physics: Conf. Series **1225** (2019) 012019 doi:10.1088/1742-6596/1225/1/012019

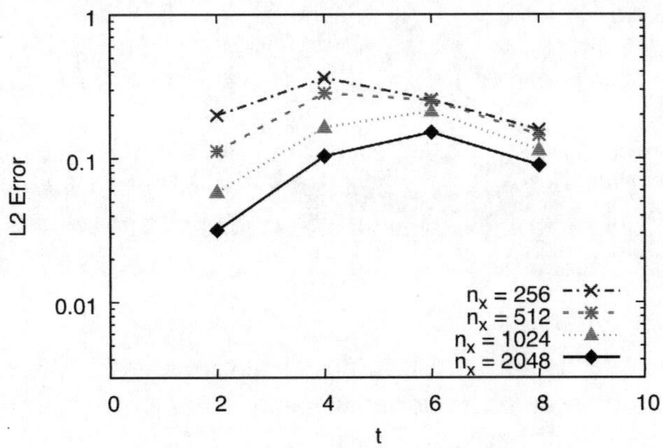

Figure 4. $\mathcal{L}_2$ errors of the SPH calculations with respect to the reference solution. The convergence is linear in resolution at $t = 2$. The convergence rate is slower in the strongly non-linear regime, though the errors continue to reduce, converging to the reference solution.

Table 1. The measured convergence rates, Γ, of the $\mathcal{L}_2$ error as fit to $\propto n_x^{-\Gamma}$.

t	Γ
2	0.89
4	0.62
6	0.26
8	0.28

by the Navier-Stokes viscosity. The convergence rate becomes sub-linear in the non-linear regime ($t \geq 4$), caused primarily by errors in the pressure gradient, which scale as $\mathcal{O}(1)$ with respect to resolution. This may be contrasted to ATHENA calculations, which have a measured rate of convergence that is between second to third order [16]. Importantly, the error of the SPH calculations is reducing at all times once the resolution is greater than $n_x = 512$ particles.

Reducing the error in the pressure gradient can be done by using higher-order smoothing kernels. For this problem, it was found that to obtain high-quality results necessitated the use of the septic spline, a high-order smoothing kernel from the B-spline family of kernels [21]. The septic spline is the high-order relative of the commonly used cubic and quintic splines. Using a high-order kernel is not an intrinsic requirement to activate the Kelvin-Helmholtz instability, but rather is needed to capture the amplitude of the initial velocity perturbation in the initial conditions and reduce spurious growth of other modes.

Figure 5 show the colour field at $t = 4$ for calculations using the cubic spline through to nonic spline. For the cubic and quartic spline, the evolution of the vortex is significantly altered with respect to the reference solution (c.f. Figure 1). The resemblance of the SPH colour fields to the reference solution improves as the quality of smoothing kernel improves. The difference between the septic and nonic spline results is negligible, thus we conclude that the kernel bias is a sub-dominant source of error when these high-order kernels are used. The calculations examined in this work have used the septic spline.

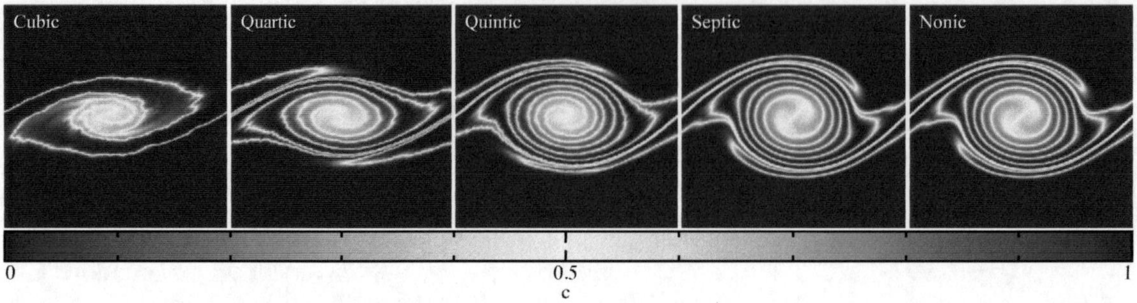

Figure 5. The colour field at $t = 4$ for SPH calculations using the cubic to nonic splines. For low-order kernels, they become the dominant source of error which degrades the quality of solution obtained. For high-order kernels (septic, nonic), the quality of smoothing kernel is no longer the dominant source of error and the solutions are indistinguishable. The calculations presented use the septic spline, which is the minimum order of B-spline that does not influence the results.

5. Summary

We have performed SPH calculations of the Kelvin-Helmholtz instability. We used the smooth, well-defined initial conditions of Lecoanet et al [16], thereby avoiding convergence issues with discontinuous initial conditions. A Navier-Stokes viscosity and thermal conductivity were included to force the evolution of the instability in the non-linear regime. A passive scalar, referred to as 'colour', was added to the two fluids to measure the degree of mixing. A high-resolution calculation using the pseudo-spectral code DEDALUS provided a reference solution to which results were compared.

The SPH results qualitatively matched the reference solution. The two fluids formed a curled vortex along the interface, which continued to spiral producing substructure and mixing of the two fluids. The substructure generated within the vortex matched between the SPH calculations and the reference solution well into the late non-linear regime.

Quantitative numerical convergence was measured for SPH on the Kelvin-Helmholtz instability, albeit with only linear convergence in the $\mathcal{L}_2$ error in the linear regime of the instability ($t = 2$), and sub-linear convergence in the strongly non-linear regime. The degree of mixing was measured by defining an entropic quantity for the 'colour' field, and the total entropy of the colour field increased monotonically in agreement with the curve from the reference solution.

Importantly, convergence has been demonstrated using the standard form of SPH with an artificial viscosity, of the kind that has been used for decades [28, 18]. No alternative SPH formulations, modifications, or hacks were employed. The only requirement to achieve the results presented here was a high-order smoothing kernel, which was needed in order to capture the initial amplitude of the velocity perturbation in the initial conditions and to curb growth of other modes from numerical noise. We conclude that the rumours of the 'fundamental flaws' of SPH have been grossly exaggerated.

Acknowledgments

This work was motivated by attendance at AstroNum 2017 in Saint-Malo, France. Thank you to Daniel Lecoanet for providing the reference DEDALUS solution, and to Daniel Price for useful discussions on physical dissipation terms in SPH.

IOP Conf. Series: Journal of Physics: Conf. Series **1225** (2019) 012019 doi:10.1088/1742-6596/1225/1/012019

References

[1] Lucy L B 1977 *AJ* **82** 1013–1024
[2] Gingold R A and Monaghan J J 1977 *MNRAS* **181** 375–389
[3] Springel V 2010 *ARA&A* **48** 391–430 (*Preprint* 1109.2219)
[4] Heß S and Springel V 2010 *MNRAS* **406** 2289–2311 (*Preprint* 0912.0629)
[5] Bauer A and Springel V 2012 *MNRAS* **423** 2558–2578 (*Preprint* 1109.4413)
[6] Sijacki D, Vogelsberger M, Kereš D, Springel V and Hernquist L 2012 *MNRAS* **424** 2999–3027 (*Preprint* 1109.3468)
[7] Hopkins P F 2013 *MNRAS* **428** 2840–2856 (*Preprint* 1206.5006)
[8] Hopkins P F 2015 *MNRAS* **450** 53–110 (*Preprint* 1409.7395)
[9] Agertz O, Moore B, Stadel J, Potter D, Miniati F, Read J, Mayer L, Gawryszczak A, Kravtsov A, Nordlund Å, Pearce F, Quilis V, Rudd D, Springel V, Stone J, Tasker E, Teyssier R, Wadsley J and Walder R 2007 *MNRAS* **380** 963–978 (*Preprint* astro-ph/0610051)
[10] Price D J 2008 *J. Comput. Phys.* **227** 10040–10057 (*Preprint* 0709.2772)
[11] Wadsley J W, Veeravalli G and Couchman H M P 2008 *MNRAS* **387** 427–438
[12] Chandrasekhar S 1961 *Hydrodynamic and hydromagnetic stability*
[13] Springel V 2010 *MNRAS* **401** 791–851 (*Preprint* 0901.4107)
[14] Robertson B E, Kravtsov A V, Gnedin N Y, Abel T and Rudd D H 2010 *MNRAS* **401** 2463–2476 (*Preprint* 0909.0513)
[15] McNally C P, Lyra W and Passy J C 2012 *ApJS* **201** 18 (*Preprint* 1111.1764)
[16] Lecoanet D, McCourt M, Quataert E, Burns K J, Vasil G M, Oishi J S, Brown B P, Stone J M and O'Leary R M 2016 *MNRAS* **455** 4274–4288 (*Preprint* 1509.03630)
[17] Stone J M, Gardiner T A, Teuben P, Hawley J F and Simon J B 2008 *ApJS* **178** 137–177 (*Preprint* 0804.0402)
[18] Monaghan J J 2005 *Rep. Progr. Phys.* **68** 1703–1759
[19] Price D J 2012 *J. Comput. Phys.* **231** 759–794 (*Preprint* 1012.1885)
[20] Morris J P and Monaghan J J 1997 *J. Comput. Phys.* **136** 41–50
[21] Schoenberg I J 1946 *Q. Appl. Math.* **4** 45–99 ISSN 0033569X, 15524485 URL http://www.jstor.org/stable/43633538
[22] Flebbe O, Muenzel S, Herold H, Riffert H and Ruder H 1994 *ApJ* **431** 754–760
[23] Watkins S J, Bhattal A S, Francis N, Turner J A and Whitworth A P 1996 *A&AS* **119** 177–187
[24] Sijacki D and Springel V 2006 *MNRAS* **371** 1025–1046 (*Preprint* astro-ph/0605301)
[25] Brookshaw L 1985 *PASA* **6** 207–210
[26] Cleary P W and Monaghan J J 1999 *J. Comput. Phys.* **148** 227–264
[27] Wang L F, Ye W H and Li Y J 2010 *Phys. Plasmas* **17** 042103
[28] Monaghan J J 1992 *ARA&A* **30** 543 574

IOP Conf. Series: Journal of Physics: Conf. Series **1225** (2019) 012020 doi:10.1088/1742-6596/1225/1/012020

Simulating 3-D Stellar Hydrodynamics using PPM and PPB Multifluid Gas Dynamics on CPU and CPU+GPU Nodes

Paul R. Woodward[1], Pei-Hung Lin[2], Huaqing Mao[1], Robert Andrassy[3] and Falk Herwig[3]

1. Minnesota Institute for Astrophysics, University of Minnesota
2. Lawrence Livermore National Laboratory, Livermore, CA
3. University of Victoria, Victoria, B. C., Canada

paul@lcse.umn.edu

Abstract. The special computational challenges of simulating 3-D hydrodynamics in deep stellar interiors are discussed, and numerical algorithmic responses described. Results of recent simulations carried out at scale on the NSF's Blue Waters machine at the University of Illinois are presented, with a special focus on the computational challenges they address. Prospects for future work using GPU-accelerated nodes such as those on the DoE's new Summit machine at Oak Ridge National Laboratory are described, with a focus on numerical algorithmic accommodations that we believe will be necessary.

1. Introduction

Our team has been simulating brief events in stars that require a 3-D treatment. These events involve the action of convection in the transport of nuclear fuel in the stellar interior. Within a convection zone, the material rapidly becomes well-mixed, but at the upper and lower boundaries of a convection zone, mixing is much more difficult to describe. In a thin region of radii, the convective motions must cease, and buoyant material located just outside the convection zone resists becoming incorporated into the convection flow but can nevertheless be pulled into it in small concentrations. Attempts to model this process of convective boundary mixing (CBM) in 1-D stellar evolution codes have difficulty establishing a convincing validity in the absence of detailed 3-D simulations. Such simulations are needed to describe the multiple nonlinear processes that cause stably stratified gas to become entrained in a turbulent convection zone. We have attempted to fill this gap in our knowledge by performing such simulations – a snap shot from one of them is shown in Fig. 1 – and by generating the averages on spherical surfaces that can inform modeling efforts. This would be a detail in stellar evolution theory if it were not the case that entrainment of stably stratified gas can have dramatic effects on the behavior of the star under a number of special conditions. Cases where this process can be pivotal occur when the boundary of a convection zone approaches a source of especially combustible fuel [1-5] (or, in a mirror circumstance, when it approaches material that, if ingested, can dampen or even extinguish a nuclear flame [6,7]).

We have given special attention to ingestion of *hydrogen* fuel, because the unusually energetic combustibility of hydrogen can give rise to especially large energy release. Even when burning hydrogen in very small concentrations, the energy produced can rival the energy generated in shells where heavier elements such as helium are consumed. The reason for this is the very strong dependence of nuclear reaction rates on the temperature. Just above the helium burning shell, in the convection zone

IOP Conf. Series: Journal of Physics: Conf. Series **1225** (2019) 012020 doi:10.1088/1742-6596/1225/1/012020

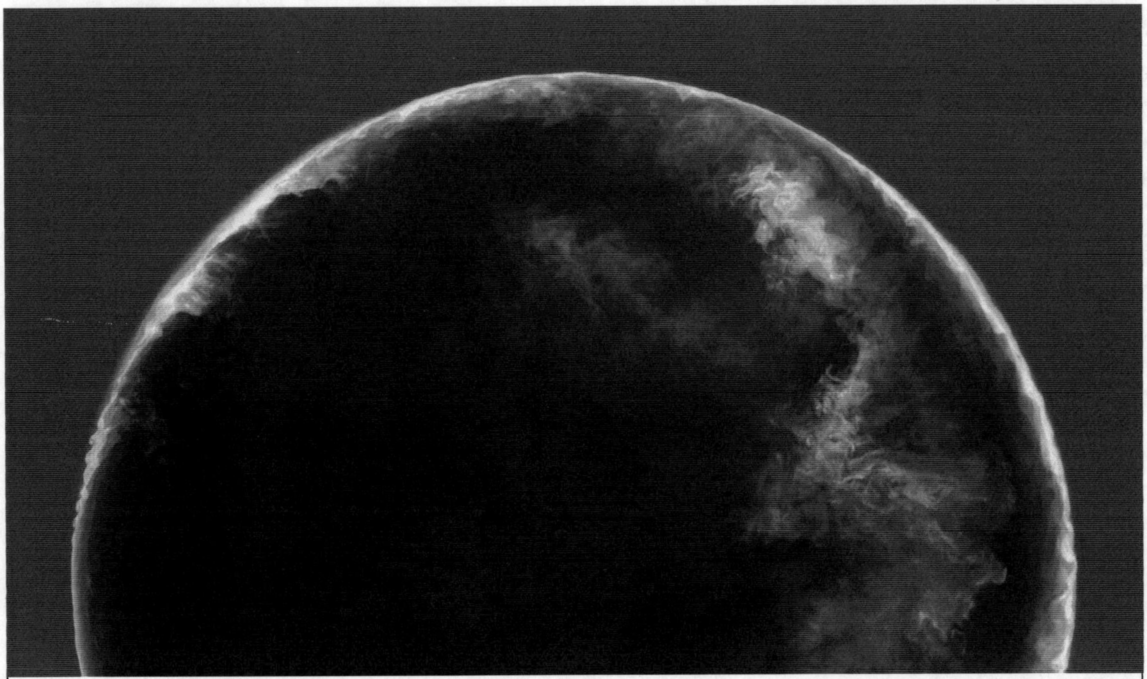

Fig. 1. A zoomed-in view of entrainment of stably stratified, relatively buoyant gas from above the top of the convection zone at the center of a 25 $M_\odot$ model star that is rotating. The rotation rate at the convection zone boundary in this slice through the equator is roughly equal to the average speed of turbulent motions in the convection zone. The log of the entrained fluid concentration is rendered here, with concentrations decreasing through red, yellow, white, aqua, blue and dark blue. This snapshot 26.2 days (for the star) into the simulation reveals a circulation flow global in scale, with the downward flow driven by the collision at the convection zone boundary of oppositely circulating large convective eddies in a roughly dipole flow pattern.

driven by this combustion, temperatures are very much higher than in the hydrogen-rich layers above the convection zone. The hydrogen is pulled downward by the convection until it reaches a level that is so hot that it burns rapidly enough to be consumed before descending still further. At these reaction rates, even small concentrations are sufficient to dramatically alter the convection flow and the overall luminosity of this region of the star.

These flows deep in stellar interiors pose several significant challenges to accurate 3-D simulation. First, the convection itself is a tiny effect, but one that has enormous consequences by moving material radially within the star. The Mach numbers range around 0.035, and in some cases can be quite a bit lower still. For the earth's atmosphere, this Mach number corresponds to a wind of about 25 mph – no hurricane, not even a squall, but enough to move things around. The entrainment of gas at the convection zone boundary is an even tinier effect. An entrained concentration as low as 1 part in 100,000 can have a significant effect, while a concentration of 0.001 can have consequences that are simply enormous. A reason for this surprising importance of tiny concentrations is the nonlinear positive feedback arising from entrainment and burning of the ingested fuel. All these unusual features of these stellar hydrodynamic problems require special design of the numerical algorithm.

The low Mach numbers demand high-order interpolations of cell interface values of variables in our explicit PPM gas dynamics scheme (see [8]). At low Mach numbers, the accuracy of the advected amounts of different quantities is dominated by the accuracy of the interpolation of the cell interface value. Our PPM scheme estimates these critical values with one order higher accuracy than it delivers for the interpolation parabola as a whole. It is also true that in this low Mach number regime, if we apply monotonicity constraints upon the interpolation parabola when they are not needed, we can destroy the benefits of the special effort we make to evaluate the cell interface values with such care.

IOP Conf. Series: Journal of Physics: Conf. Series **1225** (2019) 012020 doi:10.1088/1742-6596/1225/1/012020

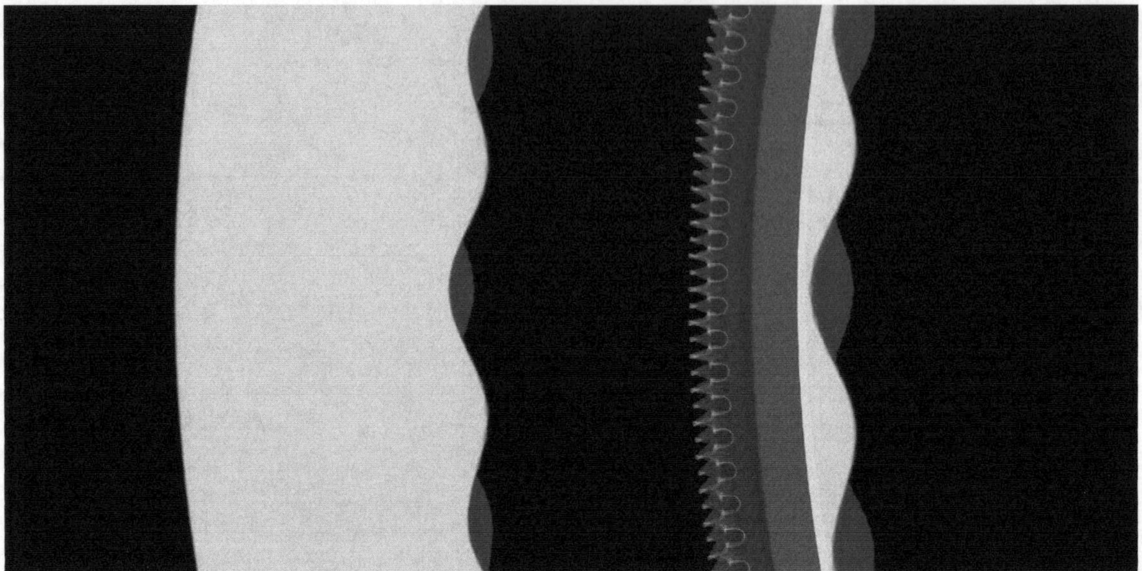

Fig. 2. Zoomed-in views, initially (left) and subsequently (right), of the dense gaseous shell from the ICF test problem study discussed in the text (see [9]). Here we inspect the small portion of the shell in which we would anticipate the most troublesome Cartesian mesh imprint phenomena to appear. No such imperfections are visible, because each surface was disturbed initially by a very high frequency and very low amplitude perturbation that only comes into view in the later image, where it has been dramatically amplified by the Rayleigh-Taylor instability of the outer surface of the shell. These images show density, and the red region has been compressed by a shock moving to the right in the image. The grid in this simulation had over a trillion cells, but the benefits of this approach can be enjoyed on any size grid capable of accurately capturing the flow features of interest. See [9] for a more complete series of images from this study.

As a result, we take the trouble to always evaluate the degree of smoothness of the function in a 5-cell portion of the grid centered on the cell of interest. When we judge the function to be smooth, we apply no monotonicity constraints. This procedure is well worth its cost, because it preserves the accuracy of the scheme as the Courant number approaches very small values. In flows where shocks can arise, we also introduce a fully 3-D dissipation inside shocks, which eliminates multiple grid imprint effects that we would otherwise observe in such flows (see [9]).

Aside from the above algorithmic accommodations, which we have made in all our codes since about 1990, we have made two other, major accommodations for our stellar hydrodynamics problems. The first is to save the initial, spherically symmetric state of the star on our grid, and then to subtract it out from the numerical scheme very carefully. We advance physical state variables that give perturbations to this initial state using nonlinear equations that are valid for perturbations of any size. Because the initial state is very rapidly varying in the radial direction, this lifts from the numerical method the burden of constantly evaluating the derivatives of this initial state with such high accuracy that small differences from those values can properly drive the dynamics of the convection flow. We take great care to explicitly cancel out all large contributions from the initial state to differences of values that are very nearly equal. This not only increases the accuracy of our computation, but it also allows that computation to be carried out with only 32-bit precision. This has an enormous benefit, not least because the speed of execution of the code is roughly doubled. This technique of subtracting out an unperturbed state was included in our earlier work on Rayleigh-Taylor instabilities [10], but the context of our present stellar hydrodynamics problems makes this subtraction much more important.

A second algorithmic accommodation we have made is to track the concentration of the entrained gas using the PPB moment-conserving advection scheme. We have described this scheme in detail elsewhere [11,8]. It is inspired by van Leer's Scheme VI from the 1970s [12], with multiple modifications that make it multi-D, robust, and efficient. We apply this scheme and our PPM gas dynamics on

IOP Conf. Series: Journal of Physics: Conf. Series **1225** (2019) 012020 doi:10.1088/1742-6596/1225/1/012020

a uniform Cartesian grid. The advantage of this grid is that, for a general flow problem, it has geometrically optimal properties that produce benefits in the ease and accuracy of directional splitting techniques as well as in MPI messaging patterns and interconnect bandwidth demands. However, our stellar problems are generally spherical, which does not match the Cartesian geometry well. Nevertheless we argue that once perturbations to our spherically symmetric initial state develop and grow into the nonlinear regime, the geometrical properties of the grid become largely irrelevant. We have addressed questions on this issue through multiple demonstrations and code comparisons. One such example is the 2-D study reported in [13]. A 3-D study, based upon a similar inertial confinement fusion (ICF) test problem, was reported in [9]. In that latter study, we specially constructed a problem involving multifluid interfaces and their instabilities in the context of a perturbed spherical implosion for which grid imprint effects arising from our Cartesian mesh would become immediately visible at a glance. We found that as long as the unstable multifluid interfaces in that ICF problem were given high frequency initial perturbations that were visibly undetectable, no grid imprints of any significance developed. In that study, we were able to increase the grid resolution to 10560^3 cells, so that convergence under grid refinement was well established. This approach to inhibiting grid imprint effects in our numerical treatment is illustrated in Fig. 2, from our study in [9]. This can be compared to the zoomed-in view of the region near the top of the convection zone in one of our recent stellar hydrodynamics simulations that is shown in Fig. 1.

To drive home the relevant numerical benefits of our PPB multifluid advection, we show results in Fig. 3 of a standard advection test problem. This test problem, which follows a multifluid boundary embedded in a rotational flow through many rotations, is quite relevant to our convection flows in stars. The main difference, however, is that this is linear advection, in which the length of the multifluid interface does not increase. It is a fundamental property of unstable interfaces that their lengths (surface areas in 3-D flows) increase. That causes a stretching of the interfaces, and this acts as a stabilizing mechanism that tends to keep our numerical representations of such interfaces as sharp as the numerical scheme allows it ever to be. This effect can be clearly seen in our study of Rayleigh-Taylor instability in [10] and [14] and is a benefit visible in action in Figs. 1 and 2. It is a feature of PPB advection that constraining the representation of the multifluid volume fraction in a cell is almost never performed, and this makes the numerical representation relatively free from glitches that can set off secondary interface instabilities, as can be seen under unfavorable circumstances with the much simpler but still highly accurate PPM advection technique.

2. Core convection in rotating main sequence stars.

We have reported work on stellar hydrodynamic problems involving ingestion of nuclear fuels at convection zone boundaries in multiple articles since 2012 [3,8,15-18,36]. This work can be compared to multiple efforts of other investigators using quite different numerical approaches [19-23, 37,38]. We believe the principal advantages of our approach described above come from our ability to embody these techniques in a code that scales to extreme levels and also runs very well on each of the thousands of multi-CPU nodes it uses. On the Blue Waters machine at NCSA, and now also on the Niagara machine on Compute-Canada's SciNet, we are therefore able to simulate the entire central portion of a model star for a run of a million time steps or more in just about a day, running at around 20 time steps per second. We use modest grids of only 1536^3 cells, and standard running configurations of 7344 nodes on Blue Waters and 1088 nodes on Niagara. The algorithm is explicit, and therefore we prefer to run problems for which the maximum Mach number encountered on the grid is at least about 0.035.

In this section, we present preliminary results of simulations on Blue Waters of core convection in massive main sequence stars. In this case, the Mach numbers are very low, but we address this issue by increasing the driving luminosity by a factor of 1000, which causes the velocities in the flow to increase by a factor of 10, bringing them into a cost-effective range for our code. We have established these scaling relationships in multiple series of simulations for different flows of this nature (see, for example, [24] and [17]). The scaling relationships can be understood by the following plausibility

IOP Conf. Series: Journal of Physics: Conf. Series **1225** (2019) 012020 doi:10.1088/1742-6596/1225/1/012020

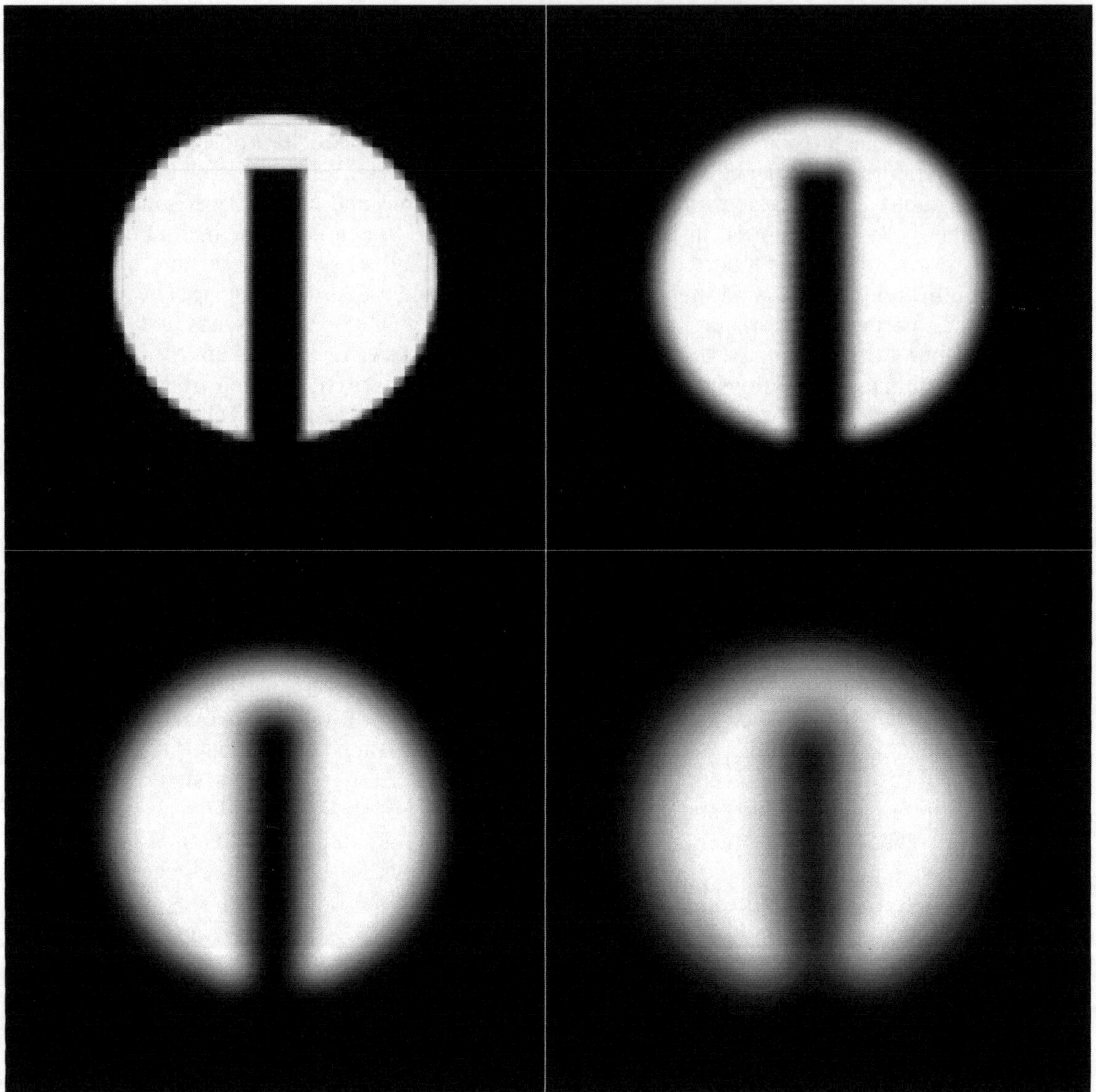

Fig. 3. Results of the standard advection problem originally devised by Zalesak. We fill a slotted circle with gas whose fractional volume in grid cells we track with our PPB advection scheme in 2D on a uniform Cartesian grid of 100×100 cells. Only the upper central 50×50 cell portion of this grid is shown. A velocity field is applied which causes the gas to rotate around the center of the entire grid as a solid body. The fractional volume of tracked gas is shown at 0, 1, 10, and 100 revolutions of the rotational flow. Each grid cell is represented by a 4×4 square of subcells in which the fractional volume is computed using the 6 low-order moments of this variable that are advanced in a conservative manner by the PPB scheme. In the initial state, at the top left, the fractional volumes and higher moments are computed exactly for a shape with boundaries of zero thickness. Grid imprints are clearly evident. By one revolution, these have vanished.

arguments. The entrainment rate scaling linearly with driving luminosity suggests that it is not the details of the flow near the convective boundary that determine the entrainment rate, but instead the energy available in the convection flow that can be used to drag down relatively buoyant gas. With more energy, we get more fluid dragged down. The argument for the scaling of the convective velocity cubed with the driving luminosity is as follows. The luminosity determines the size of the kinetic energy flux in the convection zone. Our extensive series of simulations of convection in the 1990s

IOP Conf. Series: Journal of Physics: Conf. Series **1225** (2019) 012020 doi:10.1088/1742-6596/1225/1/012020

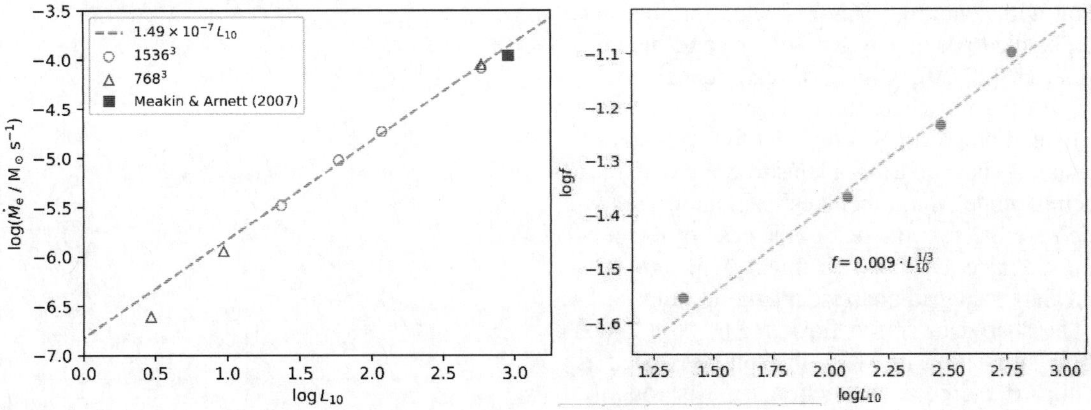

Fig. 4. Scaling laws for entrainment rate (left) and mixing region thickness (right) as determined by 3-D PPMstar simulations of convective boundary mixing at the top of the convection zone above the O-burning shell in model 25 M⊙ star. Our new code converges more rapidly than the old one as the grid is refined and gives results at low luminosities that are more in keeping with higher luminosity runs. This, we believe, is due to our subtracting out the unperturbed state of the star.

[25,26] indicate that this flux is a roughly constant fraction of the net convective energy flux. Increasing the overall energy flux in the layer by a factor of 10 would increase the kinetic energy flux by this same factor. The convection zone is fully turbulent in all our star models. Consequently, in the bulk of this zone, all 3 velocity components have, on average, the same rms value. The kinetic energy flux is proportional to the product of 3 such rms values, and therefore each of these 3 scales with the 1/3 power of the driving luminosity. The scaling relationships are observed results of our studies. We can use the above arguments to understand them, even though these arguments are not rigorous. Evidence for these relationships is shown in Figs. 4 and 5.

3. A 25 M⊙ rotating stellar model.

Exploiting the scaling relationships reported above, we consider core convection in a 25 M⊙ rotating star. We treat the rotation by adding a solid body rotational flow component to our simulation gradually over the first several thousand time steps of the run. We stop this process when we reach a desired rate of rotation at the top of the convection zone that is centered on the core of the star. Our boundary condition is that our star is enclosed in an impenetrable and immovable sphere that is inscribed in the cubical domain of our uniform Cartesian grid. We make no special effort to have this confining sphere be frictionless, since in our problems there is usually very little material, relatively speaking, located in the region of this sphere. As an additional simplifying device, we let the gravity holding the star together drop to zero a few cells before the confining sphere is reached. This allows us to use a very simple boundary condition at the sphere. The extent to which any aspect of this boundary treatment matters in the simulation should be evident from the simulation results. It is very important to note that our confining sphere is *not* holding the star together by pushing on it radially with a pressure. Instead, our star is held together by gravity, and the pressure at the location of our confining sphere is essentially negligible in most of our problems.

The main interest in studying massive main sequence stars in this way is to obtain accurate main sequence lifetimes for such stars. These stars can be observed via asteroseismology using recent satellites launched to locate exoplanets. Ingestion of hydrogen fuel from the top of the convection zone can keep the core hydrogen fire burning longer than would otherwise be possible. When the star is not rotating, we find a global convection flow that is dominated by the $l = 1$ spherical harmonic, with a dipole convection pattern. This is the same sort of flow that we found 2 decades ago in simulating the deep convective envelope of a giant star [26]. That earlier study explored the effect of the spherical geometry on convection. In the present work, we have such deep spherical convection flows inter-

IOP Conf. Series: Journal of Physics: Conf. Series **1225** (2019) 012020 doi:10.1088/1742-6596/1225/1/012020

acting with rotation. This is a situation that has been studied previously by Toomre and his collaborators [27-29], with a special focus on the case of the sun, where the convection zone does not extend downward all the way to the center of the star. They take an anelastic approach in a spectral code that includes magnetic fields, which are of great interest in the case of the sun. Here we have a simple treatment, with our uniform Cartesian grid and zero magnetic field.

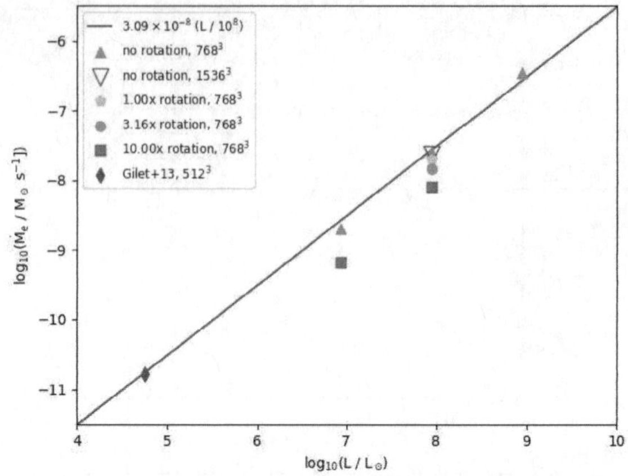

Fig. 5. Scaling of various core convection runs with luminosity. Note that the simulation of Gilet et al. [22] lies on the scaling trend, despite its separation by 2 orders of magnitude from our runs and also its use of a very different, anelastic technique.

The snap shot of the flow in Fig. 1 is taken from a simulation in which the rotation speed at the top of the core convection zone is roughly equal to the rms average convective velocity in the non-rotating case. Here the flow is modified by the rotation, but not so very much. We show results from cases rotating 3.16 times and 10 times faster in Fig. 6. These model stars rotate fast enough for the Coriolis forces to be significantly felt (although we perform the simulation in an inertial frame, where these "fictitious" forces do not apply). We can see in Fig. 5 that the entrainment of gas from above the convection zone is significantly reduced for rotating stellar models, yet the entrainment is still quite effective. Just as the convection tends to produce a well-mixed region of the star, it also transports and mixes the specific angular momentum. Although both simulations shown in Fig. 6 have been carried through multiple turn-over times for their large gyres, longer runs, ideally also carried out in rotating coordinate systems, will be necessary to study the redistribution of angular momentum by the convection in detail. Such runs are planned in the future. They should be made far more cost effective by our implementation of our code on GPU-accelerated nodes of large machines like Summit at the Oak Ridge National Laboratory. How we plan to do this is discussed in the next section.

4. Implementing the PPM+PPB algorithm for efficient execution on GPUs.

The PPM gas dynamics algorithm [30,11] in the form which we now use [31] has been augmented with PPB [11,8] advection of a fractional volume variable, f, that gives the fraction of the volume of a cell that is occupied by a special, tracked multifluid constituent. We are dealing with gases, and therefore f can either represent the concentration within the gas of the cell of a specific subspecies of gas, or it can represent the fraction of the cell volume that we regard as filled with a gas distinct from the rest of the gas in the cell. In our code, we treat both of these cases in the same way. For the purpose of solving a Riemann problem to determine a cell interface pressure and velocity, we regard the gases in the cell to be well mixed, while for the purpose of advection, computing fluxes of mass for each gas, we regard them as distinct but in a local pressure and temperature equilibrium with each other. At the level of a single cell, the two cases cannot be effectively distinguished based only on the information we store for each cell, since this is the resolution element of our simulation. However, if we look at a region of cells, it is immediately clear which of these two cases applies. We have designed the advection scheme so that it handles both cases, distinguishing one from the other automatically by exploiting the specialness of the values 0 and 1 for f.

The cell average of f and its 9 lowest-order moments, $\langle f\,\tilde{x}^l\tilde{y}^m\tilde{z}^n\rangle = \iiint_{-1/2}^{+1/2} f\,\tilde{x}^l\tilde{y}^m\tilde{z}^n\,d\tilde{x}d\tilde{y}d\tilde{z}$, with $l,m,n = 0,1,2$, are updated by the algorithm in a directionally split technique (see [8]). Here the coordinates $\tilde{x},\tilde{y},\tilde{z}$ are all scaled so that the origin is at the center of the cell and the width of the cell is unity in each coordinate. In the absence of any necessary constraints to force f to remain non-

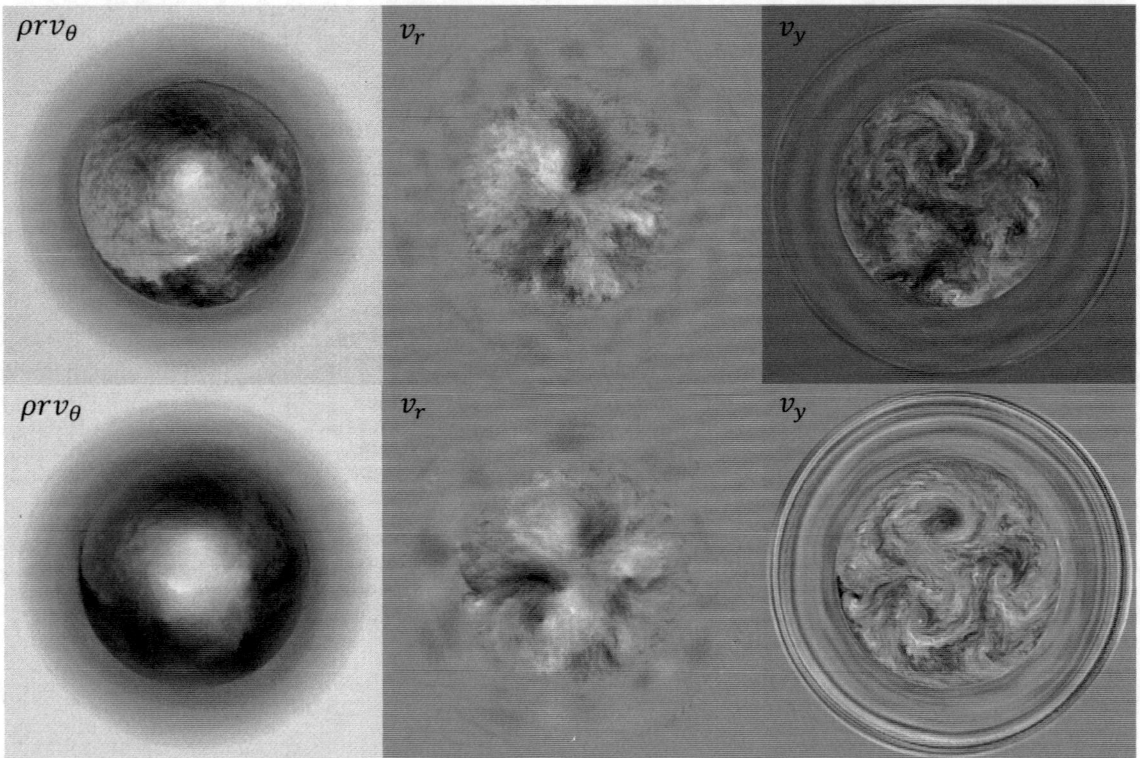

Fig. 6. Snapshots from simulations of the same 25 $M_\odot$ main sequence stellar model that is shown in Figure 1, but on the top row here the star rotates at 3.16 times the rms velocity in the convection zone and on the bottom row it rotates at 10 times that velocity. These rotation rates apply at the top of the core convection region, a boundary that is clearly seen in all these images. Further out in radius, the location of our bounding sphere, which is not rotating, is visible. The images at the left show the angular momentum in the equatorial plane. The star is rotating clockwise, so that its highest rotational angular momentum is black, while smaller values are shades of gray to white. The central images show the radial component of the velocity in the equatorial plane, with white showing outward motion and black inward. Large gyres are evident that span the convection zone between the center of the star and the convection zone top. The images at the right show the component of velocity out of the page. Here the large gyres are particularly easily made out. These have the appearance of hurricanes, and their spinning is caused by the same mechanism.

negative and less than or equal to unity, conservation laws for the 10 moments are obeyed to machine accuracy, which delivers multiple benefits. Even with 32-bit precision, a sharp transition from $f = 0$ to $f = 1$ can be maintained as this transition region is moved through the grid for a great many time steps with only very small diffusion, as can be seen from Fig. 3. Our directional splitting of this advection algorithm simplifies it enormously, especially in terms of constraints to enforce positivity and the strict upper limit of unity for f. Because advection, as contrasted with nonlinear gas dynamics, is the same whether 1-D passes are performed in the order XYZ or ZYX, the accuracy of the scheme does not appear to suffer from our directional splitting. We have even designed a version of PPB advection that explicitly handles transverse shear in each 1-D pass. This version does deliver greater accuracy, but we do not believe that extra accuracy is worth the cost it involves. We have not been able to compare this scheme directly with any of the modern and elaborate volume of fluid (VOF) techniques, but all of those techniques strike us as likely to be more expensive than our PPB implementation, and the results we have seen do not motivate us to take the substantial effort that would be needed to try them out for our gas-gas interfaces. Results of our PPB treatment have been published in [13] and [14] for DoE test problems that are simpler to set up than the problem shown in Fig. 2 and discussed in [9]; however, we have not as yet seen VOF treatments of these test problems.

IOP Conf. Series: Journal of Physics: Conf. Series **1225** (2019) 012020 doi:10.1088/1742-6596/1225/1/012020

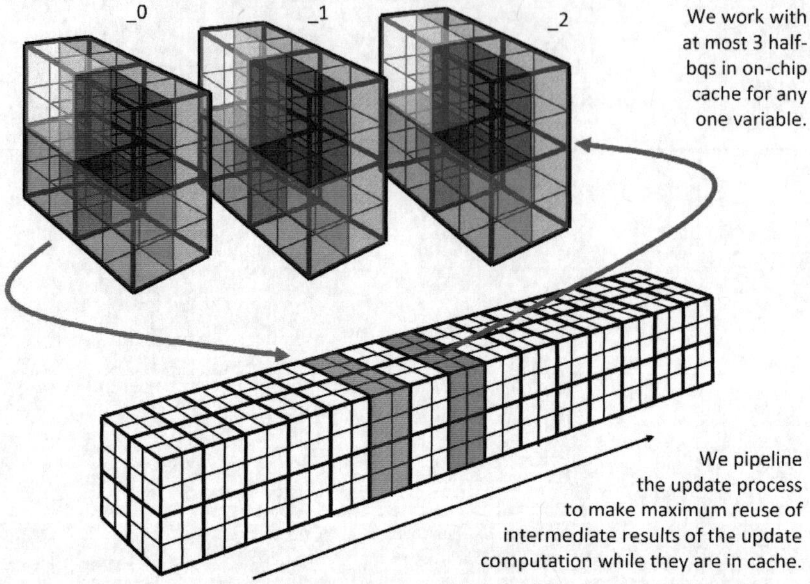

Fig. 7. Diagram of the grid pencil update process. Each half-briquette record consists of 4 mini-briquette records, as explained in the text. Execution proceeds in the direction of the arrow, with 3 half-briquettes in the process of being updated at any one time using only cached data. Each half-briquette record is 2 KB.

We will not repeat the details of the PPM and PPB algorithms here. We simply note that we compose all the steps necessary to update grid cells for a single 1-D pass into a single, massive, vectorized loop that extends for thousands of lines of Fortran code. For decades, we composed such algorithmic steps as a series of subroutines, each operating upon either single strips of grid cells or bundles of such strips, which we called *grid pencils*. In such a classic implementation, vector lengths were equal to the number of grid cells in a strip, plus a few extra cells on each end. Each subroutine consisted of a series of vector loops with progressively fewer and fewer of these extra cells contained in its vectors. Because this is a difference scheme, these vectors could not be aligned. For processing on devices manufactured after about 2000, these loops had to be "strip mined" by the compiler/hardware into a series of loops with much shorter vectors, and those vectors had each, in general, to be shifted to produce the perfect alignment required by SIMD engines, the main computational engines in all devices manufactured today. This code structure produced long vector loops containing the necessary parallelism. However, the processor cache rapidly fills up with the long vectors and vector temporaries needed to implement such loops running at a reasonable speed in the absence of sufficient memory bandwidth. Consequently, only a few of these loops can be executed out of cached data, and there must therefore be a great deal of data traveling unproductively back and forth between the processing chip and its directly connected off-chip memory to perform all the many stages of the numerical algorithm.

Our present code organization has resulted from our work with the Cell processor nodes of the Los Alamos Roadrunner machine [32-34,10]. We retain the concept of the grid pencil, but, as shown in Fig. 7, we update it progressively, one pair of 2 cross-sectional grid planes of 4×4 cells at a time. To facilitate this process, we store the physical state variables for each 2×2×2 cell mini-briquette together as a contiguous mbq-record in main memory. For an *x*-pass, with *i*, *j*, and *k* denoting *x*, *y*, and *z* indices, the mbq-record would be represented as a Fortran array indexed as mbqrec(j,k,i,iv), with the iv-index ranging over the 16 physical state variables. Such records are composed in 2×2×2 arrangements into briquette records, or bq-records, in memory so that they would be indexed in Fortran as bqrec(j,k,i,iv,jmbq,kmbq,imbq), a 7-dimensional array. Each of the 2 values of imbq, the *x* mbq-record index, corresponds to a different half-briquette, the primary data entity for our processing. The

IOP Conf. Series: Journal of Physics: Conf. Series **1225** (2019) 012020 doi:10.1088/1742-6596/1225/1/012020

Performance gains

Redundancy in computation eliminated

	Workspace / thread (KB)	flop/cell		
		Fortran-W	Pipelined	% redundancy
RK-adv	16.59	379.89	162.92	133.16
PPM-adv	19.2	454.61	273.31	66.34
tp3	208.28	5195.77	3218.67	61.43

Performance gains for PPM-adv

	Speed-up from		
	briquettes	pipelining-for-reuse & memory reduction	both
Nehalem	2x	3.33x	6.69x
Sandy Bridge	3.78x	1.66x	6.28x

Nehalem : Xeon 5570 ; Intel 9 Fortran Compiler; 16 OpenMP threads running on two sockets
Dual-socket, 4-core @ 2.93GHz, SSE-4.2 (128-bit)
Sandy Bridge : Xeon E5-2670; Intel 13 Fortran Compiler; 32 OpenMP threads running on two sockets
Dual-socket, 8-core @ 2.6GHz, AVX (256-bit)

Table 1. Results of a study of performance in 2014 by Jagan Jararaj working with our team (cf. [34]) that quantifies performance increases resulting from the 2 principal code restructuring techniques discussed here for our PPM codes. The first performance gain comes from the short, aligned vector operands provided by the briquette data structure, and the second comes from pipelining of the briquette update to remove redundant computation and reduce the size of the on-chip data workspace. No changes in recent hardware act to reduce this overall benefit of over a factor of 6. GPUs of the latest design can also benefit from these code structuring techniques, because we have reduced the on-chip data workspace requirement from 208 KB in this table to meet their limit of 32 KB plus an additional 16 KB of "shared memory" in Nvidia's Volta GPU.

composition of each bq-record in main memory out of 8 mbq-records allows us to transpose this data for processing in the y- or z-passes by transposing the contents of each mbq-record, and then transposing the arrangement of these records in the bq-record. All these transpositions can be performed while the relevant data is on the processing chip, so that the cost is small, although the code involved is significant and, unfortunately, confusing. The positions of the bq-records themselves inside the grid brick data structure remain unchanged in the transposition process.

We go to a great deal of trouble in packing our data into records so that it can be easily and efficiently prefetched, as indicated in Fig. 7, long before its presence on the chip is required. Once we demand that it must have landed, we must rearrange the contents to form 32-word vectors of individual physical state variables. We do this so that the contents of each vector line up as shown in Fig. 7, with the effective Fortran structure as in var(j,k,jmbq,kmbq,i), although of course we almost never expose all these interior dimensions of the data. We have revised our PPM+PPB algorithm so that the difference stencil extends outward only 4 cells in each direction along the x-axis for the x-pass. This means that we never need to hold in the on-chip cache more than 3 of our 32-word vectors for any quantity. With GPUs in mind, which have forced us to use 32-word rather than 16-word vectors, we do not add a final imbq-dimension. Instead, we manage separate vectors to which we append suffixes _0, _1, _2 as shown in Fig. 7. Each time we prepare to process a new half-briquette, we need to overwrite many of our vectors to accomplish a barrel shift of the data. This sounds like more trouble than it entails, and as far as our testing indicates, its cost is small. The speed of our implementation on no device now available is significantly limited by these data manipulations. Instead, it appears to be limited by the ability of the devices to perform our arithmetic at a good clip when all necessary operands are in the on-chip data store. Our codes run very well, as one can see from Tables 1 and 2, but of course they could run better. To our knowledge, our code's performance is not now limited on

IOP Conf. Series: Journal of Physics: Conf. Series **1225** (2019) 012020 doi:10.1088/1742-6596/1225/1/012020

any device by insurmountable obstacles such as the speed of light, memory bandwidth, or the size of the on-chip cache.

If we consider that at the outset of a half-briquette update we have all our 32-word vectors prepared and ready to go, then the code to perform that update should be a long, uninterrupted set of 32-wide SIMD arithmetical instructions, where all operands are immediately available on the chip. Our algorithm performs between about 1300 and 1400 such 32-wide SIMD flops to update a half-briquette, so this loop should not be outrageously long, in principle. However, for increased code readability and for convenient management of the large group of vector temporaries involved, we use subroutines that call other subroutines, so that, overall, this code module increases to 6 or 7 thousand Fortran lines. About half of this code implements computations that we do only when a data output dump will follow this particular pass. The management of temporary vectors via subroutine stacks that disappear upon exit is the largest inconvenience in this programming style. It is this technique that has enabled us to bring the size of our on-chip data workspace down from the 208 KB noted in Table 1, from 2014, to 32 KB of a GPU "register file" plus 16 KB of GPU on-chip "shared memory." This aggressive data workspace reduction is unnecessary for CPUs, which have sufficient cache storage space, but is essential for GPUs, which do not. In addition to straightforward arithmetic, we have groups of lines in which we must form new 32-word vectors consisting of the last 16 words of one vector and the first 16 of another. This is easy to express in Fortran. For CPUs, many of which process no more than 16 words at a time, these statements are unnecessary and should be removed by the compiler. For GPUs, these operations are expensive due, we think, to a special hardware feature, and they seem to benefit from doing 4 of them together, when possible. Writing a program in this style is unusual, but one can become accustomed to it, so that it is no more difficult than writing a program any other way. We have found that writing our codes this way enhances their performance on all devices we have seen.

5. Code performance experience and projections.

Because we have been running our codes almost exclusively on the Blue Waters machine at NCSA, which has only a small fraction of GPU-accelerated nodes, our experience with running our codes at scale is mainly with dual-CPU nodes. We have, however, packaged a testing version of the PPM+ PPB package, written in the style described above, that runs on either a dual CPU node or on a GPU node in a performance testing mode. The performance it achieves in this mode is born out on CPU nodes in our experience when the code is scaled up to thousands of nodes running a single problem. The GPU test does not involve any data movement back and forth between the GPU's attached memory and that of the CPU host, but in a full code implementation, we do not expect these data transfers to slow us down much, because they will only contain MPI messages, which are small relative to the overall data being updated on the GPU. Our GPU implementation was described in detail in [35]. Below, we include performance information from that study, updated to include results for more recent hardware. We have retained the numbers for the older processors, because these are found in some large systems still in service today. An evolution over 6 years is shown in the table. Improvements over this time interval of factors of 6.8 for CPUs and 8.5 for GPUs are not too far below Moore's law expectations.

In Table 2, we report delivered Gflop/s in 32-bit mode and counted in the same way for a number of CPU and GPU devices. We report the number of cores and simultaneously executing threads in each case. In the table, we have reported cores and threads using the same definition of these two terms that applies on CPUs and that has been in general usage for over 2 decades. Nvidia defines these two terms differently, and in a way that inflates the core and thread counts artificially over those that apply for a CPU. We have taken a thread to mean an independent thread of control. On the GPU, running our program, this definition of thread corresponds to 32 Nvidia threads, but those threads are not executing independent instructions, and they are better viewed as vector lanes. In that view, our threads on the CPU also have 32 SIMD vector lanes, just as on the GPU. Nvidia defines a core to be the hardware that performs the arithmetic for a single vector lane. But on a CPU, a core has one or

IOP Conf. Series: Journal of Physics: Conf. Series **1225** (2019) 012020 doi:10.1088/1742-6596/1225/1/012020

more SIMD engines and performs the arithmetic for our 32 vector lanes in each thread. In our Table 2, we therefore believe that we have counted threads and cores in the same way for all devices.

As we go to higher and higher performing devices in Table 2, it becomes progressively more questionable whether the interconnect on a large system will be able to keep up. This concern is especially acute with the latest generation of GPUs, which, according to Table 2, offer roughly double the performance of dual-CPU nodes. We can always decrease the demands on the interconnect by doubling the linear dimensions of the grid bricks that we update as uninterruptible tasks for MPI ranks in our codes. Each such doubling increases the computational work between communication events by a factor of 8, while the amount of data then communicated increases by only a factor of 4. Reducing demands on the interconnect bandwidth in this way can therefore increase the wall clock time for a run very significantly. However, if job queue wait times are long for requests involving large fractions of the machine, running the same problem on 8 times fewer nodes can turn out to be a good choice from the standpoint of higher efficiency of utilization of the requested resources and shorter wait times for the run to begin. On systems where there is higher effective interconnect bandwidth on special subsets of nodes, scaling a problem to this well-connected subset can be a good idea.

Table 2. multifluid PPM+PPB 1-D Grid Pencil Update

Device	GHz	Threads of control / cores	Gflop/s (32-bit)	Required Continuous GB/sec
dual AMD Interlagos node	2.3	32 / 32	71.7	8.30
dual Intel Sandy Bridge	2.0	32 / 16	84.7	9.80
dual Intel Haswell node	2.3	64 / 32	267	30.8
dual Intel Skylake node	**2.4**	**80 / 40**	**415**	**47.9**
Nvidia K20	0.73	168 / 14	121	38.9
Nvidia K40m	0.875	120 / 15	140	15.7
1 Nvidia K80	0.875	208 / 13	248	27.8
Intel Knights Landing node	1.4	272 / 68	260	30.0
Nvidia Titan-X Pascal GP102	1.800	448 / 56	610	68.4
Nvidia Volta GPU	**1.800**	**640 / 80**	**1033**	**115.8**

Our code is designed to give MPI messages a full grid brick update time interval to arrive at their destinations, as is described in some detail in [18]. The grid brick update consists entirely of a sequence of grid pencil updates, which are all independent in each of the three 1-D passes. We write the code so that we may begin the next 1-D pass before the present one is completed, and the dependencies between one pass and the next are minimized by the order in which the grid pencil tasks are launched. As a practical matter, we may therefore consider each grid brick update to consist of a large number of grid pencil updates that are, pretty much, independent. For a grid brick of M^3 grid briquettes, the cost of the update is $(M+2)^2\times(M+1) + (M+2)\times(M+1)\times M + (M+1)\times M^2 = (M+1)\times((M+2)\times 2\times(M+1) + M^2)$ single grid briquette updates. We have found that, as a rule of thumb, one obtains excellent performance and efficiency if one chooses M equal to the number of CPU cores on the node. From Table 2, we see that for today's hardware, a choice of M=40 or M=80 is indicated. In these two cases, the grid brick update is 206,804 or 1,594,404 briquette updates of about 83,200 flops each. This amounts to 17.2 or 132.65 Gflops, and 0.0414 or 0.1284 sec. at the computation rates given in Table 2. In this time interval, $(M+2)^3$-M^3 briquette records must both leave and come onto the node performing the grid brick update. With each briquette record being 4 KB, this is 2×40.35 MB or 2×157.5 MB. This requires a continuous, actually delivered link bandwidth to the node of 1.95 or 2.45 GB/sec. These are not impossible numbers, but they are pretty much at the margin of what is possible today. Note that we have included no consideration for finite latency of message passing. These estimates imply that although in principle better device design could deliver more performance to our codes at each node, there is no point in making such design improvements unless the interconnect bandwidth in today's machines is increased. There is a further consequence. If these grid brick sizes are close to the smallest ones that can be efficiently updated on today's equipment, with a minimum of two grid bricks assigned to each node, and if we demand that our problems scale to, say, 4000 nodes, this implies minimal grids of about $(20\times M)^3$, or either 3200^3 or 6400^3 cells. By today's standards, these are very large grids.

The scientific problems that our team is addressing have no difficulty in making good use of this new capability to attack much larger and more complex scientific problems. As a new generation of computing systems is being put into place, we will continue to adapt our code design to exploit these systems to simulate 3-D phenomena that play crucial roles in stellar evolution during relatively brief periods, such as hydrogen ingestion flashes and mergers of nuclear burning shells.

6. Acknowledgements.

Simulations and code development discussed in this article were performed on the Blue Waters machine at NCSA under support from NSF PRAC awards 1440025, 1515792, and 1713200, as well as a subcontract from the Blue Waters project at NCSA. Partial support for the code development also came from contracts from the Los Alamos and Sandia National Laboratories. The scientific research described here was supported by the NSF through CDS&E grants AST-1413548 and AST-1814181 and through subcontracts from NSF's JINA-CEE center (PHY-1430152). Simulations were also carried out on Canada's Westgrid resources and research supported by an NSERC Discovery grant.

References

[1] Fujimoto, M. Y., Ikeda, Y., & Iben, I. Jr. 2000, ApJL, 529, L25
[2] Heger, A., Woosley, S. E. 2010, ApJ, 724, 341
[3] Herwig, F., Woodward, P. R., Lin, P.-H., et al. 2014, ApJL, 792, L3
[4] Ritter, C., Andrassy, R., Côté, B., et al. 2018, MNRAS Letters, 474, L1
[5] Clarkson, O., Herwig, F., & Pignatari, M. 2018, MNRAS Letters, 474, L37
[6] Denissenkov, P. A., Herwig, F., Truran, J. W., et al. 2013, ApJ, 772, 37
[7] Lecoanet, D., Schwab, J., Quataert, E., et al. 2016, ApJ, 832, 71
[8] Woodward, P. R., Herwig, F., Lin, P.-H. 2015, ApJ, 798, 49
[9] Woodward, P. R., Jayaraj, J., Lin, P.-H., et al. 2012, Proc. NECDC 2012, Oct. 2012, Livermore, Ca., LA-UR-13-20949; available at http://permalink.lanl.gov/object/tr?what=info:lanl-repo/lareport/LA-UR-13-20949
[10] Woodward, P. R., Jayaraj, J., Lin, P.-H., et al. 2010, Proc. NECDC 2010, Oct., 2010, Los Alamos, N. M., LA-UR-11-00061; available at https://permalink.lanl.gov/object/tr?what=info:lanl-repo/lareport/LA-UR-11-00061
[11] Woodward, P. R., "Numerical Methods for Astrophysicists," in *Astrophysical Radiation Hydrodynamics*, eds. K.-H. Winkler and M. L. Norman, Reidel, 1986, pp. 245-326. Available at www.lcse.umn.edu/PPMlogo
[12] van Leer, B. 1977, J. Comput. Phys. 23, 276-299
[13] Joggerst, c. c., Nelson, A., Woodward, P. R., et al. 2014, J. Comput. Phys. 275, 154-173
[14] Ramaprabhu, P., Dimonte, G., Woodward, P. R., et al. 2012, Phys. Fluids 24, 074107
[15] Jones, S., Andrassy, R., Sandalski, S., et al. 2017, MNRAS, 465, 2991
[16] Andrassy, R., Herwig, F., Woodward, P., et al. 2018, MNRAS, submitted, arXiv:1808.04014
[17] Denissenkov, P., Herwig, F., Woodward, P., et al. 2018, submitted to MNRAS, arXiv:1809.03666
[18] Woodward, P. R., Herwig, F., and Wetherbee, T. 2018, Computing in Sci. & Eng. 20. https://ieeexplore.ieee.org/document/8452054
[19] Meakin, C. A., & Arnett, W. D. 2007, Astrophys. J. 665, 690-697
[20] Stancliffe, R. J., Dearborn, D. S. P., Lattanzio, J. C., et al. 2011, Astrophys. J. 742, 121
[21] Mocák, M., Siess, L., Müller, E. 2011, Astron. & Astrophys. 533, A53
[22] Gilet, C., Almgren, A. S., Bell, J. B., et al. 2013, Astrophys. J., 773, 137
[23] Baraffe, I., Pratt, J., Goffrey, T., et al., 2017, Astrophys. J. Lett. 845

IOP Conf. Series: Journal of Physics: Conf. Series **1225** (2019) 012020 doi:10.1088/1742-6596/1225/1/012020

[24] Jones, S., Andrassy, R., Sandalski, S., et al. 2017, MNRAS, 465, 2991
[25] Porter, D. H., and Woodward, P. R. 2000, Astrophys. J. Suppl. **127**, 159-187 (2000).
[26] Porter, D. H., Woodward, P. R., and Jacobs, M. L., "Convection in Slab and Spheroidal Geometries," Proc. 14th International Florida Workshop in Nonlinear Astronomy and Physics: Astrophysical Turbulence and Convection, Univ. of Florida, Feb., 1999; in *Annals of the New York Academy of Sciences* 898, 1-20 (2000); available at *www.lcse.umn.edu/convsph*
[27] Miesch, M. S., Elliott, J. R., Toomre, J., et al. 2000, Astrophys. J., 532, 593
[28] Brun, A. S., & Toomre, J. 2002, Astrophys. J., 570, 865
[29] Brun, A. S., Strugarek A., Jacobo, V., et al. 2017, Astrophys. J., 836, 192
[30] P. Colella, and P. R. Woodward, "The Piecewise-Parabolic Method (PPM) for Gas Dynamical Simulations," J. Comput. Phys. 54, 1984, pp. 174-201.
[31] P. R. Woodward, "The PPM Compressible Gas Dynamics Scheme," in *Implicit Large Eddy Simulation: Computing Turbulent Fluid Dynamics*, eds. F. Grinstein, L. Margolin, and W. Rider (Cambridge Univ. Press, 2007), pp. 130-146. Preprint available at www.lcse.umn.edu/ILES
[32] Woodward, P. R., J. Jayaraj, P.-H. Lin, et al., 2008, Computing in Sci. & Eng., Nov., 2008, p. 16-25. Preprint available at www.lcse.umn.edu/CiSE
[33] Woodward, P. R., J. Jayaraj, P.-H. Lin, et al. 2009, Concurrency and Computation Practice and Experience, **21,** 2160-2175
[34] Jayaraj, J., Lin, P.-H., Woodward, P. R., et al. 2014, "CFD Builder: A Library Builder for Computational Fluid Dynamics," Proceedings of 28th IEEE International Parallel & Distributed Processing Symposium (IPDPS): Programming Models, Languages and Compilers Workshop for Manycore and Heterogeneous Architectures (PLC2014), Phoenix, AZ, 2014, preprint at https://www.osti.gov/servlets/purl/1140604
[35] Lin, P.-H., and Woodward, P. R., 2016, J. Parallel & Distributed Computing 93, 56-65.
[36] Clarkson, O., Herwig, F., Andrassy, R, et al. 2018, submitted to Proc. NIC 2018, arXiv:1810.12259
[37] Müller, B., Viallet, M., Heger, A., et al. 2016, Astrophys. J. 833, 124
[38] Couch, S. M., Chatzopoulos, E., Arnett, W. D., et al. 2015, Astrophys. J. Lett. 808, L21

AUTHOR INDEX

Institute of Physics
Dirac House, Temple Back
Bristol BS1 6BE UK

ISSN: 1742-6588
ISBN 978-1-5108-8945-3

CURRAN ASSOCIATES INC.
proceedings.com